VOLUME THREE HUNDRED AND FIFTY EIGHT

INTERNATIONAL REVIEW OF CELL AND MOLECULAR BIOLOGY

Immunopathology of Celiac Disease

INTERNATIONAL REVIEW OF CELL AND MOLECULAR BIOLOGY

VOLUME THREE HUNDRED AND FIFTY EIGHT

INTERNATIONAL REVIEW OF CELL AND MOLECULAR BIOLOGY

Immunopathology of Celiac Disease

Edited by

AINARA CASTELLANOS-RUBIO
Ikerbasque, Basque Foundation for Science, Bilbao, Spain

LORENZO GALLUZZI
Weill Cornell Medical College, New York, NY, United States

ELSEVIER

ACADEMIC PRESS
An imprint of Elsevier

Academic Press is an imprint of Elsevier
50 Hampshire Street, 5th Floor, Cambridge, MA 02139, United States
525 B Street, Suite 1650, San Diego, CA 92101, United States
The Boulevard, Langford Lane, Kidlington, Oxford OX5 1GB, United Kingdom
125 London Wall, London, EC2Y 5AS, United Kingdom

First edition 2021

ISBN: 978-0-323-85311-8
ISSN: 1937-6448

For information on all Academic Press publications
visit our website at https://www.elsevier.com/books-and-journals

Publisher: Zoe Kruze
Acquisitions Editor: Ashlie M. Jackman
Editorial Project Manager: Tara Nadera
Production Project Manager: Denny Mansingh
Cover Designer: Alan Studholme

Typeset by SPi Global, India

Contents

Contributors

Eduardo Arranz
Mucosal Immunology Lab. Instituto de Biología y Genética Molecular (IBGM), Universidad de Valladolid-CSIC, Valladolid, Spain

Elisa Arribas
Mucosal Immunology Lab. Instituto de Biología y Genética Molecular (IBGM), Universidad de Valladolid-CSIC, Valladolid, Spain

Salvatore Auricchio
European Laboratory for the Investigation of Food Induced Diseases (ELFID), University Federico II, Naples, Italy

Maria Vittoria Barone
European Laboratory for the Investigation of Food Induced Diseases (ELFID); Department of Translational Medical Science, University Federico II, Naples, Italy

David Bernardo
Mucosal Immunology Lab. Instituto de Biología y Genética Molecular (IBGM), Universidad de Valladolid-CSIC, Valladolid, Spain

Govind Bhagat
Department of Pathology and Cell Biology, Columbia University Irving Medical Center, New York, NY, United States

Jose Ramon Bilbao
Department of Genetics, Physical Anthropology and Animal Physiology, Faculty of Medicine and Nursing, University of the Basque Country (UPV/EHU) and Biocruces-Bizkaia Health Research Institute, Leioa; Spanish Biomedical Research Center in Diabetes and associated Metabolic Disorders, CIBERDEM, Madrid, Spain

Ainara Castellanos-Rubio
Department of Genetics, Physical Anthropology and Animal Physiology, University of the Basque Country, UPV/EHU, Leioa; Biocruces Bizkaia Health Research Institute, Barakaldo; CIBER de Diabetes y Enfermedades Metabólicas Asociadas (CIBERDEM), Instituto de Salud Carlos III, Madrid; Ikerbasque, Basque Foundation for Science, Bilbao, Spain

Fernando Gabriel Chirdo
Departamento de Ciencias Biológicas, Facultad de Ciencias Exactas, Instituto de Estudios Inmunológicos y Fisiopatológicos—IIFP (UNLP-CONICET), La Plata, Argentina

Ariadna Cilleros-Portet
Department of Genetics, Physical Anthropology and Animal Physiology, Faculty of Medicine and Nursing, University of the Basque Country (UPV/EHU) and Biocruces-Bizkaia Health Research Institute, Leioa, Spain

Ángel De Prado
Mucosal Immunology Lab. Instituto de Biología y Genética Molecular (IBGM), Universidad de Valladolid-CSIC, Valladolid, Spain

Celia Escudero-Hernández
Department of Biomedical and Clinical Sciences (BKV), Linköping University, Linköping, Sweden

Laura Espino
Laboratorio de investigación en Genética de enfermedades complejas, Hospital Clínicos San Carlos, IdISSC, Madrid, Spain

Nora Fernandez-Jimenez
Department of Genetics, Physical Anthropology and Animal Physiology, Faculty of Medicine and Nursing, University of the Basque Country (UPV/EHU) and Biocruces-Bizkaia Health Research Institute, Leioa, Spain

Aida Fiz-López
Mucosal Immunology Lab. Instituto de Biología y Genética Molecular (IBGM), Universidad de Valladolid-CSIC, Valladolid, Spain

Iraia García-Santisteban
Department of Genetics, Physical Anthropology and Animal Physiology, Faculty of Medicine and Nursing, University of the Basque Country (UPV/EHU) and Biocruces-Bizkaia Health Research Institute, Leioa, Spain

José A. Garrote
Mucosal Immunology Lab. Instituto de Biología y Genética Molecular (IBGM), Universidad de Valladolid-CSIC; Servicio de Análisis Clínicos, Hospital Universitario Río Hortega, Valladolid, Spain

Amaia Jauregi-Miguel
Wallenberg Centre for Molecular Medicine; Department of Biomedical and Clinical Sciences, Faculty of Health Science, Linköping University, Linköping, Sweden

Concepción Núñez
Laboratorio de investigación en Genética de enfermedades complejas, Hospital Clínicos San Carlos, IdISSC, Madrid, Spain

Ane Olazagoitia-Garmendia
Department of Genetics, Physical Anthropology and Animal Physiology, University of the Basque Country, UPV/EHU, Leioa; Biocruces Bizkaia Health Research Institute, Barakaldo, Spain

Irati Romero-Garmendia
Department of Genetics, Physical Anthropology and Animal Physiology, Faculty of Medicine and Nursing, University of the Basque Country (UPV/EHU) and Biocruces-Bizkaia Health Research Institute, Leioa, Spain

Maialen Sebastian-delaCruz
Department of Genetics, Physical Anthropology and Animal Physiology, University of the Basque Country, UPV/EHU, Leioa; Biocruces Bizkaia Health Research Institute, Barakaldo, Spain

Craig R. Soderquist
Department of Pathology and Cell Biology, Columbia University Irving Medical Center, New York, NY, United States

Riccardo Troncone
European Laboratory for the Investigation of Food Induced Diseases (ELFID); Department of Translational Medical Science, University Federico II, Naples, Italy

CHAPTER ONE

Celiac disease susceptibility: The genome and beyond

Iraia García-Santisteban[a], Irati Romero-Garmendia[a], Ariadna Cilleros-Portet[a], Jose Ramon Bilbao[a,b,†], and Nora Fernandez-Jimenez[a,*,†]

[a]Department of Genetics, Physical Anthropology and Animal Physiology, Faculty of Medicine and Nursing, University of the Basque Country (UPV/EHU) and Biocruces-Bizkaia Health Research Institute, Leioa, Spain
[b]Spanish Biomedical Research Center in Diabetes and associated Metabolic Disorders, CIBERDEM, Madrid, Spain
*Corresponding author: e-mail address: nora.fernandez@ehu.eus

Contents

† Equally contributed.

International Review of Cell and Molecular Biology, Volume 358

ISSN 1937-6448
https://doi.org/10.1016/bs.ircmb.2020.10.002

Abstract

Celiac Disease (CeD) is an immune-mediated complex disease that is triggered by the ingestion of gluten and develops in genetically susceptible individuals. It has been known for a long time that the *Human Leucocyte Antigen* (*HLA*) molecules DQ2 and DQ8 are necessary, although not sufficient, for the disease development, and therefore other susceptibility genes and (epi)genetic events must participate in CeD pathogenesis. The advances in Genomics during the last 15 years have made CeD one of the immune-related disorders with the best-characterized genetic component. In the present work, we will first review the main Genome-Wide Association Studies (GWAS) carried out in the disorder, and emphasize post-GWAS discoveries, including diverse integrative strategies, SNP prioritization approaches, and insights into the Microbiome through the host Genomics. Second, we will explore CeD-related Epigenetics and Epigenomics, mostly focusing on the emerging knowledge of the celiac methylome, and the vast but yet under-explored non-coding RNA (ncRNA) landscape. We conclude that much has been done in the field although there are still completely unvisited areas in the post-Genomics of CeD. Chromatin conformation and accessibility, and Epitranscriptomics are promising domains that need to be unveiled to complete the big picture of the celiac Genome.

Abbreviations

ASM	allele-specific methylation
AUC	area under the curve
CD	Crohn's disease
CeD	Celiac disease
CGI	CpG island
CIM	CpG island methylator
DMP	differentially methylated position
DMR	differentially methylated region
DVMC	differentially variable and methylated CpG
eQTL	expression quantitative trait *loci*
GFD	gluten-free diet
GWAS	genome-wide association study
HLA	human leucocyte antigen
IBD	inflammatory bowel disease
lncRNA	long non-coding RNA
LPS	lipopolysaccharide
miRNA	micro RNA
mQTL	methylation quantitative trait *loci*
MR	Mendelian randomization
MSI	microsatellite instability
ncRNA	non-coding RNA
PBMC	peripheral blood mononuclear cell
RA	rheumatoid arthritis
SE	systemic sclerosis
SLE	systemic lupus erythematosus

SNP	single-nucleotide polymorphism
T1D	type 1 diabetes
TDT	transmission disequilibrium test
TF	transcription factor

1. Introduction

CeD is an immune-related complex disease caused by a combination of multiple genetic and environmental factors (Lindfors et al., 2019). Nowadays, the only effective treatment consists in a lifelong, strict gluten free diet (GFD) that most of the times is able to revert the principal symptoms of the disease, which include abdominal pain, diarrhea, bloating and weight loss, as well as intestinal villous atrophy together with crypt hyperplasia at a histological level.

The main genetic predisposing factors lie within the *HLA* region, located on the short arm of chromosome 6. This region harbors hundreds of genes with immunological functions that are associated not only with CeD, but also with most immune-mediated diseases. The main *HLA* variants involved in CeD are *HLA-DQ2* and *HLA-DQ8*, the former present in 90% of the patients and the latter in almost all the rest of the affected individuals (Karell et al., 2003). Remarkably, in Western countries, 40% of the population possess one or both of these variants, but the vast majority of them do not develop CeD (Liu et al., 2005). These data suggest that the *HLA-DQ2/DQ8* alleles are necessary but not sufficient for the development of the disease, and that other susceptibility factors must participate in CeD onset. In addition to *HLA-DQ2/DQ8* haplotypes, which have been estimated to account for 25–40% of the genetic susceptibility, other non-*HLA* regions have been associated with CeD. The risk effect of these non-*HLA* variants has been estimated to account for ~15% of the genetic component of the disease (Withoff et al., 2016). Taken together, all the genetic variants identified to date including both *HLA* and non-*HLA* SNPs explain more than 50% of the heritability, making CeD one of the immune-related disorders with the best-characterized genetic component.

In this review, we will discuss widely about the advances in Genomics of CeD, focusing on the GWAS carried out in the disease and more precisely, on the more recent discoveries in the post-GWAS era. Among other topics, we will address different integrative analyses and SNP prioritization strategies, and also provide insights into the celiac intestinal microbiome through the study of the Genome. Additionally, the present review also summarizes some of the most important discoveries on the field of CeD-related Epigenetics and Epigenomics. On the one hand, we will disclose the increasing knowledge on the celiac methylome and recapitulate both the genetically regulated and the environmentally shaped signatures defined for different tissues and cell types. On the other hand, we will discuss in depth the state of the art of CeD-associated ncRNAs, including both micro RNAs (miRNAs) and long non-coding RNAs (lncRNAs).

2. Genomics of CeD

During the past 15 years, GWAS have disclosed thousands of genetic polymorphisms that underlie the risk to suffer different complex traits and disorders, including CeD. During this period, a number of follow-up, prioritization and functional studies have arisen, often suggesting that the causal variants driving disease risk work through transcriptional regulation of target genes, with modest, and cell type- and/or context-specific effects. Here we revisit the Genomics of CeD from GWAS to the Microbiome regulation exerted by the Genetic landscape of the host.

2.1 Genetics of CeD before large genomic studies

The first genetic studies describing the association between CeD and *HLA* genes were carried out in the early 70's (Falchuk et al., 1972; Stokes et al., 1972), and one decade later the alleles-encoding *HLA-DQ2* were identified (Sollid et al., 1989; Tosi et al., 1983). For a long time, neither the candidate gene approaches in case-control cohorts nor the linkage studies in families were able to identify additional genetic factors outside the well-known *HLA-DQ* genes. With the completion of the Human Genome Project in 2003, there was an explosion of innovative analysis techniques including GWAS, which allow the study of the genetics of complex diseases in an unbiased manner. In GWAS, the frequency of SNPs is compared between patients and controls to identify genes and pathways associated with those complex disorders without the need of prior biological knowledge. Fig. 1 gives an overview of the GWAS and follow-up studies carried out in CeD up to date, and discussed in the present review.

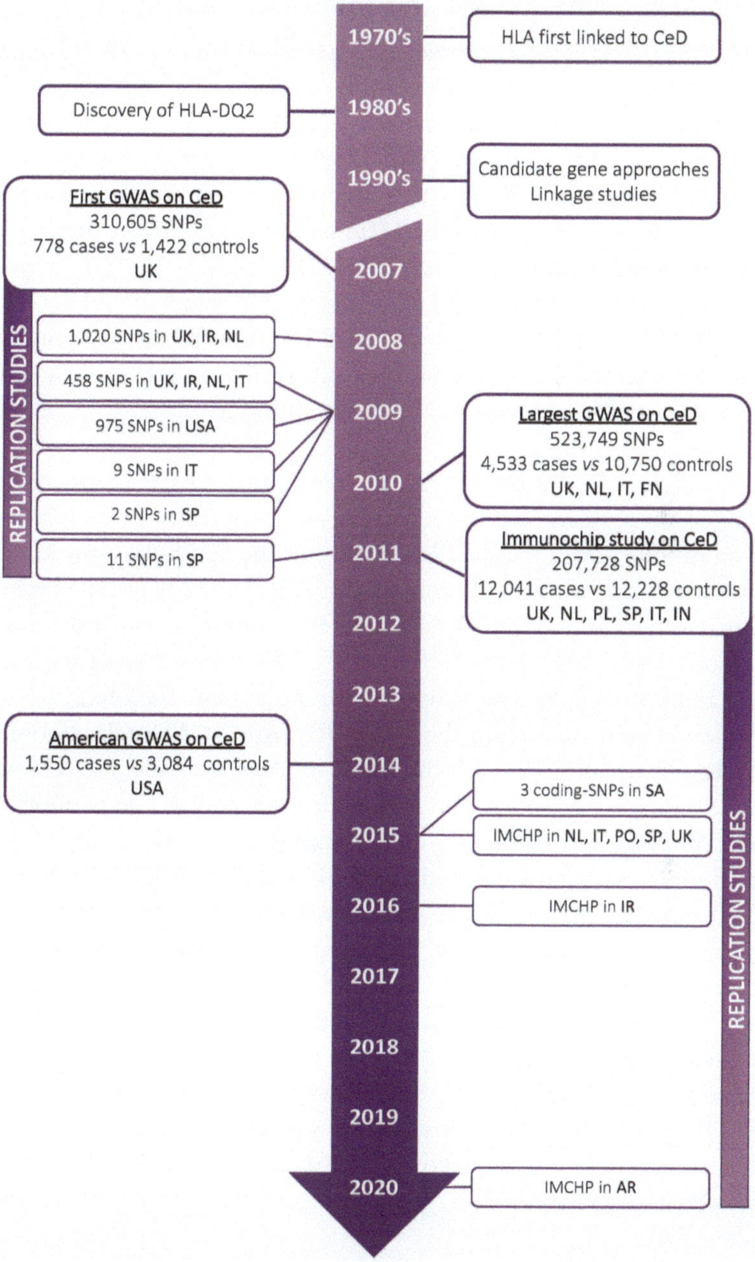

Fig. 1 Chronogram showing the GWAS and follow-up analyses performed in CeD. For each study, the number of SNPs analyzed and the origin of the cohorts is indicated (IMCHP, Immunochip; UK, United Kingdom; IR, Ireland; NL, The Netherlands; IT, Italy, SP, Spain, USA, United States; IN, India, SA, Saudi Arabia; PL, Poland; FN, Finland; AR, Argentina). In the case of GWAS, the number of cases and controls is also specified.

2.2 The first GWAS in CeD and follow-up studies

CeD was one of the earliest complex diseases subjected to a GWAS. The first one was carried out in 2007 by the group of Prof. Cisca Wijmenga in a relatively small population of the United Kingdom (UK), consisting of 778 CeD cases and 1422 controls (Van Heel et al., 2007). By genotyping 310,605 SNPs with the Illumina HumanHap300 BeadChip, besides the already known association with the *HLA locus*, this large-scale genetic study enabled the identification of the *KIAA1109-TENR-IL2-IL21* block on chromosome 4, the first non-*HLA* risk *locus* associated to CeD. Of note, the *IL1-IL21 locus*, containing genes with immune-related functions, was later associated with other immunological disorders, and is now a well-established disease-susceptibility *locus* for immune diseases (Zhernakova et al., 2009).

Afterwards, with the aim of identifying additional risk variants, the same research group performed two follow-up studies in different cohorts with a selection of SNPs from the initial GWAS. For the first follow-up study, the 1020 most strongly associated non-*HLA* markers were genotyped with the Illumina GoldenGate assay in 1643 cases and 3406 controls from the UK, Ireland and The Netherlands (Hunt et al., 2008). For the second follow-up, 458 additional SNPs that showed more modest associations in the initial GWAS study were assayed in the same UK, Irish, and Dutch cohorts, and the results were combined with the original GWAS; finally, the 6 most significant SNPs were genotyped in an independent Italian CeD cohort comprising 538 cases and 593 controls (Trynka et al., 2009). Altogether, the results of these two follow-up studies, combined with the GWAS data, presented 13 new genetic regions associated with the development of CeD in European populations. Most of these regions harbor genes that control the immune response, including cytokines or cytokine receptors, and a significant fraction of them are shared with other autoimmune diseases. Subsequently, another replication study was performed in a US population comprising 906 cases and 3819 controls with a selection of 975 out of the 1020 SNPs genotyped in the first follow-up (Garner et al., 2009). In this study, several previously detected regions were replicated, and a new association signal on chromosome 2q31, near the candidate gene *ITGA4*, was also identified.

Besides the follow-up studies performed genotyping a large number of SNPs, other replication studies were carried out with a selected number of genetic variants in several Mediterranean populations. The first one was performed in an Italian cohort composed of 538 cases and 593 controls, by genotyping 9 SNPs located in 8 previously identified CeD susceptibility *loci* (Romanos et al., 2009). This study was able to confirm the association of six out of the eight *loci*, but two of them, namely 3p21 (*CCR3*) and 2q12 (*IL18RAP*), were not replicated. Thus, in a second replication study in a Spanish population comprising 722 cases and 794 controls, those 2 SNPs were genotyped, confirming the association of the 3p21 genetic region with CeD susceptibility in this population (Dema et al., 2009). Another independent study carried out in a larger Spanish population, comprising 1094 cases and 540 controls, genotyped 11 SNPs tagging 8 GWAS regions, confirming the association of several of the SNPs analyzed (Plaza-Izurieta et al., 2011).

In addition to the follow-up studies carried out in large populations, other studies aimed at replicating GWAS regions in a family-based approach using the transmission disequilibrium test (TDT). In one of such studies 9 SNPs were genotyped in 325 Swedish-Norwegian CeD families, confirming the association between 1q31 (*RGS1*), 3p21 (*CCR1/CCR3/CCR2*), 3q25–26 (*IL12A/SCHIP1*) and 3q28 (*LPP*) as risk regions for CeD (Amundsen et al., 2010). In a similar study, 206 affected children within 97 Swedish and Norwegian families were subjected to TDT, and a new genome-wide significant risk *locus* covering the *DUSP10* gene was identified upon stratification of the affected individuals for *HLA-DQ* (Östensson et al., 2013).

2.3 Other GWAS in CeD

The largest GWAS of CeD performed up to date was carried out in 2010 by Dubois and collaborators in four well-characterized European populations from UK, Italy, Finland, and the Netherlands (Dubois et al., 2010). In a first step, 292,387 non-*HLA* SNPs from the Illumina Hap300 marker-set were tested for association in 4533 individuals with CeD and 10,750 controls; in a second step, 231,362 additional non-*HLA* markers from the Illumina Hap550 marker-set were genotyped in a subset of 3796 cases and 8154 controls. Meta-analysis identified 13 new genome-wide significant, non-*HLA* regions associated with CeD, bringing the total number to 26. In line with previous findings, the associated genetic variants identified in this work are located close to genes that are related to immunological functions, such as T- and B-cell co-stimulation or innate immune detection of viral RNA, among others. More importantly, an expression quantitative trait *loci* (*eQTL*) meta-analysis in blood samples performed by Dubois et al. pointed out that more than half of the CeD-associated SNPs correlated with the expression levels of nearby genes, highlighting the importance of the (de)regulation of gene expression in CeD pathogenesis.

Apart from the GWAS performed mainly in European cohorts, another GWAS study in a North-American population was published in 2014. In this genomic study, 1550 CeD cases and 3084 control individuals were genotyped with the Illumina Human660W-Quad array containing over 500,000 SNPs (Garner et al., 2014). In addition to replicating some of the previously identified *loci*, this GWAS provided evidence for a novel CeD *locus* in the *FRMD4B* gene on 3p14.1. After this study, no more GWAS have been performed in CeD, and research has focused on the search for the causal variants of this complex disease, using different approaches.

2.4 The immunochip and follow-up studies on CeD

One of the problems encountered by the CeD GWAS was the poor coverage of the genome at high resolution, since only 300,000–550,000 SNPs were genotyped (Dubois et al., 2010; Van Heel et al., 2007). Taking into account that 1–2% of the positions in the genome are polymorphic, and that sometimes SNPs map to regions that contain multiple genes, it is difficult to make precise associations between genetic variants and complex diseases. Another issue was that the small overlap between the arrays complicated the comparison of the results from different GWAS. To circumvent these problems, the Immunochip was developed, a low-cost Illumina Infinium High-Density array containing almost 200,000 SNPs in non-*HLA loci* that was designed to fine-map established GWAS *loci* and also to perform deep replication of 12 major autoimmune and inflammatory diseases (Cortes and Brown, 2011). In the case of CeD, in the year 2011 Trynka and collaborators used the Immunochip to genotype around 12,000 CeD individuals and a similar number of controls from seven cohorts, identifying 13 new genome-wide significant *loci*, most of which had been previously reported at lower significance levels or in other immune-mediated diseases (Trynka et al., 2011). This made a total number of 57 independent SNPs located in 39 risk *loci* outside the *HLA* region. Some years after the Immunochip study, the same group fine-mapped the *HLA* association signal in about 12,000 CeD cases and 12,000 controls of European ancestry (Gutierrez-Achury et al., 2015). This study led to the identification of 5 additional genetic variants that act independently of *HLA-DQ* alleles. These *loci*, together with the previously known 57 non-*HLA* risk *loci*, could explain up to 48% of the heritability of CeD (Gutierrez-Achury et al., 2015).

Similar to other large-scale genetic approaches, the initial Immunochip genotyping was followed by several studies in different populations. Coleman and collaborators performed a replication study in an Irish cohort comprising 425 CeD cases and 453 controls (Coleman et al., 2016). Even with the small number of individuals, this study could confirm the association of 11 out of the 57 previously identified SNPs, and found concordance in the direction of the effect of the majority of the SNPs reported in the original Immunochip study. They also meta-analyzed their data together with those reported by Trynka and collaborators, and identified two additional regions encoding *NFIA* and *ZNF335* genes that exceeded genome-wide significance. Very recently, over 4000 additional Caucasian samples from an Argentinian cohort were genotyped with the Immunochip platform, and combined with the previous data, has led to the identification of 2 novel genome-wide significant associations (Ricaño-Ponce et al., 2020). Of note, among the 57 genetic variants identified by Trynka and collaborators in the Immunochip study, 3 of them, namely rs1059702 in *IRAK1*, rs3184504 in *SH2B3*, and rs3748816 in *MMEL1*, were located in coding regions and give rise to non-synonymous mutations of the translated proteins. These three polymorphisms were genotyped in a Saudi Arabian population comprising 97 cases and 124 controls, but only the SNP located in *MMEL1* could be replicated (Saadah et al., 2015). Although these results should be interpreted with caution, given the small size of the population, they somehow stress the importance of the population ancestry to identify genetic susceptibility *loci*.

2.5 Trying to explain the missing heritability of CeD with rare genetic variants

The large-scale genomic studies performed up to date, including GWAS and the Immunochip study, have been able to explain approximately 54% of the heritability of CeD (Kumar et al., 2012). This implies that about half of the heritability of the disorder still needs to be clarified. It has been suggested that part of the missing heritability could be due to rare genetic variants with minor allele frequencies below 5–10%. Since these variants are typically excluded from GWAS, some efforts have been made to try to identify rare causal variants in CeD. One of those efforts was performed in 2013, and coding exons of 25 GWAS risk genes were sequenced in 24,892 individuals suffering from 6 autoimmune diseases (including CeD) and 17,019 controls. Among these risk genes, only two SNPs located within the gene *NCF2*, a component of the neutrophil NADPH oxidase respiratory burst complex, were shown to be associated with CeD (Hunt et al., 2013). The other study was based on exome sequencing of 75 CeD subjects from 55 European families and subsequent follow-up, but failed to identify additional CeD causing mutations (Mistry et al., 2015). Even considering that in general. Exome-sequencing studies have low statistical power, these data suggest that rare coding variants have a minor role in the pathogenesis of CeD, and that the hidden heritability of the disorder most likely lies on thousands of common genetic variants with small effect sizes.

2.6 *eQTL* and gene-expression analyses

The vast majority of genetic variants associated with CeD are located within non-coding regions, suggesting that GWAS-associated SNPs regulate gene transcription levels rather than altering the protein-coding sequence or protein structure. One approach used to identify the potential regulatory role of those genetic variants has been to perform *eQTL* mapping, where the associated SNPs are tested for their correlation with the expression of nearby genes (*cis-eQTLs*) or far-away genes (*tras-eQTLs*).

As previously mentioned, the first *eQTL* analysis was published together with the largest GWAS in CeD by Dubois and collaborators using new and published genome-wide *eQTL* datasets comprising 1469 human peripheral blood mononuclear cell (PBMC) samples. In this meta-analysis the authors reported significant *eQTLs* in 20 out of the 38 non-*HLA* CeD *loci* tested, and peak signals were located within the *MMEL1*, *NSF*, *PARK7*, *PLEK*, *TAGAP*, *RRP1*, *UBE2L3* and *ZMIZ1* genes (Dubois et al., 2010). These first *eQTL* results not only indicated that CeD risk regions are enriched in *eQTLs*, but also suggested that these risk variants might contribute to the pathogenesis of the disease by regulating gene expression.

Some years later, an integrative functional work combined the results from the *eQTL* analysis in PBMCs with co-expression analyses, obtaining a list of 49 potential CeD susceptibility genes in 39 *loci*. The expression of these genes was measured in seven different immune cell types, identifying sets of genes with higher expression level in T lymphocytes and NK cells, highlighting the importance of those cell types in the pathogenesis of the disease (Kumar et al., 2015).

Actually, an important point to consider when studying *eQTLs* is that they exhibit cell- and tissue-specific effects, implying that the genetic variants associated with complex traits like CeD are more likely to affect gene expression in a tissue-dependent manner (Fu et al., 2012). In this context, and considering the importance of T cell development in the thymus in the pathogenesis of CeD, Amundsen and collaborators performed a *cis*-

eQTL analysis of thymic tissue of 42 subjects, to evaluate the regulatory potential of 50 SNPs (Amundsen et al., 2014). Altogether, they identified 43 significant *eQTLs* in 24 out of the 39 CeD associated *loci* under study, and 14 of those *eQTLs* could represent novel thymus-specific *eQTLs*.

The main affected organ in CeD is the small intestine, and therefore it represents a very relevant tissue in the pathogenesis of the disorder. Using the small intestine of CeD patients and controls as a target tissue, our group performed several studies to measure the expression of most of the candidate genes proposed in large-scale genomic studies, in order to pinpoint their functional role. The first of those studies evaluated the expression levels of 10 candidate genes located in the regions associated with CeD in the first GWAS, and expression differences between GFD-treated CeD patients and controls were observed for several of those genes (Plaza-Izurieta et al., 2011). However, correlation between genotype and mRNA levels was only observed for *IL12A*, *LPP*, *SCHIP1*, and *SH2B3*, all of them involved in the immune response. Of note, another study by our group analyzed the expression levels of two genes located in the same GWAS peak, namely *THEMIS* and *PTPRK* (Bondar et al., 2014). Interestingly, the genotype of one of the associated SNPs in the region correlated with the expression of *THEMIS*. Additionally, we found a positive correlation between the expression levels of *THEMIS* and *PTPRK* in CeD patients but not in controls, suggesting the context-specificity of the genetic regulation and the co-expression. A third study in this topic aimed to evaluate the role of an additional set of 45 candidate genes by measuring their expression in the small intestine of CeD patients at the time of diagnosis and after 2 years on GFD, and comparing it with non-celiac controls (Plaza-Izurieta et al., 2015). Remarkably, expression levels were analyzed together with the genotypes of the main CeD-associated SNPs. In contrast to what could be expected, most of the identified *eQTLs* acted in *trans*, suggesting that the transcriptional control does not always occur at nearby *loci*. Moreover, the genotype effects discovered in this work were stimulus-dependent, since they were shown to be distinct among disease groups, confirming that a fraction of the associated genetic variants might only be relevant in a context-specific manner.

In this sense, it is important to highlight that besides tissue-specificity, *eQTLs* also exhibit context-specific behavior, being patent for instance only upon gluten ingestion. In line with this premise, in 2014 Fairfax and collaborators mapped *eQTLs* in human monocytes isolated from European individuals that were exposed to lipopolysaccharide (LPS) or interferon-γ

(IFN-γ), two molecules commonly used to trigger an inflammatory response *in vitro*. Strikingly, more than half of the *cis-eQTLs* identified were only seen in stimulated cells, supporting the idea that the disease-specific environment might be a key factor to correctly interpret the functional role of genetic variants (Fairfax et al., 2014). Of note, the stimulus-dependent *eQTLs* identified were significantly enriched for GWAS *loci*, including those from CeD (unpublished results), opening the door to novel analyses. Altogether, we can conclude that many genetic variants may have effects in a cell type-, tissue- and context-specific manner, and that the detection of these effects could provide direct insights into disease mechanisms.

2.7 Overlap between CeD and other immune-related diseases

It is well established that CeD patients show increased prevalence of other autoimmune diseases, and *vice-versa* (Lundin and Wijmenga, 2015). Since the completion of the first GWAS in 2007, we also know that the genetic variation associated with CeD overlaps with other autoimmune diseases, including type 1 diabetes (T1D), Crohn's disease (CD), ulcerative colitis (UC), rheumatoid arthritis (RA), or systemic lupus erythematosus (SLE), among others. In this sense, two approaches have been followed to pinpoint shared genetic variants between CeD and other immune-related disorders. The first approach consists of genotyping previously known risk *loci* from one disease in a case-control cohort of another disease. The second approach is based on performing a meta-analysis of publicly available GWAS in both diseases, with the aim of identifying new shared risk *loci*. An advantage of the second approach is that the statistical power is increased, and it allows the identification of novel *loci* shared between two diseases. Fig. 2 shows a Venn diagram with the shared susceptibility genes between CeD and other immune-related disorders. For a more exhaustive review, see the work by Hardy and Tye-Din (2016).

The first work that studied the overlap between CeD and T1D was performed using the first approach, and 8 known CeD-associated *loci* were genotyped in 8064 T1D patients, 9339 controls, and 3064 parent–child trios from 2828 families. At the same time, 18 known T1D-associated *loci* were genotyped in 2560 CeD patients and 9339 control subjects (Smyth et al., 2008). The study identified three CeD *loci* (*RGS1*, *IL18RAP*, and *TAGAP*) associated with T1D, and two T1D *loci* (*CCR5* and *PTPN2*) associated with CeD, indicating that the genetic susceptibility to both disorders shares common alleles.

CeD has also been associated with other intestinal disorders including Crohn's disease (CD) and ulcerative colitis (UC). The first meta-analysis looking for shared genetic *loci* between CeD and CD combined the results from the CeD GWAS from van Heel and collaborators (Van Heel et al.,

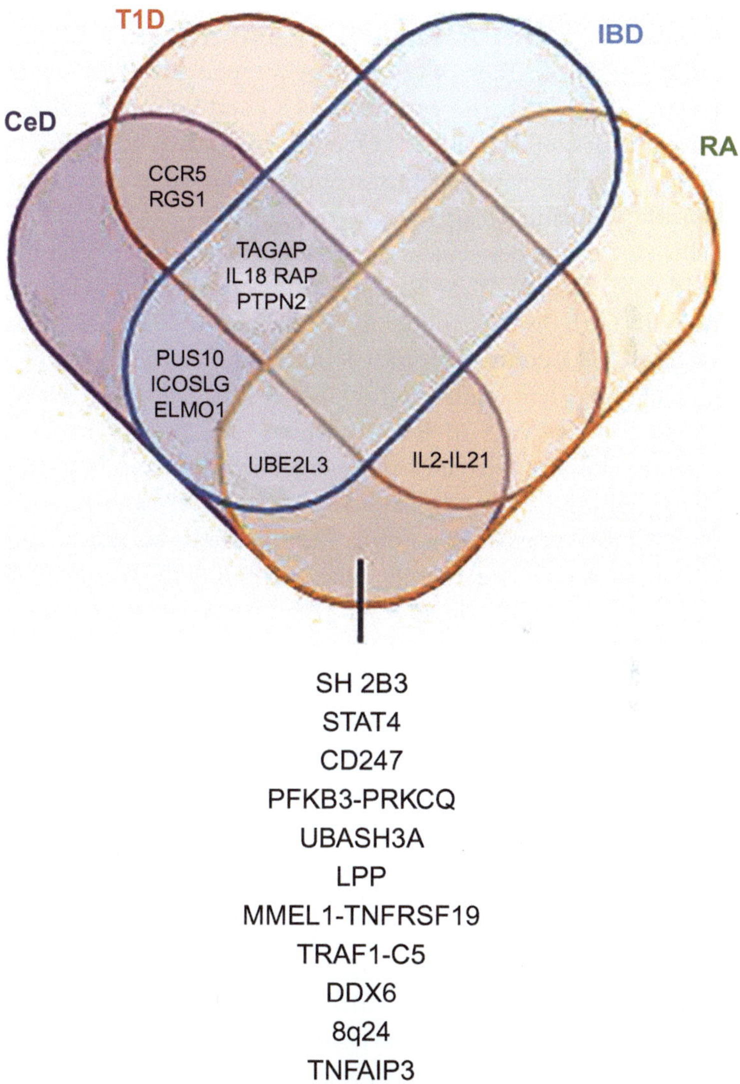

Fig. 2 Venn diagram showing the overlap between genes associated with CeD and other immune related disorders. CeD, Celiac Disease; T1D, Type 1 Diabetes; IBD, Inflammatory Bowel Disease; RA, Rheumatoid Arthritis.

2007) and a GWAS in CD published one year later (Barrett et al., 2008). The work was complemented by the replication of several new *loci* in independent cohorts for each disorder. This lead to the identification of four shared risk *loci* between CeD and CD, namely *PTPN2*, *IL18RAP*, *TAGAP*, and *PUS10* (Festen et al., 2011). One year later, another meta-analysis was carried out using the CeD GWAS data from Dubois and collaborators together with a Swedish UC data set. Additionally, 39 SNPs reported by Dubois et al. and other 6 genetic markers were also genotyped in 699 Inflammatory Bowel Disease (IBD, a disease comprising CD and UC) and 2482 control individuals (Parmar et al., 2012). Overall, this work identified several CeD risk *loci* which might confer susceptibility to IBD and/or CD/UC in the Finnish population.

Probably one of the best-characterized disorders associated with CeD is rheumatoid arthritis (RA). In 2009, Coenen and collaborators selected 11 *loci* associated with both diseases and genotyped them in 795 CeD cases, 1368 RA cases, and 1683 controls from a Dutch cohort. By combining their results in a meta-analysis with the first GWAS in CeD (Van Heel et al., 2007) and another GWAS in RA (Burton et al., 2007), they validated two previously known *loci* (including the above mentioned *IL2/IL21*) and further identified 4 new *loci* shared between CeD and RA (Coenen et al., 2009). Only 2 years later, a similar meta-analysis was carried out using public GWAS data with larger sample sizes, including the second GWAS in CeD (Dubois et al., 2010) and another GWAS in RA (Stahl et al., 2010), followed by replication in independent case-control datasets from both diseases. This study had larger statistical power, and thus was able to increase the number of shared non-*HLA* risk *loci* from 6 to 14, many of them containing genes related to T-cell signaling (Zhernakova et al., 2011). A third meta-analysis of the genetic overlap between these two immunological disorders was carried out using the CeD Immunochip data (Trynka et al., 2011) combined with the results of an independent RA cohort also genotyped with the Immunochip platform (Eyre et al., 2012). This meta-analysis was able to replicate 19 shared *loci* between CeD and RA that had previously been identified, and allowed the identification of 5 novel *loci* shared between the two disorders (Gutierrez-Achury et al., 2016).

More recently, a large meta-analysis studied the Immunochip data derived from four immune-related disorders, including CeD, T1D, RA and systemic sclerosis (SE), that together with around 22,000 healthy controls comprised almost 40,000 individuals (Márquez et al., 2018). This cross-disease meta-analysis led to the identification of 38 SNPs shared among the four diseases, 5 of which (*PADI4, NAB1, COBL, CCL21, GATA3*) were novel pleiotropic *loci*; additionally, 6 novel genome-wide associations were also identified. Interestingly, 10 of the SNPs identified showed opposite effects in different disorders, with the same allele conferring risk to one and being protective to another, adding another layer of genetic complexity to the genetic susceptibility to immune diseases.

2.8 Prioritizing GWAS hits

GWAS and follow-up replication studies have undoubtedly revolutionized our understanding of CeD Biology by successfully finding associations with multiple genetic *loci* inside and outside the *HLA* region. However, most of the identified variants are located in non-coding regions of the genome, and interpreting their functional role is a challenging task. In the last few years, computational approaches have been developed to prioritize causally associated variants and also to pinpoint the putative mechanisms through which the identified SNPs are connected to CeD pathogenesis.

In this context, in 2016 our group performed the re-analysis of the CeD Immunochip study using an approach that clusters individuals based on immunoancestry, rather than by geographical origin, before the disease association analysis (Garcia-Etxebarria et al., 2016). As a result, we proposed five novel disease-associated regions. Interestingly, *CCNT2* and *R3HDM1*, on the top novel *locus* on chromosome 2, were constitutively down-regulated in disease, even after gluten removal. One year later, another *in silico* approach was employed by Banaganapalli and collaborators (Banaganapalli et al., 2017). They carried out a systematic multidimensional analysis by first selecting CeD-linked SNPs according to their estimated deleteriousness, and then scrutinizing their function using public *eQTL* and tissue expression data, as well as effect calculations in *knockout* animal models, and network and pathway enrichment information. This systematic approach lead to the identification of *CCR2* as an immune-related candidate gene in CeD pathogenesis.

Mendelian Randomization (MR), a statistical tool used to identify causal associations between an exposure and an outcome, has also proven to be useful for the discovery of relevant genes in the context of CeD. The first MR approach in CeD was applied by our group in 2019 by means of the Summary-based Mendelian Randomization (SMR) software (Fernandez-Jimenez and Bilbao, 2019). Using publicly available gene-expression and methylation data as exposure, and CeD GWAS data as outcome, our group discovered a set of SNPs whose association with CeD could be mediated by the expression of genes in *cis* and by the methylation levels of nearby CpG positions. The most compelling result was the discovery of *UBE2L3,* as a promising diagnostic marker for CeD individuals on GFD. More recently, MR has also been applied to find causal associations between CeD and SLE, another immune-related disorder (Inamo, 2020).

Finally, several polygenic/genomic risk scores (PRS/GRS) have been proposed for CeD. PRS/GRSs aim to improve risk prediction of common complex diseases by using data from large-scale genomic studies. In 2014, Abraham and collaborators showed robust predictive models of the disease based on genome-wide SNP genotypes in a cross-validation study within six European cohorts (Area Under the Curve or AUC of 0.87–0.89) and in and independent replication across cohorts (AUC of 0.86–0.9), despite differences in ethnicity (Abraham et al., 2014). The authors of this work concluded that most of the predictive capacity was preserved in a reduced SNP set of around 200 SNPs (AUC ~0.85). Later that year, an average weighted genetic risk score was constructed including 10, 26 and 57 non-*HLA* variants (Romanos et al., 2014). In brief, the addition of 57 non-*HLA* SNPs to *HLA* haplotypes showed a significant improvement compared to scores from *HLA* only-based models (AUC of 0.854 *vs.* 0.823). In yet another study published in 2016, Coleman and collaborators found out that taking together all the CeD-associated markers from the Immunochip, they obtained an AUC of 0.83, compared to an AUC of 0.82 for *HLA* only (Coleman et al., 2016). Altogether, these studies suggest that although interesting as a potential SNP prioritization tool, PRS and similar approaches increase only moderately the prediction capacity of *HLA* testing alone.

2.9 Insights into microbiome through genomics

For a long time, a paradigm has suggested that gluten intake and genetic predisposition (understood as the presence of *HLA-DQ2* and/or *DQ8* haplotypes) were sufficient for the development of CeD. Nevertheless, mounting evidence indicates that this paradigm is unrealistic. First, we know that CeD onset can occur years after the first gluten intake. Second, as previously mentioned, 40% of the population in western countries carry the *HLA-DQ2/ DQ8* alleles, but only around 1% of them will ever develop the disease (Liu et al., 2005). Third, despite having a very strong genetic component (virtually all the CeD patients carry *HLA* risk alleles), CeD prevalence does not show a 100% concordance in monozygotic twins (Greco et al., 2002). All these data strongly suggests that although gluten intake and genetic predisposition seem to be necessary, they are insufficient to develop the disease. In this context, gut microbiota has arisen as a key environmental factor that might be mediating CeD onset and development.

The gut microbiota is composed of millions of microbes that colonize the human intestine and exert beneficial functions for the host, including the digestion of certain nutrients, the production of vitamins, and the modulation of metabolic and immunological programs. Importantly, in a mouse model of CeD, gliadin-induced damage could either be reduced or intensified by the gut microbiota, and this was related to severe intestinal damage caused by an increase of the phylum Proteobacteria (Galipeau et al., 2015). In light of such crucial roles, it is not surprising that alterations in gut microbiota composition and function have been linked to CeD. Genomic approaches based on targeted sequencing of the 16S rRNA, as well as whole-genome shotgun sequencing have permitted to compare the microbial communities of CeD patients with those of healthy controls. The cross-sectional studies performed in the small intestine of adult and child cohorts have revealed that microbial diversity differs between CeD patients and controls, with a decrease in *Bifidobacteria* and an increase in *Bacteroides* as a common feature of CeD (Bodkhe et al., 2019; D'Argenio et al., 2016; De Palma et al., 2010b; Nadal et al., 2007). However, those data must be interpreted with caution, given the inherent variability of the microbiota among individuals, the limited sample sizes (especially in the control groups), and the distinct sequencing approaches used in each study. Another limitation of those studies is that the genetic makeup of the host had not been considered.

Prospective studies have compared the composition of the gut microbiota in infants with a genetic predisposition for CeD carrying the *HLA* risk-alleles, with those who do not carry those alleles, and have followed-up cohorts over time. In line with previous reports, there were significant differences between the groups, where high-risk children displayed higher abundance of *Firmicutes* and *Proteobacteria*. Moreover, the microbiota of children positive for *HLA-DQ2/DQ8* did not stabilize at age 1 year, and remained so also at 2 years of age (Olivares et al., 2018a,b; Sánchez et al., 2011; Sellitto et al., 2012). As with cross-sectional studies, one of the limitations of the prospective studies performed so far is the limited sample size. To try to circumvent this problem, the CDGEMM (Celiac Disease, Genomic, Environmental, Microbiome and Metabolomic Study) has enrolled 500 infants aged 0–6 months with a first-degree family member with CeD in a prospective longitudinal, multi-center study ongoing in the USA, Italy and Spain since 2015 (Leonard et al., 2015). One of the aims of this project is to identify early changes in the gut microbiome of infants genetically predisposed to CeD, based on the presence/absence of *HLA-DQ2/DQ8* alleles. This information will be combined with other omics data (microbiome, transcriptomics, proteomics, metabolomics...) in order to create novel integrative networks, and to develop a personalized causal predictive model for the prevention of CeD in genetically predisposed individuals.

All the data from prospective studies suggest that *HLA-DQ* alleles might be somehow influencing the gut microbiome. Some available evidence suggests that the *HLA-DQ* genotype can affect early gut microbiota composition (De Palma et al., 2010a). This finding is not restricted to CeD, as in the last years several genomic studies have reported the influence of host genetics on the gut microbiome (Bonder et al., 2016; Turpin et al., 2016; Wang et al., 2016). Using data of one of those GWAS, our group has very recently published the first genomic study that examines the interplay between host genetics, gut microbiota, and CeD. In this study, we performed a two-sample Mendelian Randomization (2SMR) analysis using the gut microbiome data from Bonder and collaborators as exposure, and the large-scale genomic studies on CeD from Dubois (Dubois et al., 2010) or Trynka (Trynka et al., 2011) as outcomes. We identified a number of genetic variants linked to the phyla Firmicutes and Proteobacteria, bacteria whose abundance is altered in those CeD patients that harbor *HLA* risk alleles; in addition, we linked several SNPs to bacterial metabolic pathways that are related to CeD pathogenesis (García-Santisteban et al., 2020).

Unfortunately, the genomic studies on gut microbiome published up to date examine a limited number of individuals, meaning that we are still missing most of the genetic associations. In this sense, the MiBioGen initiative will definitely help, as it will not only meta-analyze the genetics and microbiome composition of approximately 20,000 individuals, but also collect disease-related phenotypic data (Wang et al., 2018).

3. DNA methylation in CeD

DNA methylation is a process by which methyl groups bind to the DNA molecule in a cytosine-guanine dinucleotide (CpG) context. Although it can change the function of a given DNA fragment without altering the sequence, methylation is a mitotically heritable modification.

Additionally, it plays an important role in the regulation of gene expression as well as in embryonic development and cell differentiation. Thus, DNA methylation is highly tissue and cell type-specific and is very often influenced both by the genotype and by the exposure to a number of internal/external factors (Teschendorff and Relton, 2018).

The epigenetic make-up has emerged as a relevant causal contributor to phenotypic variation and complex traits. In fact, DNA methylation has classically been shown as a discrete hyper- or hypomethylation coupled with a clear on/off switch of genes in imprinted genomic regions (Li et al., 1993) and different cancers (Jones and Baylin, 2002). However, more recently, the so-called "subtle change" paradigm has been adopted and is becoming the epigenetic hallmark of complex, non-malignant disorders (Leenen et al., 2016). According to the "subtle change" hypothesis, intermediary DNA methylation levels are fine-tuned, often influenced by the environment. Therefore, the association of DNA methylation with complex traits happens through slight differences in methylation levels as small as 1–5%, in very short stretches of DNA or even at single CpG positions. Over the last decade, several immune-related disorders, including IBD and T1D, as well as a number of malignancies, have been reported to display disease-specific genome-wide DNA methylation patterns (Agardh et al., 2015; Jones and Baylin, 2002; McDermott et al., 2016). However, few studies have been conducted specifically in CeD. Here, we present a detailed review of those (Fig. 3).

3.1 DNA methylation in CeD-related cancers

As far as we know, the first mention to CeD-related DNA methylation dates back to 2010, when Diosdado and collaborators characterized 48 small bowel adenocarcinomas, 15 of which were reported to be associated with CeD (Diosdado et al., 2010). In this study, chromosomal aberrations, microsatellite instability (MSI) and *APC* promoter methylation and mutation status were interrogated. Promoter hypermethylation of the *APC* gene was reported in 48% and 73% of CeD-unrelated and CeD-related small bowel adenocarcinomas, respectively, suggesting that *APC* promoter methylation could be a relatively frequent event in CeD-associated malignancies of the small intestine. Later that year, a high level CpG island methylator (CIM)/MSI phenotype was found in CeD-associated small bowel carcinoma, together with transcriptional repression of the tumor suppressor gene *MLH1* due to hypermethylation of its promoter (Bergmann et al., 2010).

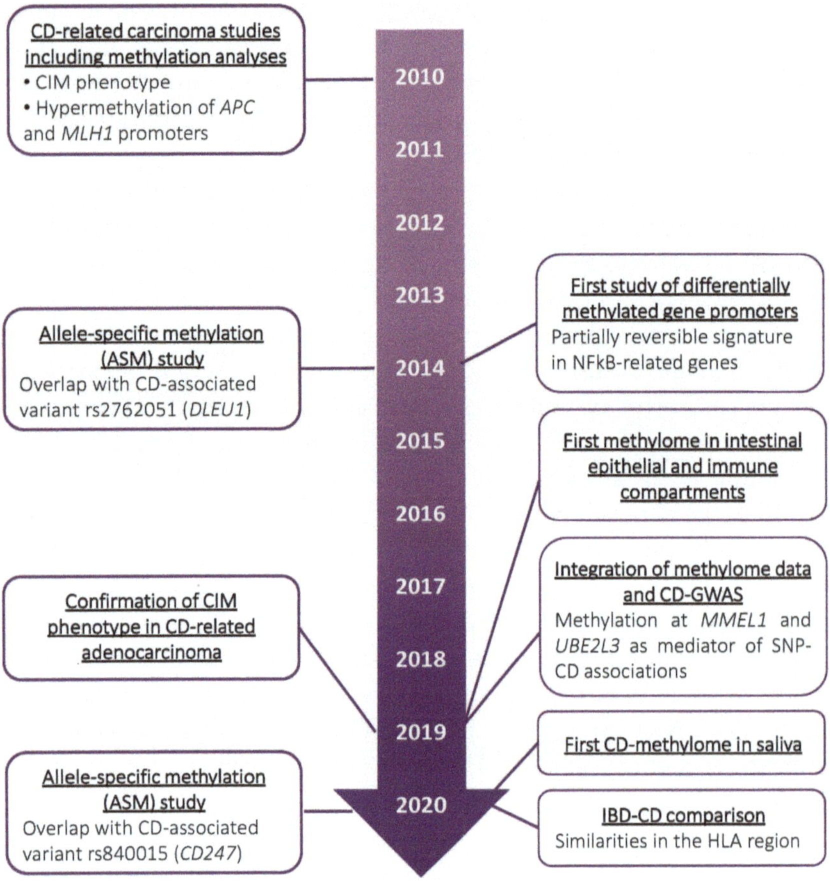

Fig. 3 Chronogram showing the findings related to DNA methylation in CeD.

More recently, 13 CeD-associated small bowel carcinomas were transcriptionally profiled and classified according to the Consensus Molecular Subtype classifier of colorectal cancer (Rizzo et al., 2019). Additionally, their CIM phenotype was evaluated using a methylation-sensitive amplification approach. Consequently, CeD-related carcinoma showed predominantly a CIM/MSI phenotype, as well as high CD3+ and CD8+ T cell infiltration and up-regulation of apoptosis, cell cycle and proteasome pathways. Altogether, those works strongly suggest the implication of DNA methylation in CeD-associated epithelial malignancies, although only small sample sets have been characterized and further studies will be needed in order to determine more precisely the underlying mechanisms.

3.2 First study on differentially methylated promoters

The first study to report differentially methylated regions (DMRs) in the celiac gut was published by our group in 2014 and investigated several NFκB pathway-related gene promoters in intestinal tissue from CeD patients at diagnosis, CeD patients after at least 2 years on a strict GFD, and non-CeD control individuals (Fernandez-jimenez et al., 2014). In summary, we found that *MALT1*, *MAP3K7* and *TRADD* presented subtle but significantly higher methylation levels in active CeD than in non-CeD controls, while *RELA* showed the opposite trend that remained significant even after GFD treatment. In contrast, LINE-1 elements did not appear to be differentially methylated in any case. Interestingly, the changes were to some extent maintained in patients even after complete adherence to a GFD during more than 2 years. However, differences with controls were less pronounced than in active CeD, suggesting that CeD-related methylation alterations might be only partially reversible or that alternatively, the duration of the treatment had not been long enough for changes to return to baseline.

On the other hand, another remarkable discovery of this work relies on the fact that, regardless of the stage of the disease, CeD patients show consistent correlations among the methylation levels of a number of NFκB-related genes, whereas non-CeD controls do not. Besides, the correlation of methylation levels (or co-methylation events) shown specifically for CeD patients appeared to be associated with the disturbance of co-expression of those same gene-pairs. We speculated that disease could coordinately alter the methylation patterns of functionally related genes to subsequently disrupt the regulatory equilibrium of a given pathway at the transcriptional level. In line with these observations, other studies have shown that gene pairs presenting strong co-methylation very often share functional similarities, but low correlation between expression levels (Akulenko and Helms, 2013).

3.3 Allele-specific methylation

Also in 2014, a study by Hutchinson and collaborators pointed to allele-specific methylation (ASM) as one of the causal contributors that could explain the phenotypic effects exerted by the variants identified in GWAS (Hutchinson et al., 2014). They screened 380,678 CpG sites within the genome, 5% of which showed evidence of ASM. Many of the sites presenting ASM in *cis* were also *eQTLs* in PBMCs and monocytes and additionally were in high linkage-disequilibrium with SNPs associated to complex diseases. Among other variants, rs2762051 (in *DLEU1*) was proposed to be implicated specifically in CeD through ASM. The study concluded that *cis*-regulated ASM may link the non-coding genetic variation with diverse phenotypic variation, and provide a way to integrate DNA methylation and GWAS findings (Hutchinson et al., 2014).

Very recently, this idea has been not only confirmed but extended since ASM has been found to help identify previously unknown regulatory SNPs, and characterized as an increased phenomenon in cancers, often through the disruption of CTCF and transcription factor binding sites (Do et al., 2020). Finally, the approach defined in this work also revealed ASM events in linkage disequilibrium with complex disease-associated variants, such as rs840015 (in *CD247*), directly highlighted for CeD. In summary, although theoretically interesting and relevant, ASM studies have disclosed moderate knowledge on the implication of DNA methylation in the Genetics of CeD, and pointed out the need of more disease-focused investigation.

3.4 First genome-wide methylation study in intestinal epithelial and immune cell fractions

The first genome-wide methylation study on CeD was published in the beginning of 2019 by our group and consisted of a set of Illumina Infinium HumanMethylation450 arrays run on sorted cells from intestinal bulk tissue of active CeD patients and non-CeD controls (Fernandez-Jimenez et al., 2019). We found 43 and 310 differentially methylated positions (DMPs) in the epithelial and immune fractions, respectively (mostly hypermethylated, FDR-corrected and with absolute methylation differences between groups >10%). In line with our prior knowledge of the disease, type 1 diabetes and adhesion molecules were the most enriched pathways among the genes the closest to the epithelial cell-specific DMPs. In turn, in the immune fraction, the most represented pathway was apoptosis. Expression analyses of the genes situated close to the cell type-specific DMPs demonstrated that methylation changes are in general more strongly correlated with expression levels in the epithelial cell compartment than in the immune, revealing a greater functional impact of the former.

Differentially variable and methylated CpGs (DVMCs), as defined by Teschendorff and collaborators, are based on the idea that differential variability between disease groups can be as informative as traditional studies comparing methylation means (Teschendorff et al., 2016). In fact, several works have proven chronic inflammatory and precancerous stages more prone to methylation variation than non-inflamed tissues or even cancer

(Teschendorff et al., 2014). In this context, the 2019 study by our group identified 560 DVMCs in the epithelial cell population, out of which 93.6% were more variable in active CeD, in line with the high variability reported for inflammatory and premalignant conditions. In contrast, we found 851 CeD-associated DVMCs in the immune cell fraction, only 26% of them being more variable in CeD and less correlated to the expression levels of adjacent genes than in the case of the epithelial compartment. A plausible explanation for the observed decrease of variability in the celiac immune fraction relies on the fact that CeD is characterized by a limited number of expanded clones dominating the T-cell repertoire (Dahal-Koirala et al., 2016), that would consequently reduce the (epi)genetic heterogeneity.

Finally, and given that DNA methylation of adjacent CpGs is often highly correlated, a dimension reduction strategy was applied in order to find CeD-related CpG clusters or DMRs (Fernandez-Jimenez et al., 2019). Remarkably, many of the identified DMRs exhibited loss of CpG island (CGI) boundaries (or blurring of the lowly methylated genomic stretches as defined by the CGIs), an event commonly observed in cancer (Timp and Feinberg, 2013). Additionally, many of the DMRs, both in the epithelial and the immune fractions, but especially in the former, were located in the *HLA* region on chromosome 6, those situated outside being predominantly related to the immune response (Table 1).

Importantly, we observed methylation changes in the *HLA* region that were associated with CeD in a cell population-specific manner and therefore

Table 1 DMRs identified in the intestinal epithelial fraction when compared to non-CeD controls lacking intestinal inflammation at the time of recruitment.

Gene	hg19 coordinates	# of probes	Mean *P*-value	Max. change (%)
TRIM15	chr6:30131458 – 30132471	19	3.23E – 25	26.13
TAP1	chr6:32819858 – 32820249	9	3.84E – 24	–36.45
HLA-B	chr6:31322766 – 31323506	8	2.75E – 23	–47.94
HLA-DPB1	chr6:33048086 – 33048706	16	1.35E – 23	36.17
NLRC5	chr16:57022486 – 57022486	1	7.15E – 23	–59.0
TMEM105	chr17:79304594 – 79304683	6	3.20E – 22	23.27

"# of probes" refers to the number of CpG positions taking part in each of the DMRs, while "max change (%)" refers to the maximum methylation percentage change between CeD patients and controls observed for each region.

were, at least partially, genotype-independent (Fernandez-Jimenez et al., 2019). This finding is controversial since it is widely accepted that the *HLA* region presents strong linkage disequilibrium and provides a major contribution to CeD susceptibility. Finally, we confirmed the differential methylation of two different gene promoters in the region (namely *TAP1* and *HLA-B*) in an independent set of biopsies from patients on GFD. Altogether, an at least partially irreversible, cell type-specific epigenetic signature was proposed for CeD. The celiac intestinal epithelium evoked the precancerous stages of solid tumors, with loss of CGI boundaries and higher epigenetic variability.

3.5 Other context-specific methylation studies

Later in 2019, a candidate gene study on the methylation patterns of the intestinal epithelial compartment in CeD was published (Cielo et al., 2019). In general, inverse modulation of gene expression and methylation was shown for *IL21* and *SH2B3*. Additionally, *cREL* and *TNFAIP3* were shown to be differentially methylated in the disease. The authors suggested that the genes linked to inflammatory processes could be generally up-regulated in CeD, while genes involved in the maintenance of intestinal barrier integrity could be down-regulated. More recently, Illumina InfiniumMethylation450 arrays have been used to analyze CeD methylation patterns in saliva, in a set of samples from patients that had been on GFD during at least 2 years and were asymptomatic at the time of recruitment (Hearn et al., 2020). Several DMRs were retested through bisulfite pyrosequencing in two additional independent cohorts of CeD and non-CeD individuals. Consistent with the previous studies, trends of differential methylation were shown for the *HLA* region, although in this case, differences could not be validated. On the other hand, differences were also detected near *SLC17A3,* but validation yielded contradictory results. Altogether, these investigations highlight the wide epigenetic variability across tissues and cell types while demonstrate the vast work yet to be done to decipher the celiac methylome.

3.6 Integrative OMIC analyses involving methylation

In recent years, more and more studies have been carried out to screen for genes related to complex diseases by combining methylation data with GWAS results, in order to identify methylation quantitative trait *loci* or *mQTLs*. In this regard, one of the strengths of our investigation published in 2019 was that we had both methylome and genome-wide genotype data from the same individuals (Fernandez-Jimenez et al., 2019). This enabled us to identify, despite the limited sample set, 1423 and 1457 *mQTLs* for the epithelial and the immune compartments, respectively. In order to ascertain the proportion of CeD-associated SNPs that participate to cell type-specific *mQTLs*, we searched for overlaps between the *mQTL*-SNPs and the top CeD-associated SNPs from the Immunochip Project (Fernandez-Jimenez et al., 2019). We found that 363 epithelial and 311 immune *mQTL*-SNPs overlapped with the 7144 most associated SNPs ($P < 1e-4$), resulting in very significant enrichments in both fractions. Remarkably, 178 *mQTL*-SNPs were common to both cell populations. As expected, most of them mapped to the *HLA* region; the only exception consisted of 13 SNPs that correlated with the methylation of a single CpG position (namely cg21209485) and were located in *MMEL1* on chromosome 1. These results strongly support the very plausible mediation of DNA methylation in the association of the region with CeD. Finally, it is important to note that none of the DMPs/DMRs identified in this study were among the *mQTL*-CpGs, suggesting again at least a partial genotype-independence of the proposed epigenetic signatures, even within the very controversial *HLA* region.

On the other hand, as previously stated, in yet a more recent study by our group we used MR to infer causal relationships between either gene expression or DNA methylation, and CeD as an outcome (Fernandez-Jimenez and Bilbao, 2019). Regarding the part of this work in which we aimed to ascertain which CpG positions mediated the SNP-CeD associations detected by GWAS, we identified four different CpGs that were mediating between rs3748816 and *MMEL1* on chromosome 1, and six CpGs as intermediaries of rs5754217 and *UBE2L3* on chromosome 22. In the case of *MMEL1*, this was the second time that completely independent (epi)genome-wide data pointed almost exclusively to the same region of the genome as a DNA methylation mediator in CeD genetic susceptibility.

Briefly, in this review we have listed four SNPs, rs2762051 (Hutchinson et al., 2014), rs840015 (Do et al., 2020), rs3748816 (Fernandez-Jimenez et al., 2019; Fernandez-Jimenez and Bilbao, 2019) and rs5754217 (Fernandez-Jimenez and Bilbao, 2019), that according to the literature are both associated with CeD and related to the methylation levels of adjacent CpG positions. In general, those variants have been proposed by different authors as regulatory SNPs that could relate to CeD through the modification of the surrounding methylation landscape. In a very recent publication by Airaksinen and collaborators comparing familial and sporadic CeD, the authors concluded that for example familial CeD usually surfaces in screenings of relatives of CeD patients and that affected individuals are more prone to carry *HLA-DQ* risk alleles (Airaksinen et al., 2020). The authors also discovered that four out of the 39 best-known CeD-related non-*HLA* SNPs are associated with familial CeD, including rs2762051 (located in the long non-coding RNA *DLEU1*) and rs3748816 in *MMEL1*. Surprisingly, those two SNPs are included in the aforementioned list. The fact that those SNPs seem both to be associated with CeD through methylation and to provide susceptibility to a familial form of the disease deserves further investigation and opens the door to predisposing, genetically-determined epigenetic marks.

3.7 Comparison with inflammatory bowel disease

Finally, the celiac methylation signature has been recently compared to the IBD methylome (Agliata et al., 2020). In the study by Agliata and collaborators, previously published DNA methylation data of the intestinal mucosa from IBD patients and controls was jointly analyzed and compared to the cell-type specific CeD methylome published in 2019 (Fernandez-Jimenez et al., 2019). Interestingly, 8 and 31 out of the 4205 IBD-DMPs resulting from the meta-analysis were among the 43 epithelial and the 310 immune DMPs described in CeD, respectively, pointing to a significant overlap between CeD and IBD DMPs, and suggesting a common methylation signature. All common DMPs were concordant, following the same trend (*i.e.*, hypo- or hypermethylation) in both disorders. Nevertheless, methylation changes (compared to controls) were more pronounced in CeD, probably due to the fact that CeD DMPs were identified in sorted cell populations, while methylation in IBD was assessed in whole intestinal tissue, potentially blurring cell type-specific signatures (Agliata et al., 2020).

On the other hand, it is important to note that 5/8 DMPs shared between CeD epithelial fraction and IBD were located on the *HLA* region. One of the main claims of the 2019 celiac methylome consisted on the discovery of a genotype-independent methylation signature that included *loci* in the *HLA* region. These patterns probably responded to environmental factors rather than genetic. It is known that the association of *HLA* with IBD is weaker than with CeD and besides, is explained by different *HLA* haplotypes (Goyette et al., 2015). In this context, the fact that the overlap between the two disease-specific methylation profiles occurs in the *HLA* supports the idea of the genotype-independence and the inflammatory environment as a not negligible driving force in the region.

In conclusion, the celiac methylation landscape arises from the equilibrium between both genetic predisposition and environmental factors and presents a dual signature, with a complex mixture of both genetically determined and environmentally acquired patterns.

4. Non-coding RNAs in CeD

Our current knowledge on the human genome indicates that most of it is transcribed not only to produce protein-coding mRNAs but also a large variety of ncRNAs that make up more than 90% of all transcripts produced; these include, among others, miRNAs and lncRNAs, that have regulatory roles and can modulate the expression of protein-coding genes.

4.1 Micro RNAs in CeD

MiRNAs are short (18–22 nucleotides long) ncRNAs that regulate gene expression at the post-transcriptional level and target mRNAs. These small ncRNAs are predicted to regulate the translation of up to 60% of protein-coding genes (Esteller, 2011) and they are able to regulate transcription factor (TF) activity, affecting several intracellular pathways.

MiRNAs have been shown to be involved in several diseases such as cancer, neurological disorders, cardiovascular disorders or inflammatory diseases (Esteller, 2011). Alterations in miRNA regulation have also been related to the development of immune dysfunctions and autoimmunity. Several studies have focused on the role of miRNAs in autoimmune diseases and different expression profiles have been identified as biomarkers of certain autoimmune conditions such as SLE, RA and Sjögren's syndrome (Chen et al., 2016).

In CeD, a number of studies have investigated miRNA expression at the site of CeD pathology, analyzing intestinal biopsies. One of the first studies was carried out by Capuano and collaborators and the authors evaluated miRNA expression in the small intestine of children affected by active CeD, children with CeD on GFD and non-celiac children as controls (Capuano et al., 2011). They observed that children with CeD have a different miRNA expression pattern as compared to controls. In particular, miR-449a showed very high expression in both active and GFD-treated celiac children (Fig. 4). They observed that *NOTCH1* and *KLF4*, which are involved in intestinal epithelial homeostasis and differentiation and maturation of goblet cells, are putative target genes of miR-449a. These genes were deregulated in small intestinal biopsies of CeD children, suggesting that

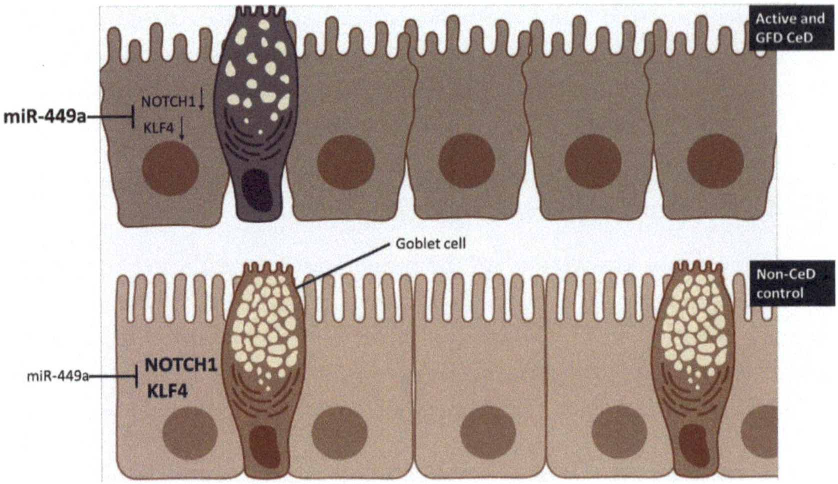

Fig. 4 Potential mode of action of a CeD-related miRNA. The elevated expression of miR-449a in celiac children (above) leads to the down-regulation of target genes *NOTCH1* and *KLF4*, altering epithelial proliferation and globet cell differentiation in the intestine, while normal levels of miR-449a maintain the expression of its targets and intestinal homeostasis.

miRNAs could drive the increased proliferation and the decreased differentiation of intestinal cells in CeD.

In another study, Vaira and collaborators analyzed miRNA expression in the duodenal mucosa of adult patients with active CeD, GFD-treated patients (with or without duodenal normalization), and controls (Vaira et al., 2014). Seven miRNAs (miR-31-5p, miR-192-3p, miR-194-5p, miR-551a, miR-551b-5p, miR-638 and miR-1290) were found to be downregulated in CeD patients with different clinical phenotypes when compared to controls. The expression of these miRNAs was studied in duodenal fibroblasts obtained from CeD patients and controls treated *in vitro* with gliadin peptides (13- and 33-mer). miR-192-3p (linked to extracellular matrix remodeling) was down-modulated by gliadin peptides in fibroblasts derived from CeD patients, whereas its expression remained unchanged in controls. The authors concluded that the analysis of miRNAs could contribute to the treatment and management of CeD.

In a study carried out by Magni and collaborators, it was observed that deregulation of miRNAs in the duodenal mucosa of CeD patients could alter the expression of molecules involved in the immune response (Magni et al., 2014). In particular, seven miRNAs (miR-638, miR-192-5p, miR-483-3p, miR-31-5p, miR-517c, miR-338-3p, and miR-197) were found to be down-regulated in the duodenum of adult CeD patients when compared to controls. *CXCL2* and *NOD2*, possible targets of miR-192-5p involved in the innate and adaptive immunity, were up-regulated in patients with intestinal atrophy and showed an inverse correlation with the decreased expression of the regulatory miR-192-5p. *FOXP3*, *RUNX1*, and *interleukin-18* (targets of miR-31-5p, miR-338-3p, and miR-197, respectively) were also up-regulated in CeD patients. In addition, gliadin was able to trigger the expression of *CXCL2* and *NOD2*, *FOXP3*, miR-192-5p, and miR-31-5p in CeD patients.

In a more recent work, autophagy-related genes and their regulatory miRNAs, particularly *BECN1* and miR-30a, and *ATG7* and miR-17 were investigated in blood and in intestinal biopsies of pediatric patients with active CeD and controls (Comincini et al., 2017). The expression of *BECN1* was altered in the blood of CeD patients and all of the investigated targets showed significant differences between CeD and control subjects in intestinal biopsies. They showed that the autophagic status of the cell was altered when modulating miR-17 levels.

Buoli Comani and collaborators showed that the expression of some miRNAs and their target genes presented different expression patterns in the duodenal mucosa of pediatric and adult CeD patients (Buoli Comani et al., 2015). In particular, the authors observed a decreased miR-192-5p expression in pediatric CeD patients, but no alterations were observed in *NOD2* and *CXCL2*, targets that had been previously identified in adults (Magni et al., 2014). On the other hand, the expression of another regulatory miRNA, miR-486-5p, was up-regulated in pediatric CeD patients, but not in adults. Moreover, another miR-192-5p target, *MAD2L1* (related to cell cycle control), was significantly increased in pediatric CeD patients but not in adults, supporting the presence of different miRNA-regulatory patterns.

Another aspect that must be taken into consideration when addressing the potential alterations in gene expression is the need for non-invasive biomarkers that could replace endoscopic biopsies. In this sense, circulating miRNA expression has been explored as potential biomarkers of CeD.

Buoli Comani and collaborators claimed that the expression of circulating miRNAs could be used as a diagnostic tool, and they analyzed the expression of miRNAs in plasma (Buoli Comani et al., 2015). Results were consistent with what they had observed in duodenal biopsies; miR-192-5p and miR-31-5p were down-regulated in CeD patients compared to controls, while miR-21-5p and miR-21-3p increased their expression. miR-192-5p remained down-regulated in presence of GFD, whereas miR-

31-5p, 21-5p and 21-3p in GFD returned to controls levels (Vaira et al., 2014). Finally, the authors suggested that a wider panel of plasmatic miRNAs could reduce the need for invasive procedures.

Amr and collaborators investigated the possibility of using circulating miRNAs as non-invasive biomarkers in pediatric patients with CeD (Amr et al., 2019). They demonstrated an up-regulation of miR-21 and down-regulation of miR-31 expression in active CeD patients in comparison with the treated CeD patients and controls. Moreover, they found that there was a positive correlation between miR-21 expression and IgA autoantibodies against tissue transglutaminase (tTG), a major auto-antigen in CeD. The authors concluded that both miRNAs could be useful as potential biomarkers for pediatric CeD patients.

In a recent study, they assessed the expression of the inflammatory miRNAs miRNA-146a, miRNA-155, miRNA-21, and miRNA-125b in PBMCs, monocytes, intestinal mucosa, and plasma, in adult CeD patients (Bascuñán et al., 2020). They found an increase in the expression of miRNA-146a, miRNA-155, miRNA-21 in PBMCs, miR-155 and miRNA-146a in monocytes, and miR-155, miR-21, and miR-125b in plasma in both untreated and treated CeD patients when compared to controls. Gliadin stimulation of PBMCs and monocytes lead to an up-regulation of miR-146a and miR-155 in both CeD groups. In intestinal biopsies, expression of those miRNAs did not change. The authors suggest that miRNA-146a and miRNA-155 should be evaluated as potential biomarkers for CeD due to their specificity for the presence of CeD.

In conclusion, according to the different studies analyzing miRNA expression in CeD, both in intestinal biopsies and circulating miRNAs, there is no doubt that miRNA expression is altered in the disease and that these molecules could be involved in multiple pathways participating in the pathogenesis. These studies do not only show the involvement of miRNAs in CeD, but they also open a window for new possible biomarkers for diagnosis. Expression analysis of circulating miRNAs could be used as a non-invasive biomarker tool for the diagnosis of pediatric CeD patients. To our opinion, further studies are required to improve the diagnosis through miRNA expression.

4.2 Long non-coding RNAs in CeD

Long non-coding RNAs are the largest class of non-coding RNA molecules, with more than 16,000 lncRNA genes annotated in the GRCh38.p13 assembly. These transcripts are >200 nucleotides long and resemble protein-coding genes in that they display an exon-intron-exon structure, but do not harbor open-reading frames and are not translated. As in the case of protein coding genes, lncRNA are expressed in a cell- and tissue-specific manner and respond to environmental stimuli. Although the particular functions and mechanisms of action of the majority of lncRNAs remains to be elucidated, the paradigm in force holds that they play essential roles in the regulation of the different levels of the process of gene expression, including transcription activation and silencing through effects on chromatin structure or allocation of RNA synthesis machinery; mRNA maturation, transport and degradation, as well as protein synthesis (Wang and Chang, 2011).

On the other hand, the large number of GWAS performed in different complex diseases have made it clear that not more than 10% of disease-associated SNPs map to the open reading frames of protein coding genes, so it can be assumed that the vast majority of the associated variants modulate the genetic risk to a particular disease through allele-specific regulatory effects. In fact, many disease-associated SNPs are located in, or nearby, lncRNA genes, suggesting that lncRNAs might be necessary for the correct understanding of GWAS results. It is plausible to speculate that certain SNPs in regulatory regions can affect lncRNA expression levels, or that while those that modify the lncRNA sequence itself can modify the structure and function of lncRNAs, and finally affect their regulatory capacity. Several of those lncRNA SNPs have been associated with different immune-mediated diseases, including CeD, but the mechanism by which they participate in the pathogenesis of disease is in most of the cases far from being clarified (Hrdlickova et al., 2014).

A systematic analysis of the association of SNPs with expression levels of lncRNAs in different human tissues discovered a strong, tissue-dependent correlation between SNP genotypes and the expression of nearby lncRNAs (*cis-eQTL*) that was generally specific to the non-coding gene and did not affect neighboring protein-coding genes. However, considering that lncRNAs can regulate the expression of protein-coding genes located elsewhere in the genome, it is possible that these SNPs also influence the function of those protein-coding genes through the action of the lncRNA. Additionally, they observed that many of the lncRNA *cis-eQTLs* are disease-associated SNPs, suggesting that disease-associated, non-coding SNPs that are *eQTLs* of lncRNAs in turn affect protein-coding genes, and could explain part of the GWAS associations with the human diseases (Kumar et al., 2013). In recent years, several lncRNAs have been shown to be involved in CeD pathogenesis, in some cases through the effect of non-coding GWAS SNPs.

In a gene expression study of the CeD-associated genomic regions in duodenal tissue, our group identified a subset of genes whose expression seemed to be regulated by three SNPs, including rs1018326 on chromosome 2q31.3, located between *UBE2E3* and *ITGA4*, on top of lncRNA AC104820.2, a gene that is strongly expressed in CD8+ T-cells and is thought to influence T-cell proliferation (Plaza-Izurieta et al., 2015). The expression of AC104820.2 was up-regulated in biopsy samples from active CeD patients compared to biopsies taken from the same patients after more than 2 years on GFD. Another study confirmed that the genotype of SNP rs1018326 affects the expression of *AC104820.2* (Ricaño-Ponce et al., 2016). In the same work, the authors showed that the expression of lncRNA *AP002954.4* on chromosome 11q23.3 was significantly affected by the genotype of CeD-associated SNP rs10892258. This non-coding RNA is expressed specifically in monocytes and participates in the regulation of cytokine responses and defense against fungal infection.

The first experimental demonstration of the role of a GWAS-SNP on a lncRNA in the pathogenesis of a complex human disease was the discovery of a novel lncRNA gene, named *lnc13* and the demonstration of the effect of rs917997 genotype in the disease process (Castellanos-Rubio et al., 2016). The expression of *lnc13* was reduced in the duodenum of CeD patients, and it was observed that *in vitro* reduction of *lnc13* resulted in the up-regulation of the NFkB pathway, a situation that is found constitutively in the celiac gut (Fig. 5). The authors also showed that *lnc13* interacts with chromatin and the multifunctional protein hnRNPD and is necessary to form a complex involved in the repression of several pro-inflammatory genes that participate in CeD. Moreover, they demonstrated that the binding occurs in an allele-specific manner, with the CeD-associated rs917997 risk allele bound to hnRNPD less efficiently, resulting in the up-regulation of downstream pro-inflammatory genes.

As mentioned above, the expression of certain lncRNAs can be modified by external signals. The expression of *Carlr* increases upon NFkB signaling in macrophages, and is necessary for the expression of downstream NFkB pathway genes and for the interaction between macrophages and intestinal cells in an inflammatory environment (Castellanos-Rubio et al., 2017). *Carlr* was found to be augmented in the cytoplasm of CeD intestinal samples together with the overexpression of NFkB pathway genes.

Apart from the implication of lncRNAs in the regulation of the immune response, there are two works that focus on the participation of the epithelial compartment of the gut in the development of CeD and the implication of

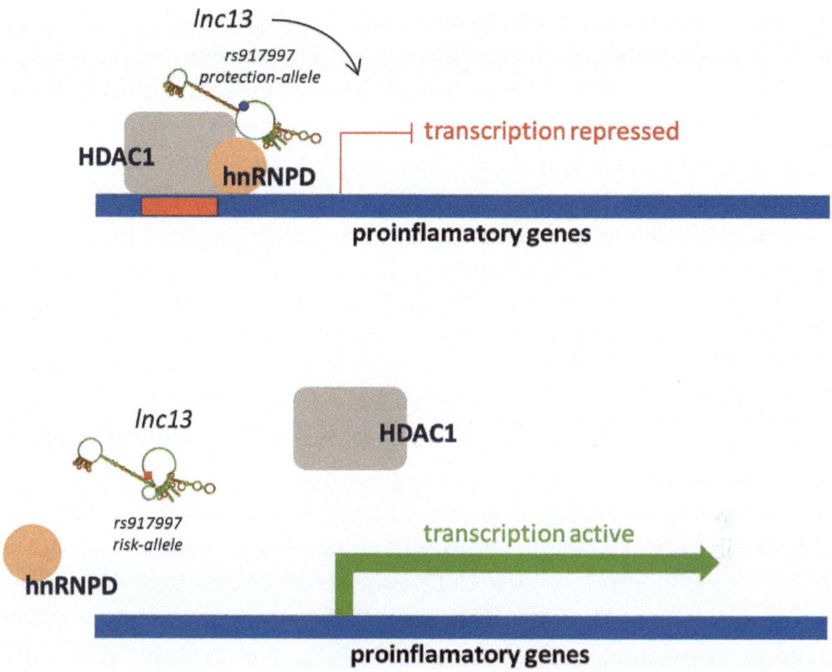

Fig. 5 The first experimental demonstration of the implication of a CeD-associated variant (rs917997) in the pathogenesis of the disease through lnc13, a regulatory lncRNA. This lncRNA is down-regulated in the celiac duodenum and interacts with hnRNPD, necessary to form a complex that includes HDAC1 and that when assembled to chromatin, is able to repress the expression of several pro-inflammatory genes. Moreover, the risk allele of rs917997 binds to hnRNPD less efficiently, resulting in the overexpression of the target genes.

epithelial lncRNAs. In a fine mapping SNP genotyping of the *HLA* region comparing *HLA-DR3-DQ2* CeD and non-celiac individuals, Santin and collaborators identified rs3135316, a SNP that confers susceptibility to CeD independently from the *HLA-DQ* genotype (Santin et al., 2018). It is located in the intron *HCG14* (*HLA* complex group 14), a lncRNA whose expression is down-regulated in epithelial cells from duodenal biopsies of CeD patients. The authors also showed that the expression of *NOD1*, a candidate gene for several inflammatory diseases, is modulated by *HCG14* in intestinal cells. A study by Jauregi-Miguel and collaborators was able to replicate the association of rs6962966 with CeD, a SNP located in an intron of *MAGI2*, a gene that is part of the Tight-Junction pathway and is involved in maintaining the integrity of the intestinal barrier (Jauregi-Miguel et al., 2019). In fact, a defect in intestinal permeability is characteristic of the celiac gut. The *MAGI* intron harbors a non-coding transcript named *RP4-587D13.2*, a lncRNA of unknown function. The authors showed that both genes are co-regulated and constitutively down-regulated in the CeD intestine, and conclude that *RP4-587D13.2* regulates the expression of *MAGI2* in a genotype-dependent manner. They suggest that the constitutive down-regulation of both transcripts in CeD patients could affect the Tight-Junction pathway and influence the response to the ingested gluten.

5. Conclusions

Overall, it is clear that a large amount of knowledge on Genomics of CeD has accumulated in the last decade. However, there are still completely unexplored post-Genomics domains in the disease, including uncovered Epigenetics layers such as chromatin conformation and accessibility. Additionally, further investigations will be needed in order to explore the very diverse contexts in which CeD-associated variants exert their regulatory effects, if any. Alternatively, unrevealed (epi)genetic events, probably linked to non-causative CeD-associated SNPs, will probably arise in the near future. Finally, Epitranscriptomics is a promising domain that although yet not completely understood, is being investigated by groups like ours. In fact, unpublished data suggest that different SNP-CeD associations could be mediated by m6A methylation of RNA, leading to the regulation of proteins that are relevant to the disease pathogenesis. These and other mysteries need to be solved in the following years in order to have the global picture of the celiac Genome.

References

Abraham, G., Tye-Din, J.A., Bhalala, O.G., Kowalczyk, A., Zobel, J., Inouye, M., 2014. Accurate and robust genomic prediction of celiac disease using statistical learning. PLoS Genet. 10, 1004137.

Agardh, E., Lundstig, A., Perfilyev, A., Volkov, P., Freiburghaus, T., Lindholm, E., Rönn, T., Agardh, C.D., Ling, C., 2015. Genome-wide analysis of DNA methylation in subjects with type 1 diabetes identifies epigenetic modifications associated with proliferative diabetic retinopathy. BMC Med. 13, 182. BioMed Central Ltd.

Agliata, I., Fernandez-Jimenez, N., Goldsmith, C., Marie, J.C., Bilbao, J.R., Dante, R., Hernandez-Vargas, H., 2020. The DNA methylome of inflammatory bowel disease (IBD) reflects intrinsic and extrinsic factors in intestinal mucosal cells. Epigenetics 15, 1068–1082. Taylor and Francis Inc.

Airaksinen, L., Myllymäki, L., Kaukinen, K., Saavalainen, P., Huhtala, H., Lindfors, K., Kurppa, K., 2020. Differences between familial and sporadic celiac disease. Dig. Dis. Sci. 1, 3. Springer.

Akulenko, R., Helms, V., 2013. DNA co-methylation analysis suggests novel functional associations between gene pairs in breast cancer samples. Hum. Mol. Genet. 22, 3016–3022. Oxford Academic.

Amr, K.S., Bayoumi, F.S., Eissa, E., Abu-Zekry, M., 2019. Circulating microRNAs as potential non-invasive biomarkers in pediatric patients with celiac disease. Eur. Ann. Allergy Clin. Immunol. 51, 159–164.

Amundsen, S.S., Rundberg, J., Adamovic, S., Gudjónsdóttir, A.H., Ascher, H., Ek, J., Nilsson, S., Lie, B.A., Naluai, Å., Sollid, L.M., 2010. Four novel coeliac disease regions replicated in an association study of a Swedish-Norwegian family cohort. Genes Immun. 11, 79–86.

Amundsen, S.S., Viken, M.K., Sollid, L.M., Lie, B.A., 2014. Coeliac disease-associated polymorphisms influence thymic gene expression. Genes Immun. 15, 355–360.

Banaganapalli, B., Rashidi, O., Saadah, O.I., Wang, J., Khan, I.A., Al-Aama, J.Y., Shaik, N.A., Elango, R., 2017. Comprehensive computational analysis of GWAS loci identifies CCR2 as a candidate gene for celiac disease pathogenesis. J. Cell. Biochem. 118, 2193–2207. Wiley-Liss Inc.

Barrett, J.C., Hansoul, S., Nicolae, D.L., Cho, J.H., Duerr, R.H., Rioux, J.D., Brant, S.R., Silverberg, M.S., Taylor, K.D., Barmada, M.M., Bitton, A., Dassopoulos, T., Datta, L.W., Green, T., Griffiths, A.M., Kistner, E.O., Murtha, M.T., Regueiro, M.D., Rotter, J.I., Schumm, L.P., Steinhart, A.H., Targan, S.R., Xavier, R.J., Libioulle, C., Sandor, C., Lathrop, M., Belaiche, J., Dewit, O., Gut, I., Heath, S., Laukens, D., Mni, M., Rutgeerts, P., Van Gossum, A., Zelenika, D., Franchimont, D., Hugot, J.P., De Vos, M., Vermeire, S., Louis, E., Cardon, L.R., Anderson, C.A., Drummond, H., Nimmo, E., Ahmad, T., Prescott, N.J., Onnie, C. M., Fisher, S.A., Marchini, J., Ghori, J., Bumpstead, S., Gwilliam, R., Tremelling, M., Deloukas, P., Mansfield, J., Jewell, D., Satsangi, J., Mathew, C.G., Parkes, M., Georges, M., Daly, M.J., 2008. Genome-wide association defines more than 30 distinct susceptibility loci for Crohn's disease. Nat. Genet. 40, 955–962.

Bascuñán, K.A., Pérez-Bravo, F., Gaudioso, G., Vaira, V., Roncoroni, L., Elli, L., Monguzzi, E., Araya, M., 2020. A miRNA-based blood and mucosal approach for detecting and monitoring celiac disease. Dig. Dis. Sci. 65, 1982–1991. Springer.

Bergmann, F., Singh, S., Michel, S., Kahlert, C., Schirmacher, P., Helmke, B., von Knebel Doeberitz, M., Kloor, M., Bläker, H., 2010. Small bowel adenocarcinomas in celiac disease follow the CIM-MSI pathway. Oncol. Rep. 24, 1535–1539. Spandidos Publications.

Bodkhe, R., Shetty, S.A., Dhotre, D.P., Verma, A.K., Bhatia, K., Mishra, A., Kaur, G., Pande, P., Bangarusamy, D.K., Santosh, B.P., Perumal, R.C., Ahuja, V., Shouche, Y.S., Makharia, G.K., 2019. Comparison of small gut and whole gut microbiota of first-degree relatives with adult celiac disease patients and controls. Front. Microbiol. 10, 137–140. Frontiers Media S.A.

Bondar, C., Plaza-Izurieta, L., Fernandez-Jimenez, N., Irastorza, I., Withoff, S., Wijmenga, C., Chirdo, F., Bilbao, J.R., 2014. THEMIS and PTPRK in celiac intestinal mucosa: coexpression in disease and after in vitro gliadin challenge. Eur. J. Hum. Genet. 22, 358–362.

Bonder, M.J., Kurilshikov, A., Tigchelaar, E.F., Mujagic, Z., Imhann, F., Vila, A.V., Deelen, P., Vatanen, T., Schirmer, M., Smeekens, S.P., Zhernakova, D.V., Jankipersadsing, S.A., Jaeger, M., Oosting, M., Cenit, M.C., Masclee, A.A.M., Swertz, M.A., Li, Y., Kumar, V., Joosten, L., Harmsen, H., Weersma, R.K., Franke, L., Hofker, M.H., Xavier, R.J., Jonkers, D., Netea, M.G., Wijmenga, C., Fu, J., Zhernakova, A., 2016. The effect of host genetics on the gut microbiome. Nat. Genet. 48, 1407–1412.

Buoli Comani, G., Panceri, R., Dinelli, M., Biondi, A., Mancuso, C., Meneveri, R., Barisani, D., 2015. miRNA-regulated gene expression differs in celiac disease patients according to the age of presentation. Genes Nutr. 10, 1–12. Springer Verlag.

Burton, P.R., Clayton, D.G., Cardon, L.R., Craddock, N., Deloukas, P., Duncanson, A., Kwiatkowski, D.P., McCarthy, M.I., Ouwehand, W.H., Samani, N.J., Todd, J.A., Donnelly, P., Barrett, J.C., Davison, D., Easton, D., Evans, D., Leung, H.T., Marchini, J.L., Morris, A.P., Spencer, C.C.A., Tobin, M.D., Attwood, A.P., Boorman, J.P., Cant, B., Everson, U., Hussey, J.M., Jolley, J.D., Knight, A.S., Koch, K., Meech, E., Nutland, S., Prowse, C.V., Stevens, H.E., Taylor, N.C., Walters, G.R., Walker, N.M., Watkins, N.A., Winzer, T., Jones, R.W., McArdle, W.L., Ring, S.M., Strachan, D.P., Pembrey, M., Breen, G., St. Clair, D., Caesar, S., Gordon-Smith, K., Jones, L., Fraser, C., Green, E.K., Grozeva, D., Hamshere, M.L., Holmans, P.A., Jones, I.R., Kirov, G., Moskvina, V., Nikolov, I., O'Donovan, M.C., Owen, M.J., Collier, D.A., Elkin, A., Farmer, A., Williamson, R., McGuffin, P., Young, A.H., Ferrier, I.N., Ball, S.G., Balmforth, A.J., Barrett, J.H., Bishop, D.T., Iles, M.M., Maqbool, A., Yuldasheva, N., Hall, A.S., Braund, P.S., Dixon, R.J., Mangino, M., Stevens, S., Thompson, J.R., Bredin, F., Tremelling, M., Parkes, M., Drummond, H., Lees, C.W., Nimmo, E.R., Satsangi, J., Fisher, S.A., Forbes, A., Lewis, C.M., Onnie, C.M., Prescott, N.J., Sanderson, J., Mathew, C.G., Barbour, J., Mohiuddin, M.K., Todhunter, C.E., Mansfield, J.C., Ahmad, T., et al., 2007. Genome-wide association study of 14,000 cases of seven common diseases and 3,000 shared controls. Nature 447, 661–678. Nature Publishing Group.

Capuano, M., Iaffaldano, L., Tinto, N., Montanaro, D., Capobianco, V., Izzo, V., Tucci, F., Troncone, G., Greco, L., Sacchetti, L., 2011. MicroRNA-449a overexpression, reduced NOTCH1 signals and scarce goblet cells characterize the small intestine of celiac patients. PLoS One 6, e29094. Public Library of Science.

Castellanos-Rubio, A., Fernandez-Jimenez, N., Kratchmarov, R., Luo, X., Bhagat, G., Green, P.H.R., Schneider, R., Kiledjian, M., Bilbao, J.R., Ghosh, S., 2016. A long non-coding RNA associated with susceptibility to celiac disease. Science 352, 91–95. American Association for the Advancement of Science.

Castellanos-Rubio, A., Kratchmarov, R., Sebastian, M., Garcia-Etxebarria, K., Garcia, L., Irastorza, I., Ghosh, S., 2017. Cytoplasmic form of carlr lncRNA facilitates inflammatory gene expression upon NF-κB activation. J. Immunol. 199, 581–588. The American Association of Immunologists.

Chen, J.Q., Papp, G., Szodoray, P., Zeher, M., 2016. The role of microRNAs in the pathogenesis of autoimmune diseases. Elsevier B.V, Autoimmun. Rev. 15, 1171–1180.

Cielo, D., Galatola, M., Fernandez-Jimenez, N., De Leo, L., Garcia-Etxebarria, K., Loganes, C., Tommasini, A., Not, T., Auricchio, R., Greco, L., Bilbao, J.R., 2019. Combined analysis of methylation and gene expression profiles in separate compartments of small bowel mucosa identified celiac disease patients' signatures. Sci. Rep. 9, 10020.

Coenen, M.J.H., Trynka, G., Heskamp, S., Franke, B., van Diemen, C.C., Smolonska, J., van Leeuwen, M., Brouwer, E., Boezen, M.H., Postma, D.S., Platteel, M., Zanen, P., Lammers, J.-W.W.J., Groen, H.J.M., Mali, W.P.T.M., Mulder, C.J., Tack, G.J., Verbeek, W.H.M., Wolters, V.M., Houwen, R.H.J., Mearin, M.L., van Heel, D.A., Radstake, T.R.D.J., van Riel, P.L.C.M., Wijmenga, C., Barrera, P., Zhernakova, A., 2009. Common and different genetic background for rheumatoid arthritis and coeliac disease. Hum. Mol. Genet. 18, 4195–4203.

Coleman, C., Quinn, E.M., Ryan, A.W., Conroy, J., Trimble, V., Mahmud, N., Kennedy, N., Corvin, A.P., Morris, D.W., Donohoe, G., O'Morain, C., MacMathuna, P., Byrnes, V., Kiat, C., Trynka, G., Wijmenga, C., Kelleher, D., Ennis, S., Anney, R.J., McManus, R., 2016. Common polygenic variation in coeliac disease and confirmation of ZNF335 and NIFA as disease susceptibility loci. Eur. J. Hum. Genet. 24, 291–297.

Comincini, S., Manai, F., Meazza, C., Pagani, S., Martinelli, C., Pasqua, N., Pelizzo, G., Biggiogera, M., Bozzola, M., 2017. Identification of autophagy-related genes and their regulatory miRNAs associated with celiac disease in children. Int. J. Mol. Sci. 18, 391. MDPI AG.

Cortes, A., Brown, M.A., 2011. Promise and pitfalls of the immunochip. BioMed Central, Arthritis Res. Ther. 13 (1), 101.

Dahal-Koirala, S., Risnes, L.F., Christophersen, A., Sarna, V.K., Lundin, K.E., Sollid, L.M., Qiao, S.W., 2016. TCR sequencing of single cells reactive to DQ2.5-glia-α2 and DQ2.5-glia-ω2 reveals clonal expansion and epitope-specific V-gene usage. Mucosal Immunol. 9, 587–596. Nature Publishing Group.

D'Argenio, V., Casaburi, G., Precone, V., Pagliuca, C., Colicchio, R., Sarnataro, D., Discepolo, V., Kim, S.M., Russo, I., Del Vecchio Blanco, G., Horner, D.S., Chiara, M., Pesole, G., Salvatore, P., Monteleone, G., Ciacci, C., Caporaso, G.J., Jabrì, B., Salvatore, F., Sacchetti, L., 2016. Metagenomics reveals dysbiosis and a potentially pathogenic *N. flavescens* strain in duodenum of adult celiac patients. Am. J. Gastroenterol. 111, 879–890.

De Palma, G., Capilla, A., Nadal, I., Nova, E., Pozo, T., Varea, V., Polanco, I., Castillejo, G., López, A., Garrote, J.A., Calvo, C., García-Novo, M.D., Cilleruelo, M.L., Ribes-Koninckx, C., Palau, F., Sanz, Y., 2010a. Interplay between human leukocyte antigen genes and the microbial colonization process of the newborn intestine. Curr. Issues Mol. Biol. 12, 1–10.

De Palma, G., Nadal, I., Medina, M., Donat, E., Ribes-Koninckx, C., Calabuig, M., Sanz, Y., 2010b. Intestinal dysbiosis and reduced immunoglobulin-coated bacteria associated with coeliac disease in children. BMC Microbiol. 10, 63.

Dema, B., Martínez, A., Fernández-Arquero, M., Maluenda, C., Polanco, I., De La Concha, E.G., Urcelay, E., Núñez, C., 2009. Association of IL18RAP and CCR3 with coeliac disease in the Spanish population. J. Med. Genet. 46, 617–619. BMJ Publishing Group Ltd.

Diosdado, B., Buffart, T.E., Watkins, R., Carvalho, B., Ylstra, B., Tijssen, M., Bolijn, A.S., Lewis, F., Maude, K., Verbeke, C., Nagtegaal, I.D., Grabsch, H., Mulder, C.J.J., Quirke, P., Howdle, P., Meijer, G.A., 2010. High-resolution array comparative genomic hybridization in sporadic and celiac disease-related small bowel adenocarcinomas. Clin. Cancer Res. 16, 1391–1401. American Association for Cancer Research.

Do, C., Do, C., Dumont, E.L.P., Dumont, E.L.P., Salas, M., Salas, M., Castano, A., Castano, A., Mujahed, H., Maldonado, L., Singh, A., Dasilva-Arnold, S.C., Bhagat, G., Bhagat, G., Lehman, S., Christiano, A.M., Madhavan, S., Nagy, P.L., Green, P.H.R., Feinman, R., Feinman, R., Feinman, R., Trimble, C., Illsley, N.P., Marder, K., Marder, K., Honig, L., Honig, L., Monk, C., Goy, A., Goy, A., Goy, A., Chow, K., Chow, K., Chow, K., Goldlust, S., Goldlust, S., Kaptain, G., Kaptain, G., Siegel, D., Siegel, D., Siegel, D., Tycko, B., Tycko, B., Tycko, B., 2020. Allele-specific DNA methylation is increased in cancers and its dense mapping in normal plus neoplastic cells increases the yield of disease-associated regulatory SNPs. Genome Biol. 21, 153. BioMed Central Ltd.

Dubois, P.C.A., Trynka, G., Franke, L., Hunt, K.A., Romanos, J., Curtotti, A., Zhernakova, A., Heap, G.A.R., Ádány, R., Aromaa, A., Bardella, M.T., Van Den Berg, L.H., Bockett, N.A., De La Concha, E.G., Dema, B., Fehrmann, R.S.N., Fernández-Arquero, M., Fiatal, S., Grandone, E., Green, P.M., Groen, H.J.M., Gwilliam, R., Houwen, R.H.J., Hunt, S.E., Kaukinen, K., Kelleher, D., Korponay-Szabo, I., Kurppa, K., MacMathuna, P., Mäki, M., Mazzilli, M.C., McCann, O.T., Mearin, M.L., Mein, C.A., Mirza, M.M., Mistry, V., Mora, B., Morley, K.I., Mulder, C.J., Murray, J.A., Ñez, C., Oosterom, E., Ophoff, R.A., Polanco, I., Peltonen, L., Platteel, M., Rybak, A., Salomaa, V., Schweizer, J.J., Sperandeo, M.P.,

Tack, G.J., Turner, G., Veldink, J.H., Verbeek, W.H.M., Weersma, R.K., Wolters, V.M., Urcelay, E., Cukrowska, B., Greco, L., Neuhausen, S.L., McManus, R., Barisani, D., Deloukas, P., Barrett, J.C., Saavalainen, P., Wijmenga, C., Van Heel, D.A., 2010. Multiple common variants for celiac disease influencing immune gene expression. Nat. Genet. 42, 295–302.

Esteller, M., 2011. Non-coding RNAs in human disease. Nature Publishing Group, Nat. Rev. Genet. 12 (12), 861–874. <www.nature.com/reviews/genetics>. 2 September 2020.

Eyre, S., Bowes, J., Diogo, D., Lee, A., Barton, A., Martin, P., Zhernakova, A., Stahl, E., Viatte, S., McAllister, K., Amos, C.I., Padyukov, L., Toes, R.E.M., Huizinga, T.W.J., Wijmenga, C., Trynka, G., Franke, L., Westra, H.J., Alfredsson, L., Hu, X., Sandor, C., De Bakker, P.I.W., Davila, S., Khor, C.C., Heng, K.K., Andrews, R., Edkins, S., Hunt, S.E., Langford, C., Symmons, D., Concannon, P., Onengut-Gumuscu, S., Rich, S.S., Deloukas, P., Gonzalez-Gay, M.A., Rodriguez-Rodriguez, L., Ärlsetig, L., Martin, J., Rantapää-Dahlqvist, S., Plenge, R.M., Raychaudhuri, S., Klareskog, L., Gregersen, P.K., Worthington, J., Moitra, R.K., Prouse, P.J., Shawe, D.J., Crisp, A.J., Gaston, J.S.H., Hall, F.C., Hazleman, B.L., Jenner, J.R., Lillicrap, M.S., Ostor, A., Silverman, B., Speed, C., Bruce, I.N., Hyrich, K., Ho, P., Gorodkin, R., Armstrong, D., Chuck, A.J., Hailwood, S., Kumar, N., Badcock, L.J., Deighton, C.M., O'Reilly, S.C., Raj, N., Regan, M.R., Summers, G.D., Williams, R.A., Lambert, J.R., Stevens, R., Wilkinson, C., Hamilton, J., Heycock, C.R., Kelly, C.A., Saravanan, V., Rees, D.H., Williams, R.B., Bingham, S., Emery, P., Morgan, A., Bird, H.A., Conaghan, P.G., Pease, C.T., Wakefield, R.J., Chalam, S.V., Mulherin, D., Price, T., Sheeran, T., Venkatachalam, S., Gaffney, K., Macgregor, A.J., Marshall, T., Merry, P., Scott, D.G.I., Birrell, F.N., et al., 2012. High-density genetic mapping identifies new susceptibility loci for rheumatoid arthritis. Nat. Genet. 44, 1336–1340.

Fairfax, B.P., Humburg, P., Makino, S., Naranbhai, V., Wong, D., Lau, E., Jostins, L., Plant, K., Andrews, R., McGee, C., Knight, J.C., 2014. Innate immune activity conditions the effect of regulatory variants upon monocyte gene expression. Science 343 (6175), 1246949. American Association for the Advancement of Science.

Falchuk, Z.M., Rogentine, G.N., Strober, W., 1972. Predominance of histocompatibility antigen HL-A8 in patients with gluten-sensitive enteropathy. J. Clin. Investig. 51, 1602–1605. American Society for Clinical Investigation.

Fernandez-Jimenez, N., Bilbao, J.R., 2019. Mendelian randomization analysis of celiac GWAS reveals a blood expression signature with diagnostic potential in absence of gluten consumption. Hum. Mol. Genet. 28, 3037–3042.

Fernandez-jimenez, N., Castellanos-rubio, A., Plaza-izurieta, L., Irastorza, I., Elcoroaristizabal, X., Jauregi-miguel, A., Lopez-euba, T., Tutau, C., De pancorbo, M.M., Vitoria, J.C., Bilbao, J.R., 2014. Coregulation and modulation of NFκB-related genes in celiac disease: uncovered aspects of gut mucosal inflammation. Hum. Mol. Genet. 23, 1298–1310. Oxford Academic.

Fernandez-Jimenez, N., Garcia-Etxebarria, K., Plaza-Izurieta, L., Romero-Garmendia, I., Jauregi-Miguel, A., Legarda, M., Ecsedi, S., Castellanos-Rubio, A., Cahais, V., Cuenin, C., Degli Esposti, D., Irastorza, I., Hernandez-Vargas, H., Herceg, Z., Bilbao, J.R., 2019. The methylome of the celiac intestinal epithelium harbours genotype-independent alterations in the HLA region. Sci. Rep. 9, 1298. Nature Publishing Group.

Festen, E.A.M., Goyette, P., Green, T., Boucher, G., Beauchamp, C., Trynka, G., Dubois, P.C., Lagacé, C., Stokkers, P.C.F., Hommes, D.W., Barisani, D., Palmieri, O., Annese, V., van Heel, D.A., Weersma, R.K., Daly, M.J., Wijmenga, C., Rioux, J.D., 2011. A meta-analysis of genome-wide association scans identifies IL18RAP, PTPN2, TAGAP, and PUS10 as shared risk loci for crohn's disease and celiac disease. PLoS Genet. 7 (1), e1001283.

Fu, J., Wolfs, M.G.M., Deelen, P., Westra, H.J., Fehrmann, R.S.N., te Meerman, G.J., Buurman, W.A., Rensen, S.S.M., Groen, H.J.M., Weersma, R.K., van den Berg, L.H., Veldink, J., Ophoff, R.A., Snieder, H., van Heel, D., Jansen, R.C., Hofker, M.H., Wijmenga, C., Franke, L., 2012. Unraveling the regulatory mechanisms underlying tissue-dependent genetic variation of gene expression. PLoS Genet. 8, 1002431.

Galipeau, H.J., McCarville, J.L., Huebener, S., Litwin, O., Meisel, M., Jabri, B., Sanz, Y., Murray, J.A., Jordana, M., Alaedini, A., Chirdo, F.G., Verdu, E.F., 2015. Intestinal microbiota modulates gluten-induced immunopathology in humanized mice. Am. J. Pathol. 185, 2969–2982.

Garcia-Etxebarria, K., Jauregi-Miguel, A., Romero-Garmendia, I., Plaza-Izurieta, L., Legarda, M., Irastorza, I., Bilbao, J.R., 2016. Ancestry-based stratified analysis of Immunochip data identifies novel associations with celiac disease. Eur. J. Hum. Genet. 24, 1831–1834.

García-Santisteban, I., Cilleros-Portet, A., Moyua-Ormazabal, E., Kurilshikov, A., Zhernakova, A., Garcia-Etxebarria, K., Fernandez-Jimenez, N., Bilbao, J.R., 2020. A two-sample mendelian randomization analysis investigates associations between gut microbiota and celiac disease. Nutrients 12, 1420. Multidisciplinary Digital Publishing Institute.

Garner, C.P., Murray, J.A., Ding, Y.C., Tien, Z., van Heel, D.A., Neuhausen, S.L., 2009. Replication of celiac disease UK genome-wide association study results in a US population. Hum. Mol. Genet. 18, 4219–4225.

Garner, C., Ahn, R., Ding, Y.C., Steele, L., Stoven, S., Green, P.H., Fasano, A., Murray, J.A., Neuhausen, S.L., 2014. Genome-wide association study of celiac disease in North America confirms FRMD4B as new celiac locus. PLoS One 9.

Goyette, P., Boucher, G., Mallon, D., Ellinghaus, E., Jostins, L., Huang, H., Ripke, S., Gusareva, E.S., Annese, V., Hauser, S.L., Oksenberg, J.R., Thomsen, I., Leslie, S., Daly, M.J., Van Steen, K., Duerr, R.H., Barrett, J.C., McGovern, D.P.B., Schumm, L.P., Traherne, J.A., Carrington, M.N., Kosmoliaptsis, V., Karlsen, T.H., Franke, A., Rioux, J.D., Abraham, C., Achkar, J.P., Ahmad, T., Amininejad, L., Ananthakrishnan, A.N., Andersen, V., Anderson, C.A., Andrews, J.M., Aumais, G., Baidoo, L., Baldassano, R.N., Balschun, T., Bampton, P.A., Barclay, M., Bayless, T.M., Bethge, J., Bis, J.C., Bitton, A., Brand, S., Brant, S.R., Buning, C., Chew, A., Cho, J.H., Cleynen, I., Cohain, A., Croft, A., D'Amato, M., Danese, S., De Jong, D., De Vos, M., Denapiene, G., Denson, L.A., Devaney, K.L., Dewit, O., D'Inca, R., Dubinsky, M., Edwards, C., Ellinghaus, D., Essers, J., Ferguson, L.R., Festen, E.A., Fleshner, P., Florin, T., Franchimont, D., Fransen, K., Gearry, R., Georges, M., Gieger, C., Glas, J., Green, T., Griffiths, A.M., Guthery, S.L., Hakonarson, H., Halfvarson, J., Hanigan, K., Haritunians, T., Hart, A., Hawkey, C., Hayward, N.K., Hedl, M., Henderson, P., Hu, X., Hui, K.Y., Imielinski, M., Ippoliti, A., Jonaitis, L., Kennedy, N.A., Khan, M.A., Kiudelis, G., Kugathasan, S., Kupcinskas, L., Latiano, A., Laukens, D., et al., 2015. High-density mapping of the MHC identifies a shared role for HLA-DRB1*01:03 in inflammatory bowel diseases and heterozygous advantage in ulcerative colitis. Nat. Genet. 47, 172–179. Nature Publishing Group.

Greco, L., Romino, R., Coto, I., Di Cosmo, N., Percopo, S., Maglio, M., Paparo, F., Gasperi, V., Limongelli, M.G., Cotichini, R., D'Agate, C., Tinto, N., Sacchetti, L., Tosi, R., Stazi, M.A., 2002. The first large population based twin study of coeliac disease. Gut 50, 624–628. BMJ Publishing Group.

Gutierrez-Achury, J., Zhernakova, A., Pulit, S.L., Trynka, G., Hunt, K.A., Romanos, J., Raychaudhuri, S., Van Heel, D.A., Wijmenga, C., De Bakker, P.I.W., 2015. Fine

mapping in the MHC region accounts for 18% additional genetic risk for celiac disease. Nat. Genet. 47, 577–578.
Gutierrez-Achury, J., Zorro, M.M., Ricaño-Ponce, I., Zhernakova, D.V., Diogo, D., Raychaudhuri, S., Franke, L., Trynka, G., Wijmenga, C., Zhernakova, A., 2016. Functional implications of disease-specific variants in loci jointly associated with coeliac disease and rheumatoid arthritis. Hum. Mol. Genet. 25, 180–190.
Hardy, M.Y., Tye-Din, J.A., 2016. Coeliac disease: a unique model for investigating broken tolerance in autoimmunity. Clin. Transl. Immunol. 5, e112.
Hearn, N.L., Chiu, C.L., Lind, J.M., 2020. Comparison of DNA methylation profiles from saliva in coeliac disease and non-coeliac disease individuals. BMC Med. Genomics 13, 16. BioMed Central Ltd.
Hrdlickova, B., Kumar, V., Kanduri, K., Zhernakova, D.V., Tripathi, S., Karjalainen, J., Lund, R.J., Li, Y., Ullah, U., Modderman, R., Abdulahad, W., Lähdesmäki, H., Franke, L., Lahesmaa, R., Wijmenga, C., Withoff, S., 2014. Expression profiles of long non-coding RNAs located in autoimmune disease-associated regions reveal immune cell-type specificity. Genome Med. 6, 88. BioMed Central Ltd.
Hunt, K.A., Zhernakova, A., Turner, G., Heap, G.A.R., Franke, L., Bruinenberg, M., Romanos, J., Dinesen, L.C., Ryan, A.W., Panesar, D., Gwilliam, R., Takeuchi, F., McLaren, W.M., Holmes, G.K.T., Howdle, P.D., Walters, J.R.F., Sanders, D.S., Playford, R.J., Trynka, G., Mulder, C.J.J., Mearin, M.L., Verbeek, W.H.M., Trimble, V., Stevens, F.M., O'Morain, C., Kennedy, N.P., Kelleher, D., Pennington, D.J., Strachan, D.P., McArdle, W.L., Mein, C.A., Wapenaar, M.C., Deloukas, P., McGinnis, R., McManus, R., Wijmenga, C., Van Heel, D.A., 2008. Newly identified genetic risk variants for celiac disease related to the immune response. Nat. Genet. 40, 395–402.
Hunt, K.A., Mistry, V., Bockett, N.A., Ahmad, T., Ban, M., Barker, J.N., Barrett, J.C., Blackburn, H., Brand, O., Burren, O., Capon, F., Compston, A., Gough, S.C.L., Jostins, L., Kong, Y., Lee, J.C., Lek, M., MacArthur, D.G., Mansfield, J.C., Mathew, C.G., Mein, C.A., Mirza, M., Nutland, S., Onengut-Gumuscu, S., Papouli, E., Parkes, M., Rich, S.S., Sawcer, S., Satsangi, J., Simmonds, M.J., Trembath, R.C., Walker, N.M., Wozniak, E., Todd, J.A., Simpson, M.A., Plagnol, V., Van Heel, D.A., 2013. Negligible impact of rare autoimmune-locus coding-region variants on missing heritability. Nature 498, 232–235.
Hutchinson, J.N., Raj, T., Fagerness, J., Stahl, E., Viloria, F.T., Gimelbrant, A., Seddon, J., Daly, M., Chess, A., Plenge, R., 2014. Allele-specific methylation occurs at genetic variants associated with complex disease. PLoS One 9 (6), e98464.
Inamo, J., 2020. Association between celiac disease and systemic lupus erythematosus: a Mendelian randomization study. Rheumatology 59 (9), 2642–2644.
Jauregi-Miguel, A., Santin, I., Garcia-Etxebarria, K., Olazagoitia-Garmendia, A., Romero-Garmendia, I., Sebastian-delaCruz, M., Irastorza, I., Castellanos-Rubio, A., Bilbao, J.R., 2019. MAGI2 gene region and celiac disease. Front. Nutr. 6, 187. Frontiers Media S.A.
Jones, P.A., Baylin, S.B., 2002. The fundamental role of epigenetic events in cancer. Nature Publishing Group, Nat. Rev. Genet. 3 (6), 415–428. <www.nature.com/reviews/genetics>. 1 September 2020.
Karell, K., Louka, A.S., Moodie, S.J., Ascher, H., Clot, F., Greco, L., Ciclitira, P.J., Sollid, L.M., Partanen, J., 2003. HLA types in celiac disease patients not carrying the DQA1 *05-DQB1 *02 (DQ2) heterodimer: results from the European genetics cluster on celiac disease. Hum. Immunol. 64, 469–477. Elsevier Inc.
Kumar, V., Wijmenga, C., Withoff, S., 2012. From genome-wide association studies to disease mechanisms: celiac disease as a model for autoimmune diseases. Semin. Immunopathol. 34 (4), 567–580.

Kumar, V., Westra, H.J., Karjalainen, J., Zhernakova, D.V., Esko, T., Hrdlickova, B., Almeida, R., Zhernakova, A., Reinmaa, E., Võsa, U., Hofker, M.H., Fehrmann, R.S.N., Fu, J., Withoff, S., Metspalu, A., Franke, L., Wijmenga, C., 2013. Human disease-associated genetic variation impacts large intergenic non-coding rna expression. PLoS Genet. 9, e1003201. Public Library of Science.

Kumar, V., Gutierrez-Achury, J., Kanduri, K., Almeida, R., Hrdlickova, B., Zhernakova, D.V., Westra, H.J., Karjalainen, J., Ricaño-Ponce, I., Li, Y., Stachurska, A., Tigchelaar, E.F., Abdulahad, W.H., Lähdesmäki, H., Hofker, M.H., Zhernakova, A., Franke, L., Lahesmaa, R., Wijmenga, C., Withoff, S., 2015. Systematic annotation of celiac disease loci refines pathological pathways and suggests a genetic explanation for increased interferon-gamma levels. Hum. Mol. Genet. 24, 397–409.

Leenen, F.A.D., Muller, C.P., Turner, J.D., 2016. DNA methylation: conducting the orchestra from exposure to phenotype? Springer Verlag, Clin. Epigenet. 8 (1), 92. <http://creativecommons.org/publicdomain/zero/1.0/>. 1 September 2020.

Leonard, M.M., Camhi, S., Huedo-Medina, T.B., Fasano, A., 2015. Celiac disease genomic, environmental, microbiome, and metabolomic (CDGEMM) study design: approach to the future of personalized prevention of celiac disease. Nutrients 7, 9325–9336.

Li, E., Beard, C., Jaenisch, R., 1993. Role for DNA methylation in genomic imprinting. Nature 366, 362–365. Nature Publishing Group.

Lindfors, K., Ciacci, C., Kurppa, K., Lundin, K.E.A., Makharia, G.K., Mearin, M.L., Murray, J.A., Verdu, E.F., Kaukinen, K., 2019. Coeliac disease. Nature Publishing Group, Nat. Rev. Dis. Primers 5, 3.

Liu, E., Rewers, M., Eisenbarth, G.S., 2005. Genetic testing: who should do the testing and what is the role of genetic testing in the setting of celiac disease? Gastroenterology 128, S33–S37.

Lundin, K.E.A., Wijmenga, C., 2015. Coeliac disease and autoimmune disease—genetic overlap and screening. Nature Publishing Group, Nat. Rev. Gastroenterol. Hepatol. 12, 507–515. <https://www.nature.com/articles/nrgastro.2015.136>. 25 June 2020.

Magni, S., Comani, G.B., Elli, L., Vanessi, S., Ballarini, E., Nicolini, G., Rusconi, M., Castoldi, M., Meneveri, R., Muckenthaler, M.U., Bardella, M.T., Barisani, D., 2014. miRNAs affect the expression of innate and adaptive immunity proteins in celiac disease. Am. J. Gastroenterol. 109, 1662. Nature Publishing Group.

Márquez, A., Kerick, M., Zhernakova, A., Gutierrez-Achury, J., Chen, W.M., Onengut-Gumuscu, S., González-Álvaro, I., Rodriguez-Rodriguez, L., Rios-Fernández, R., González-Gay, M.A., Mayes, M.D., Raychaudhuri, S., Rich, S.S., Wijmenga, C., Martín, J., 2018. Meta-analysis of Immunochip data of four autoimmune diseases reveals novel single-disease and cross-phenotype associations. Genome Med. 10 (1), 97.

McDermott, E., Ryan, E.J., Tosetto, M., Gibson, D., Burrage, J., Keegan, D., Byrne, K., Crowe, E., Sexton, G., Malone, K., Harris, R.A., Kellermayer, R., Mill, J., Cullen, G., Doherty, G.A., Mulcahy, H., Murphy, T.M., 2016. DNA methylation profiling in inflammatory bowel disease provides new insights into disease pathogenesis. J. Crohns Colitis 10, 77–86. Oxford University Press.

Mistry, V., Bockett, N.A., Levine, A.P., Mirza, M.M., Hunt, K.A., Ciclitira, P.J., Hummerich, H., Neuhausen, S.L., Simpson, M.A., Plagnol, V., Van Heel, D.A., 2015. Exome sequencing of 75 individuals from multiply affected coeliac families and large scale resequencing follow up. PLoS One 10, 116845.

Nadal, I., Donant, E., Ribes-Koninckx, C., Calabuig, M., Sanz, Y., 2007. Imbalance in the composition of the duodenal microbiota of children with coeliac disease. J. Med. Microbiol. 56, 1669–1674.

Olivares, M., Benítez-Páez, A., de Palma, G., Capilla, A., Nova, E., Castillejo, G., Varea, V., Marcos, A., Garrote, J.A., Polanco, I., Donat, E., Ribes-Koninckx, C., Calvo, C., Ortigosa, L., Palau, F., Sanz, Y., 2018b. Increased prevalence of pathogenic bacteria in the gut microbiota of infants at risk of developing celiac disease: the PROFICEL study. Gut Microbes 9, 551–558.

Olivares, M., Walker, A.W., Capilla, A., Benítez-Páez, A., Palau, F., Parkhill, J., Castillejo, G., Sanz, Y., 2018. Gut microbiota trajectory in early life may predict development of celiac disease. Microbiome 6 (1), 36.

Östensson, M., Montén, C., Bacelis, J., Gudjonsdottir, A.H., Adamovic, S., Ek, J., Ascher, H., Pollak, E., Arnell, H., Browaldh, L., Agardh, D., Wahlström, J., Nilsson, S., Torinsson-Naluai, Å., 2013. A possible mechanism behind autoimmune disorders discovered by genome-wide linkage and association analysis in celiac disease. PLoS One 8, 70174.

Parmar, A.S., Lappalainen, M., Paavola-Sakki, P., Halme, L., Färkkilä, M., Turunen, U., Kontula, K., Aromaa, A., Salomaa, V., Peltonen, L., Halfvarson, J., Törkvist, L., D'Amato, M., Saavalainen, P., Einarsdottir, E., 2012. Association of celiac disease genes with inflammatory bowel disease in Finnish and Swedish patients. Genes Immun. 13, 474–480.

Plaza-Izurieta, L., Castellanos-Rubio, A., Irastorza, I., Fernández-Jimenez, N., Gutierrez, G., Bilbao, J.R., Alsina, M., Esteve, M., Pardo, L., Castaño, L., Oliver, P., Vitoria, J.C., Rodrigo, L., López-Larrea, C., López-Vázquez, A., Hualde, I., García, Z., Hernández, B., Ramos-Arroyo, M.A., Sánchez-Valverde, F., Farré, C., Marqués, T., Vilar, P., Arranz, E., Garrote, J.A., 2011. Revisiting genome wide association studies (GWAS) in coeliac disease: replication study in Spanish population and expression analysis of candidate genes. J. Med. Genet. 48, 493–496. BMJ Publishing Group Ltd.

Plaza-Izurieta, L., Fernandez-Jimenez, N., Irastorza, I., Jauregi-Miguel, A., Romero-Garmendia, I., Vitoria, J.C., Bilbao, J.R., 2015. Expression analysis in intestinal mucosa reveals complex relations among genes under the association peaks in celiac disease. Eur. J. Hum. Genet. 23, 1100–1105.

Ricaño-Ponce, I., Zhernakova, D.V., Deelen, P., Luo, O., Li, X., Isaacs, A., Karjalainen, J., Di Tommaso, J., Borek, Z.A., Zorro, M.M., Gutierrez-Achury, J., Uitterlinden, A.G., Hofman, A., van Meurs, J., Netea, M.G., Jonkers, I.H., Withoff, S., van Duijn, C.M., Li, Y., Ruan, Y., Franke, L., Wijmenga, C., Kumar, V., BIOS Consortium, 2016. Refined mapping of autoimmune disease associated genetic variants with gene expression suggests an important role for non-coding RNAs. J. Autoimmun. 68, 62–74. Academic Press.

Ricaño-Ponce, I., Gutierrez-Achury, J., Costa, A.F., Deelen, P., Kurilshikov, A., Zorro, M.M., Platteel, M., van der Graaf, A., Sugai, E., Moreno, M.L., Vázquez, H., Barrientos, E., Mauriño, E., Talamo, M., Villarruel, A., Sanna, S., Daffra, O., Zhernakova, A., Fu, J., Trynka, G., Smecuol, E., Niveloni, S.I., Bai, J.C., Kumar, V., Wijmenga, C., 2020. Immunochip meta-analysis in European and Argentinian populations identifies two novel genetic loci associated with celiac disease. Eur. J. Hum. Genet. 28, 313–323. Springer Nature.

Rizzo, F., Vanoli, A., Sahnane, N., Cerutti, R., Trapani, D., Rinaldi, A., Sellitto, A., Ciacci, C., Volta, U., Villanacci, V., Calabrò, A., Arpa, G., Luinetti, O., Paulli, M., Solcia, E., Di Sabatino, A., Sessa, F., Weisz, A., Furlan, D., 2019. Small-bowel carcinomas associated with celiac disease: transcriptomic profiling shows predominance of microsatellite instability-immune and mesenchymal subtypes. Virchows Arch. 476, 711–723. Springer Verlag.

Romanos, J., Barisani, D., Trynka, G., Zhernakova, A., Bardella, M.T., Wijmenga, C., 2009. Six new coeliac disease loci replicated in an Italian population confirm association with coeliac disease. J. Med. Genet. 46, 60–63. BMJ Publishing Group Ltd.

Romanos, J., Rosén, A., Kumar, V., Trynka, G., Franke, L., Szperl, A., Gutierrez-Achury, J., Van Diemen, C.C., Kanninga, R., Jankipersadsing, S.A., Steck, A., Eisenbarth, G., Van Heel, D.A., Cukrowska, B., Bruno, V., Mazzilli, M.C., Núñez, C., Bilbao, J.R., Mearin, M.L., Barisani, D., Rewers, M., Norris, J.M., Ivarsson, A., Boezen, H.M., Liu, E., Wijmenga, C., 2014. Improving coeliac disease risk prediction by testing non-HLA variants additional to HLA variants. Gut 63, 415–422. BMJ Publishing Group.

Saadah, O.I., Shaik, N.A., Banaganapalli, B., Salama, M.A., Al-Harthi, S.E., Wang, J., Shawoosh, H.A., Alghamdi, S.A., Bin-Taleb, Y.Y., Alhussaini, B.H., Elango, R., Al-Aama, J.Y., 2015. Replication of GWAS Coding SNPs Implicates MMEL1 as a potential susceptibility locus among saudi arabian celiac disease patients. Dis. Markers 2015, 351673.

Sánchez, E., De Palma, G., Capilla, A., Nova, E., Pozo, T., Castillejo, G., Varea, V., Marcos, A., Garrote, J.A., Polanco, I., López, A., Ribes-Koninckx, C., García-Novo, M.D., Calvo, C., Ortigosa, L., Palau, F., Sanz, Y., 2011. Influence of environmental and genetic factors linked to celiac disease risk on infant gut colonization by Bacteroides species. Appl. Environ. Microbiol. 77, 5316–5323.

Santin, I., Jauregi-Miguel, A., Velayos, T., Castellanos-Rubio, A., Garcia-Etxebarria, K., Romero-Garmendia, I., Fernandez-Jimenez, N., Irastorza, I., Castaño, L., Bilbao, J.R., 2018. Celiac diasease–associated lncRNA named HCG14 regulates NOD1 expression in intestinal cells. J. Pediatr. Gastroenterol. Nutr. 67, 225–231. Lippincott Williams and Wilkins.

Sellitto, M., Bai, G., Serena, G., Fricke, W.F., Sturgeon, C., Gajer, P., White, J.R., Koenig, S.S.K., Sakamoto, J., Boothe, D., Gicquelais, R., Kryszak, D., Puppa, E., Catassi, C., Ravel, J., Fasano, A., 2012. Proof of concept of microbiome-metabolome analysis and delayed gluten exposure on celiac disease autoimmunity in genetically at-risk infants. PLoS One 7, 33387.

Smyth, D.J., Plagnol, V., Walker, N.M., Cooper, J.D., Downes, K., Yang, J.H., Howson, J.M., Stevens, H., McManus, R., Wijmenga, C., Heap, G.A., Dubois, P.C., Clayton, D.G., Hunt, K.A., Van Heel, D.A., Todd, J.A., 2008. Shared and distinct genetic variants in type 1 diabetes and celiac disease. N. Engl. J. Med. 359, 2767–2777.

Sollid, L.M., Markussen, G., Ek, J., Gjerde, H., Vartdal, F., Thorsby, E., 1989. Evidence for a primary association of celiac disease to a particular HLA-DQ α/β heterodimer. J. Exp. Med. 169, 345–350.

Stahl, E.A., Raychaudhuri, S., Remmers, E.F., Xie, G., Eyre, S., Thomson, B.P., Li, Y., Kurreeman, F.A.S., Zhernakova, A., Hinks, A., Guiducci, C., Chen, R., Alfredsson, L., Amos, C.I., Ardlie, K.G., Barton, A., Bowes, J., Brouwer, E., Burtt, N.P., Catanese, J.J., Coblyn, J., Coenen, M.J.H., Costenbader, K.H., Criswell, L.A., Crusius, J.B.A., Cui, J., De Bakker, P.I.W., De Jager, P.L., Ding, B., Emery, P., Flynn, E., Harrison, P., Hocking, L.J., Huizinga, T.W.J., Kastner, D.L., Ke, X., Lee, A.T., Liu, X., Martin, P., Morgan, A.W., Padyukov, L., Posthumus, M.D., Radstake, T.R.D.J., Reid, D.M., Seielstad, M., Seldin, M.F., Shadick, N.A., Steer, S., Tak, P.P., Thomson, W., Van Der Helm-Van Mil, A.H.M., Van Der Horst-Bruinsma, I.E., Van Der Schoot, C.E., Van Riel, P.L.C.M., Weinblatt, M.E., Wilson, A.G., Wolbink, G.J., Wordsworth, B.P., Wijmenga, C., Karlson, E.W., Toes, R.E.M., De Vries, N., Begovich, A.B., Worthington, J., Siminovitch, K.A., Gregersen, P.K., Klareskog, L., Plenge, R.M., 2010. Genome-wide association study meta-analysis identifies seven new rheumatoid arthritis risk loci. Nat. Genet. 42, 508–514. Peter K Gregersen 21.

Stokes, P.L., Holmes, G.K.T., Asquith, P., Mackintosh, P., Cooke, W.T., 1972. Histocompatibility antigens associated with adult cœliac disease. Lancet 300, 162–164. Elsevier.

Teschendorff, A.E., Relton, C.L., 2018. Statistical and integrative system-level analysis of DNA methylation data. Nature Publishing Group, Nat. Rev. Genet. 19 (3), 129–147. <www.nature.com/nrg>. 1 September 2020.

Teschendorff, A.E., Liu, X., Caren, H., Pollard, S.M., Beck, S., Widschwendter, M., Chen, L., 2014. The dynamics of DNA methylation covariation patterns in carcinogenesis. PLoS Comput. Biol. 10, 1003709.

Teschendorff, A.E., Gao, Y., Jones, A., Ruebner, M., Beckmann, M.W., Wachter, D.L., Fasching, P.A., Widschwendter, M., 2016. DNA methylation outliers in normal breast tissue identify field defects that are enriched in cancer. Nat. Commun. 7, 1–12. Nature Publishing Group.

Timp, W., Feinberg, A.P., 2013. Cancer as a dysregulated epigenome allowing cellular growth advantage at the expense of the host. Nature Publishing Group, Nat. Rev. Cancer 13 (7), 497–510. <https://www.nature.com/articles/nrc3486>. 1 September 2020.

Tosi, R., Vismara, D., Tanigaki, N., Ferrara, G.B., Cicimarra, F., Buffolano, W., Follo, D., Auricchio, S., 1983. Evidence that celiac disease is primarily associated with a DC locus allelic specificity. Clin. Immunol. Immunopathol. 28, 395–404. Academic Press.

Trynka, G., Zhernakova, A., Romanos, J., Franke, L., Hunt, K.A., Turner, G., Bruinenberg, M., Heap, G.A., Platteel, M., Ryan, A.W., De Kovel, C., Holmes, G.K.T., Howdle, P.D., Walters, J.R.F., Sanders, D.S., Mulder, C.J.J., Mearin, M.L., Verbeek, W.H.M., Trimble, V., Stevens, F.M., Kelleher, D., Barisani, D., Bardella, M.T., McManus, R., Van Heel, D.A., Wijmenga, C., 2009. Coeliac disease-associated risk variants in TNFAIP3 and REL implicate altered NF-κB signalling. Gut 58, 1078–1083. BMJ Publishing Group.

Trynka, G., Hunt, K.A., Bockett, N.A., Romanos, J., Mistry, V., Szperl, A., Bakker, S.F., Bardella, M.T., Bhaw-Rosun, L., Castillejo, G., De La Concha, E.G., De Almeida, R.C., Dias, K.R.M., Van Diemen, C.C., Dubois, P.C.A., Duerr, R.H., Edkins, S., Franke, L., Fransen, K., Gutierrez, J., Heap, G.A.R., Hrdlickova, B., Hunt, S., Izurieta, L.P., Izzo, V., Joosten, L.A.B., Langford, C., Mazzilli, M.C., Mein, C.A., Midah, V., Mitrovic, M., Mora, B., Morelli, M., Nutland, S., Núñez, C., Onengut-Gumuscu, S., Pearce, K., Platteel, M., Polanco, I., Potter, S., Ribes-Koninckx, C., Ricaño-Ponce, I., Rich, S.S., Rybak, A., Santiago, J.L., Senapati, S., Sood, A., Szajewska, H., Troncone, R., Varadé, J., Wallace, C., Wolters, V.M., Zhernakova, A., Thelma, B.K., Cukrowska, B., Urcelay, E., Bilbao, J.R., Mearin, M.L., Barisani, D., Barrett, J.C., Plagnol, V., Deloukas, P., Wijmenga, C., Van Heel, D.A., 2011. Dense genotyping identifies and localizes multiple common and rare variant association signals in celiac disease. Nat. Genet. 43, 1193–1201.

Turpin, W., Espin-Garcia, O., Xu, W., Silverberg, M.S., Kevans, D., Smith, M.I., Guttman, D.S., Griffiths, A., Panaccione, R., Otley, A., Xu, L., Shestopaloff, K., Moreno-Hagelsieb, G., Paterson, A.D., Croitoru, K., 2016. Association of host genome with intestinal microbial composition in a large healthy cohort. Nat. Genet. 48, 1413–1417.

Vaira, V., Roncoroni, L., Barisani, D., Gaudioso, G., Bosari, S., Bulfamante, G., Doneda, L., Conte, D., Tomba, C., Bardella, M.T., Ferrero, S., Locatelli, M., Elli, L., 2014. microRNA profiles in coeliac patients distinguish different clinical phenotypes and are modulated by gliadin peptides in primary duodenal fibroblasts. Clin. Sci. 126, 417–423. Portland Press.

Van Heel, D.A., Franke, L., Hunt, K.A., Gwilliam, R., Zhernakova, A., Inouye, M., Wapenaar, M.C., Barnardo, M.C.N.M., Bethel, G., Holmes, G.K.T., Feighery, C., Jewell, D., Kelleher, D., Kumar, P., Travis, S., Walters, J.R., Sanders, D.S., Howdle, P., Swift, J., Playford, R.J., McLaren, W.M., Mearin, M.L., Mulder, C.J., McManus, R., McGinnis, R., Cardon, L.R., Deloukas, P., Wijmenga, C., 2007.

A genome-wide association study for celiac disease identifies risk variants in the region harboring IL2 and IL21. Nat. Genet. 39, 827–829.
Wang, K.C., Chang, H.Y., 2011. Molecular mechanisms of long noncoding RNAs. Cell Press, Mol. Cell 43 (6), 904–914.
Wang, J., Thingholm, L.B., Skiecevičie, J., Rausch, P., Kummen, M., Hov, J.R., Degenhardt, F., Heinsen, F.A., Rühlemann, M.C., Szymczak, S., Holm, K., Esko, T., Sun, J., Pricop-Jeckstadt, M., Al-Dury, S., Bohov, P., Bethune, J., Sommer, F., Ellinghaus, D., Berge, R.K., Hübenthal, M., Koch, M., Schwarz, K., Rimbach, G., Hübbe, P., Pan, W.H., Sheibani-Tezerji, R., Häsler, R., Rosenstiel, P., D'Amato, M., Cloppenborg-Schmidt, K., Künzel, S., Laudes, M., Marschall, H.U., Lieb, W., Nöthlings, U., Karlsen, T.H., Baines, J.F., Franke, A., 2016. Genome-wide association analysis identifies variation in Vitamin D receptor and other host factors influencing the gut microbiota. Nat. Genet. 48, 1396–1406.
Wang, J., Kurilshikov, A., Radjabzadeh, D., Turpin, W., Croitoru, K., Bonder, M.J., Jackson, M.A., Medina-Gomez, C., Frost, F., Homuth, G., Rühlemann, M., Hughes, D., Kim, H.N., Spector, T.D., Bell, J.T., Steves, C.J., Timpson, N., Franke, A., Wijmenga, C., Meyer, K., Kacprowski, T., Franke, L., Paterson, A.D., Raes, J., Kraaij, R., Zhernakova, A., 2018. Meta-analysis of human genome-microbiome association studies: the MiBioGen consortium initiative. Microbiome 6 (1), 101.
Withoff, S., Li, Y., Jonkers, I., Wijmenga, C., 2016. Understanding celiac disease by genomics. Elsevier Ltd, Trends Genet. 32 (5), 295–308.
Zhernakova, A., Van Diemen, C.C., Wijmenga, C., 2009. Detecting shared pathogenesis from the shared genetics of immune-related diseases. Nature Publishing Group, Nat. Rev. Genet. 10 (1), 43–55.
Zhernakova, A., Stahl, E.A., Trynka, G., Raychaudhuri, S., Festen, E.A., Franke, L., Westra, H.J., Fehrmann, R.S.N., Kurreeman, F.A.S., Thomson, B., Gupta, N., Romanos, J., McManus, R., Ryan, A.W., Turner, G., Brouwer, E., Posthumus, M.D., Remmers, E.F., Tucci, F., Toes, R., Grandone, E., Mazzilli, M.C., Rybak, A., Cukrowska, B., Coenen, M.J.H., Radstake, T.R.D.J., van Riel, P.L.C.M., Li, Y., de Bakker, P.I.W., Gregersen, P.K., Worthington, J., Siminovitch, K.A., Klareskog, L., Huizinga, T.W.J., Wijmenga, C., Plenge, R.M., 2011. Meta-analysis of genome-wide association studies in celiac disease and rheumatoid arthritis identifies fourteen non-HLA shared loci. PLoS Genet. 7, 1002004.

CHAPTER TWO

The HLA complex and coeliac disease

Laura Espino and Concepción Núñez*
Laboratorio de investigación en Genética de enfermedades complejas, Hospital Clínicos San Carlos, IdISSC, Madrid, Spain
*Corresponding author: e-mail address: conchita.npardo@gmail.com

Contents

Abstract

The Human Leukocyte Antigen (HLA) has a crucial role in the development and pathogenesis of coeliac disease (CD). The genes *HLA-DQA1* and *HLA-DQB1*, both lying in this region and encoding the HLA-DQ heterodimer, are the main genetic predisposing factors to CD. Approximately 90% of CD patients carry the heterodimer HLA-DQ2.5, leaving only a small proportion of patients with lower risk heterodimers (HLA-DQ8, HLA-DQ2.2 or HLA-DQ7.5). These HLA-DQ molecules act as receptors present in the surface of antigen presenting cells and show high affinity for deamidated gluten peptides, which bind and present to $CD4^+$ T cells. This triggers the immunological

International Review of Cell and Molecular Biology, Volume 358
ISSN 1937-6448
https://doi.org/10.1016/bs.ircmb.2020.09.009

reaction that evolves into CD. Since specific HLA genetics is present in almost the totality of CD patients, HLA typing has a very high negative predictive value, and it can be used to support diagnosis in specific scenarios. HLA risk has been associated to different CD-related features, such as age at onset, clinical outcomes, antibody levels and grade of histological lesion; but further research is needed. HLA-DQ genotypes have been also suggested to modulate the composition of the gut microbiota.

1. Introduction

The Major Histocompatibility Complex (MHC) was discovered in the early 1900s when studying tumor transplantation in mice. Years later, it was established that genes with similar functions were present in all mammals, playing a major role in the immune response against protein antigens. In humans, it received the name of Human Leukocyte Antigen (HLA). The HLA complex is located in the 6p21 chromosomal region and contains more than 200 genes that are grouped in three main classes: I, II and III. Besides the huge gene density, HLA is mainly characterized by the extremely high polymorphism and by the codominant inheritance, this means the expression of the protein products of both alleles. Both characteristics ensure a high molecular diversity and increase the ability of the immune system to respond to multitude of pathogens and foreign proteins. The extensive linkage disequilibrium, i.e., the non-random association between alleles at different loci, is also a notorious feature of the HLA. Despite the high genetic variability of this complex, the allele distribution is not as wide as expected, being common the predominance of specific combinations of alleles at different loci that are passed together to the offspring and are designated as haplotypes (Ahmad et al., 2003). Notably, the pattern of linkage disequilibrium varies between regions and populations and this has a high impact in disease susceptibility.

2. Genetic susceptibility

Genetic studies have described HLA loci related to susceptibility to numerous diseases (Karnes et al., 2017). In CD, the genetic influence of the HLA was first discovered in the 1970s by linkage studies, which designated this locus as *CELIAC1* (Liu et al., 2002). Subsequent genetic association studies showed that the main genetic predisposing factors lie specifically on the HLA class II region, pinpointing *HLA-DQA1* and *HLA-DQB1* (Fig. 1). These genes encode the α and β chains, respectively,

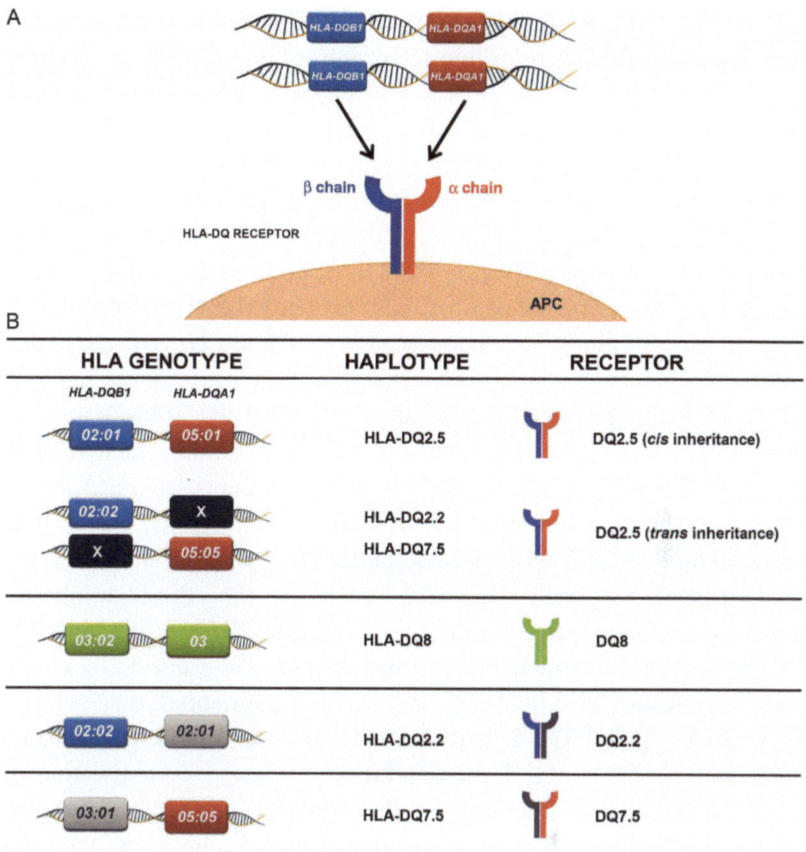

Fig. 1 HLA genetic susceptibility to celiac disease (CD). (A) *HLA-DQA1* and *HLA-DQB1* genes are the main genetic factors associated to CD risk. They encode the α and β chains, respectively, that conform the HLA-DQ molecule, present in the surface of antigen presenting cells (APC). (B) The specific HLA-DQ allele combinations inherited from each progenitor (haplotypes) determine the different HLA-DQ receptors that are formed, which receive the same name than its encoding haplotype with the exception of HLA-DQ2.5 that can be encoded also in *trans* by inheriting HLA-DQ2.2 and HLA-DQ7.5. The alleles and haplotypes in the Figure are the most common ones causing risk, but *HLA-DQA1*03* refers to any allele of that family, being *HLA-DQA1*03:01* and *HLA-DQA1*03:02* the most frequent; and other alleles of the *HLA-DQA1*05* and *HLA-DQB1*02* family can be also present.

that conform the heterodimeric HLA-DQ molecule or receptor (Sollid et al., 1989). Due to the extensive linkage disequilibrium in the region, other loci such as *HLA-A*, *HLA-D* or *HLA-DRB1* were initially associated with CD susceptibility (Falchuk et al., 1972; Keuning et al., 1976; Mearin et al., 1983).

Alleles encoding HLA-DQ2 and HLA-DQ8 are the main genetic risk variants for CD. The heterodimeric receptor HLA-DQ2.5 is considered to be present in around 90–95% of CD patients, who carry any allele of the family *HLA-DQA1*05* and *HLA-DQB1*02*, being the most frequent the ones cited below. These alleles can be inherited in *cis* or *trans* configuration. The HLA-DQ2.5 molecules encoded by those two configurations differ in residues that do not involve the site of antigenic union and therefore do not modify the risk to develop CD. In *cis* configuration both alleles are inherited from the same progenitor, who presents the HLA-DQ2.5 haplotype, with the specific alleles *HLA-DQA1*05:01* and *HLA-DQB1*02:01*. HLA-DQ2.5 can be also encoded in *trans* configuration by the haplotypes HLA-DQ2.2 (characterized by the allele *HLA-DQB1*02:02*) and HLA-DQ7.5 (characterized by the allele *HLA-DQA1*05:05*), each haplotype received from one progenitor. Almost all the patients lacking the HLA-DQ2.5 receptor are considered to carry HLA-DQ8, which is encoded by the alleles *HLA-DQA1*03* (any allele of the *HLA-DQA1*03* family, being *HLA-DQA1*03:01* and *HLA-DQA1*03:02* the most frequent in populations of European ancestry) and *HLA-DQB1*03:02*, both alleles always present in the same haplotype and therefore transmitted by one progenitor (Spurkland et al., 1992). It is well documented that some CD patients lack HLA-DQ2.5 and HLA-DQ8 heterodimers (Fernandez-Banares et al., 2017; Karell et al., 2003). In these patients, the allele *HLA-DQB1*02*, encoding one of the chains required to form HLA-DQ2.5, is predominant and in some populations it reaches a frequency quite similar to that of HLA-DQ8 (Delgado et al., 2014; Martinez-Ojinaga et al., 2018). It is followed by the allele encoding the other HLA-DQ chain: *HLA-DQA1*05*. Although at an extremely low frequency, some patients do not carry any known HLA risk allele.

When looking for specific HLA data, some precautions are needed to confront the mass of published results. Many of them include low numbers of CD patients, which may lead to biased results. Percentages of the HLA-DQ2.5 heterodimer higher than those present may be found, but also the finding by chance of one CD patient with non-permissive HLA genetics in small total numbers will contribute to obtain an artificially high percentage for this group. Husby et al. showed HLA data obtained in 55 studies, but only 15 (27%) included more than 100 CD patients (Husby et al., 2012). This problem applies to works including European populations, but mostly affects to those studying populations of a different ethnicity, none of them reaching such a total number. Selecting those works with $N > 100$,

European populations show a percentage of HLA-DQ2.5 receptors ranging from 83% to 95%, HLA-DQ8 from 2% to 8% and HLA-DQ2 or HLA-DQ8 between 89% and 99%.

Specifically regarding non-HLA-DQ2.5/DQ8 genetics, the most complete study dates back to 2003 and the results are concordant with that remaining 1–11% (Karell et al., 2003). That study was performed by the European Genetics cluster on CD by including data of 1008 CD patients recruited from six countries. The frequency of non-HLA-DQ2.5/DQ8 CD showed a mean value of 6%, with variations among countries: 10.5% in Italy, 6.5% in France, 4.2% in United Kingdom, 4% in Finland and 3.3% in Norway and Sweden. In the non-HLA-DQ2.5/DQ8 group, 93% of them carried one of the alleles encoding HLA-DQ2.5: mostly *HLA-DQB1*02* (67%), followed by *HLA-DQA1*05* (26%). This left 7% of non-HLA-DQ2.5/DQ8 CD without any known HLA risk allele, with some variability across European countries: 12.5% in United Kingdom, 11.5% in Italy and zero in Norway/Sweden, France and Finland. No clear association was found between CD and any previously unknown HLA risk allele. In 2016, a systematic review and a meta-analysis were performed to investigate the presence of non-HLA-DQ2.5/DQ8 CD in Spain (Fernandez-Banares et al., 2017). A total of 2963 Spanish CD patients were included and 3% of them were observed to be negative for HLA-DQ2.5/DQ8, most of them also carrying *HLA-DQB1*02* (55%). But, in contrast with the previous study, 20% of the negative HLA-DQ2.5/DQ8 CD patients in Spain were negative for the two alleles of HLA-DQ2.5 and for HLA-DQ8 (Fig. 2). All these CD patients were diagnosed based on positive serology and atrophy, and according to the data presented further in this chapter, it would be interesting to investigate the percentage of non-HLA-DQ2.5/DQ8 genetics in seronegative patients and those with low-grade enteropathy.

Very few studies have included populations with no European ancestry, and they showed data of low numbers of CD patients. Although the compatible genetics is expected to predispose to CD independently of the ethnicity, and it is corroborated in those studies, the possibility of existing additional HLA susceptibility factors at low frequency has not been sufficiently explored. In China, the haplotype *HLA-DQA1*03-DQB1*03:03* (HLA-DQ9.3) can reach considerable frequency and it has been described as a CD susceptibility factor (Wang et al., 2015). Noticeably, a role of this haplotype in CD had been previously pointed out by Bodd et al. after finding HLA-DQ9.3-restricted gluten-reactive T cells in the small intestine of a CD patient (Bodd et al., 2012b).

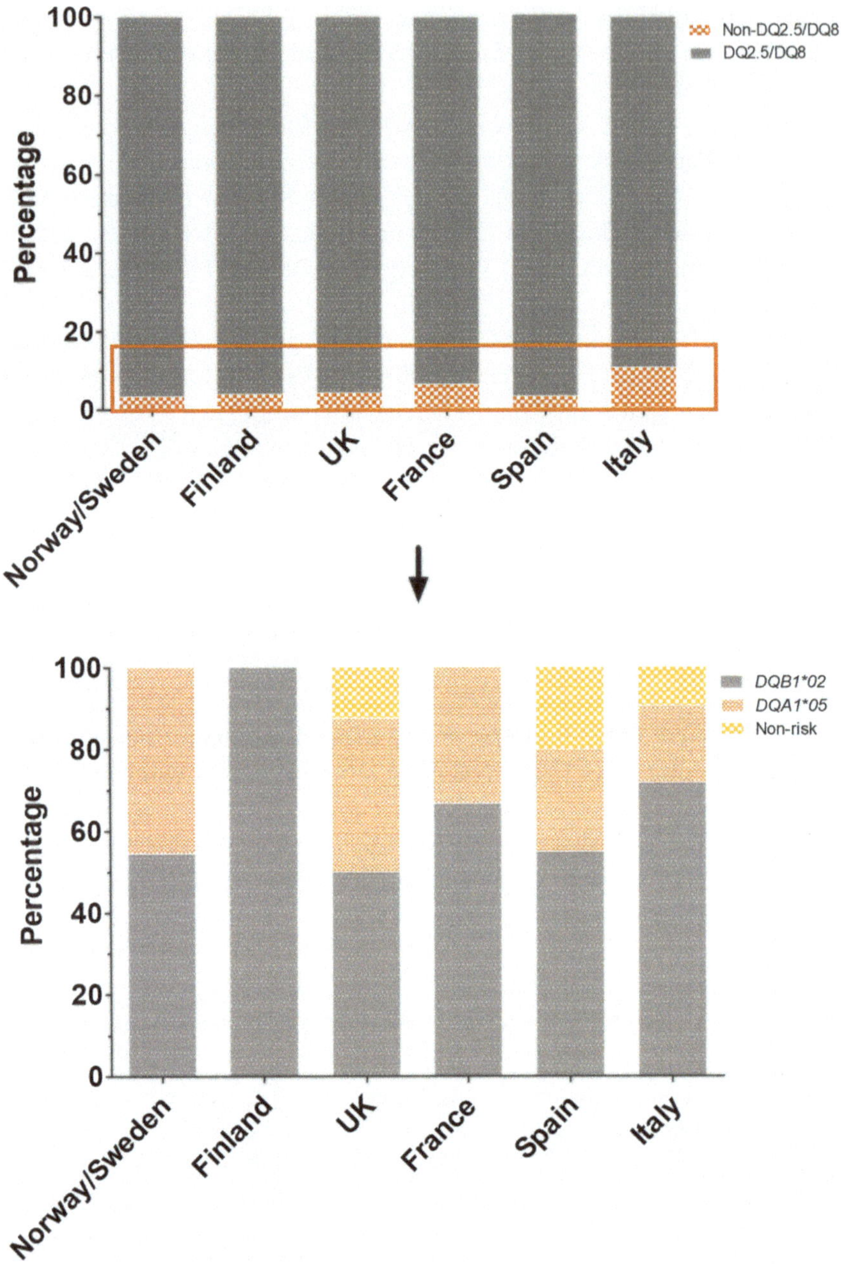

Fig. 2 Frequency and allele distribution of non-HLA-DQ2.5/DQ8 celiac disease in European countries. Upper graph shows the reported frequency of non-HLA-DQ2.5/DQ8 and HLA-DQ2.5/DQ8 genetics in several countries. The graph below shows the distribution of the specific alleles observed in the group with non-HLA-DQ2.5/DQ8 genetics in each of those countries.

3. Risk gradient

Besides the presence of specific alleles, CD risk depends on the gene dosage (Fig. 3). Individuals carrying the HLA-DQ2.5 heterodimer hold the highest risk to develop CD when they carry two copies of the *HLA-DQB1*02* allele (Demarchi et al., 1983; Louka et al., 2002; Ploski et al., 1993; van Belzen et al., 2004). The next risk category is confirmed by HLA-DQ2.5 subjects with one copy of *HLA-DQB1*02*, i.e., those with one HLA-DQ2.5 haplotype or those with the HLA-DQ2.5 receptor encoded in *trans*. Lower risk exist when only HLA-DQ8 or HLA-DQ2.2 are present. A gene dosage effect for HLA-DQ8 has been also proposed, and HLA-DQ8 homozygotes would be moved to higher risk categories (Karell et al., 2003; Martinez-Ojinaga et al., 2018). The presence of HLA-DQ7.5 constitutes the lowest genetic risk category. HLA-DQ7.5

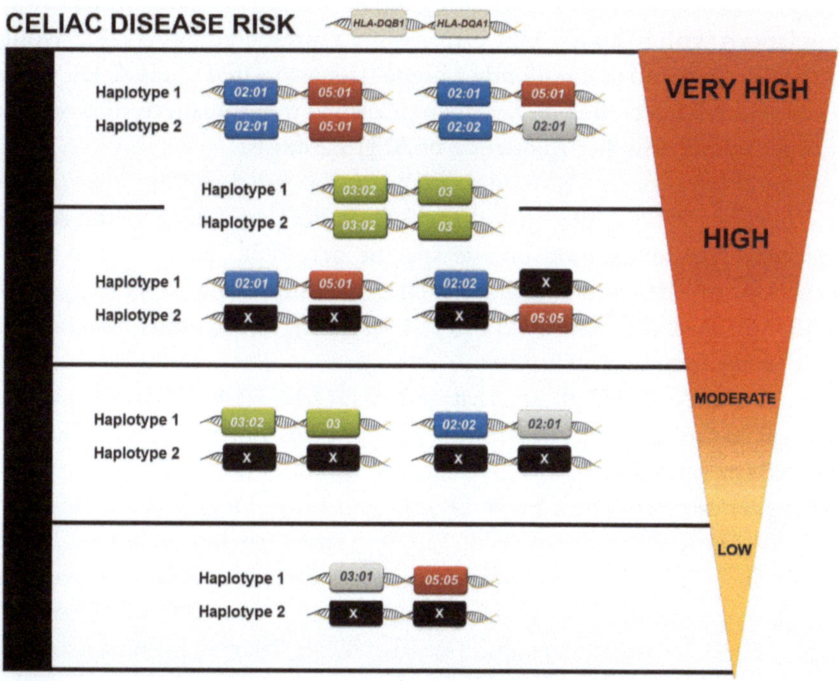

Fig. 3 Gene dosage effect on celiac disease (CD) risk. The risk to develop CD depends on the *HLA-DQA1* and *HLA-DQB1* alleles, and can be graded from very high to low. More studies are needed to know the category risk of individuals who are homozygotes for HLA-DQ8 (*HLA-DQA1*03 HLA-DQB1*03:02*).

appears at very low frequency in CD and a significant association between this haplotype and CD susceptibility is not detected, but its presence in CD individuals lacking all the known HLA risk variants denote it as an haplotype compatible with CD development.

Some authors have established several subdivisions in the high and moderate risk categories. As a matter of fact, a slightly higher risk for individuals carrying HLA-DQ2.5 in *trans* vs those with *cis* inheritance has been described in some European populations (Margaritte-Jeannin et al., 2004; Medrano et al., 2012). This could be explained by the presence of an additional risk factor in the HLA region. The haplotype HLA-DQ2.5 can be extended to different *HLA-DRB1*03-HLA-DQA1*05-HLA-DQB1*02* haplotypes, and it is present in two ancestral haplotypes (AH): AH8.1 and AH18.2. A higher risk to CD has been associated to AH8.1 (Medrano et al., 2012), haplotype that receives its name by the presence of the *HLA-B*08* allele. Interestingly, *HLA-B*08* has been associated to CD susceptibility (Gutierrez-Achury et al., 2015). Additional risk factors could be also present in individuals inheriting HLA-DQ2.5 in *trans*. In this way, populations with HLA-DQ2.5 *in cis* mainly represented by AH18.2 would carry lower risk and consequently a higher risk associated to HLA-DQ2.5 in *trans* could be detected. This has been described in populations of Northern Europe, where a higher frequency of AH18.2 exists.

There have been also some attempts for subclassifying the risk of heterozygous HLA-DQ2.5 (with *cis* inheritance) attending to the second haplotype, i.e., depending on the specific genotype. However, with the exception of the well-recognized effect of supplying a second *HLA-DQB1*02* allele (HLA-DQ2.5 or HLA-DQ2.2), a clear effect has not been consistently found across studies (Table 1).

It must be highlighted that *HLA-DQA1-HLA-DQB1* haplotype frequencies follow a north-south gradient in Europe, with southern populations showing lower frequencies of the HLA-DQ2.5 haplotype and higher frequencies of HLA-DQ2.2 and HLA-DQ7.5 (supplementary fig. 1 in Gutierrez-Achury et al., 2015; Margaritte-Jeannin et al., 2004) (www.allefrequencies.net). Differences across European countries exist also for the two extended ancestral haplotypes AH8.1 and AH18.2 (Demarchi et al., 1979; Margaritte-Jeannin et al., 2004; Martinez-Ojinaga et al., 2018). Considering that HLA-DQ frequencies in CD patients are concordant with the HLA composition of the studied population, this could generate slight risk variations among countries. Consistent with that, AH8.1 is less frequent in northern European populations, being those where

Table 1 Celiac disease risk according to the HLA-DQ genotype observed in different studies.

HLA-DQ genotype	**Odds**				
	Italy (Megiorni et al., 2009)	Italy (Piccini et al., 2012)	Spain (Ruiz-Ortiz et al., 2014)	Morocco (Piacantelli et al., 2017)	Spain (Martinez-Ojinaga et al., 2018)
DQ2.5/DQ2.5	1:10	1:7	1:12	1:14	1:12
DQ2.5/DQ2.2					1:12
DQ8/DQ8	–	1:43[a]	–	–	1:25
DQ2.2/DQ7.5[b]	1:35	1:47	1:64	1:10	1:35
DQ2.5/DQX				1:150	1:42
DQ2.5/DQ7.5				1:60	1:60
DQ2.5/DQ8	1:7	1:41	1:22	1:170	1:72
DQ8/DQ7.5	–	–	–	–	1:605
DQ8/DQ2.2	1:24	1:43[a]	1:22	1:120	1:681
DQ2.2/DQ2.2	1:26	1:45	1:1063	1:59	1:929[c]
DQ2.2/DQX	1:210	1:75		1:205	
DQ8/DQX	1:89[d]	1:85[d]	1:265[d]	1:200[d]	
DQ7.5/DQX	1:1842	1:1818	–	1:209	1:1135

[a]Includes DQ8/DQ8 and DQ8/DQ2.2.
[b]HLA-DQ2.5 in *trans*.
[c]No patients carry DQ2.2/DQ2.2.
[d]DQ8/DQX also includes DQ8/DQ7.5.
Calculations are based on a prevalence for celiac disease of 1%. DQX indicates "non-risk."

higher risk to HLA-DQ2.5 inherited in *trans* has been described, as previously cited. The lower risk haplotypes HLA-DQ2.2 and HLA-DQ7.5 appear at higher frequency in populations of southern Europe. This could explain the high proportion of HLA-DQ7.5 observed in CD in Italy, where a frequency of 28% was reported for that haplotype (Tinto et al., 2015).

When trying to establish risk categories valid in most countries, we consider that the four categories here established best fit to all the populations of European descent (Fig. 3, Table 2). The suitable specific haplotype combinations in each risk category are specifically stated in Table 2.

Table 2 Specific haplotype combinations present in the different genetic risk categories for celiac disease.

Category risk	HLA genotypes	Description
Very high	DQ2.5/DQ2.5 DQ2.5/DQ2.2	HLA-DQ2.5 carriers with two copies of *HLA-DQB1*02*
High	DQ2.2/DQ7.5[a] DQ2.5/DQ8 DQ2.5/DQ7.5 DQ2.5/DQX DQ8/DQ8	HLA-DQ2.5 carriers with one copy of *HLA-DQB1*02* or HLA-DQ8/HLA-DQ8 individuals
Moderate	DQ8/DQ7.5 DQ8/DQ2.2 DQ8/DQX DQ2.2/DQ2.2 DQ2.2/DQX	HLA-DQ8 heterozygous carriers (non-HLA-DQ2.5) or HLA-DQ2.2 carriers (non-HLA-DQ2.5)
Low	DQ7.5/DQ7.5 DQ7.5/DQX	HLA-DQ7.5 carriers (non-HLA-DQ2.5)

[a]HLA-DQ2.5 in *trans*.
DQX indicates "non-risk."

In 2015, after performing a fine-mapping study of the HLA region and correcting for the collective effects of these classical HLA-DQ risk variants, five further independent variants were identified in this region: *HLA-B* (including the already cited *HLA-B*08*); HLA-DPβ1 (position 9); and the SNPs rs1611710, which exerts a *cis*-eQTL effect on *HLA-F* expression, and rs2301226, a *cis*-eQTL for *B3GALT4* and *HLA-DPB1* (Gutierrez-Achury et al., 2015).

4. Heritability

The classical HLA associated variants account for 22% of CD heritability (Gutierrez-Achury et al., 2015). The new HLA factors described in 2015 were first reported to explain an additional 18% of the disease heritability, but this value was corrected by the authors to 2.5–3% (A. Zhernakova, personal communication at the ICDS, 2015). Taking together, MHC risk variants account for 25% of CD heritability.

Twin studies were conducted in order to disentangle the effect of genetic and environmental factors in CD development. HLA appeared, as already known, as a condition necessary for CD development, but some authors suggested that it has minor effects on CD heritability. Kuja-Halkola et al. studied the concordance rate (probability of a subject to develop CD if their twin has CD) and the heritability of CD in a population-based study on Swedish twins. They showed a concordance ratio of only 49% between monozygotic twins vs 10% in dizygotic ones (Kuja-Halkola et al., 2016). Their estimated heritability for the atrophy-based CD was 75%, being 68% due to non-HLA genetic factors and leaving a quite similar proportion to shared and non-shared environmental factors (17% and 15%, respectively). This left only moderate impact to HLA. Those values strongly contrast with the ones previously published in Italy (Greco et al., 2002; Nistico et al., 2006). Greco et al. described 86% of concordance between monozygotic twin pairs and 20% between dizygotic pairs by performing CD screening in 47 twin pairs (20 monozygotic twins), and years later corroborated very similar values (83% and 17% for monozygotic and dizygotic twin pairs, respectively) after adding 26 new twin pairs (3 new monozygotic twin pairs). It is noteworthy that Kuja-Halkola et al. did not investigate CD in all the included cases, i.e., they studied diagnosed CD, which most probably led to underestimate the real number of CD cases. However, all these studies lead to the same conclusion: HLA seems to be a necessary condition, but once the permissive HLA is present, the specific HLA alleles do not influence CD concordance. When considering different HLA risk categories, CD concordance was always higher in monozygotic pairs compared with dizygotic ones sharing identical HLA chromosomes. Moreover, the dizygotic twins/siblings ratio of CD prevalence was close to 1. Both considerations suggest that non-HLA susceptibility factors have a relevant effect on CD pathogenesis.

5. Familial risk

CD shows a strong genetic component. As a consequence, the risk of having CD in the relatives of patients with CD is higher than the 1% accepted in the general population. In a meta-analysis published in 2015, CD prevalence was estimated in 7.5% in first-degree relatives of CD patients and 2.3% in second-degree relatives, but differences exist depending of the kinship: 9% in siblings, 8% in offspring and 3% in parents; and also on sex, with sisters and daughters of index patients showing the highest values (Singh et al., 2015).

It must be considered that CD development needs a compatible HLA genetics and this is more commonly found in first-degree relatives of CD patients. Lionetti et al. studied the prevalence of CD autoimmunity and of overt CD in 861 newborns with first-degree relatives with CD (Lionetti et al., 2014). The first outstanding observation was that 82% of those newborns were HLA-DQ2/8, and similar values were observed in other studies (Wessels et al., 2018). After a 15-month follow-up, none children with no HLA risk had developed CD. The authors excluded those patients with no HLA risk and observed 16.5% of children with CD autoimmunity (positive antibodies) and 13% with overt CD (positive antibodies and Marsh2 or Marsh 3). They performed stratified analyses according to the presence of high HLA risk (HLA-DQ2.5/HLA-DQ2.5 or HLA-DQ2.5/HLA-DQ2.2) or standard HLA risk (HLA-DQ2.5 with one copy of *HLA-DQB1*02*, HLA-DQ8 or HLA-DQ2.2). They found a significantly increased prevalence in the high HLA risk group for both conditions after a follow-up of at least 5 years (mean 7.9 years, range 5.2–10.6): 38% vs 19% for CD autoimmunity and 26% vs 16% for overt CD. Thus, high risk HLA was significantly associated with increased development of CD autoimmunity and overt CD in first-degree relatives, but other factors studied such as type of relative with CD, number of affected relatives, dietary pattern and early intestinal infections did not show association.

We can compare those data with the ones obtained in a similar study but performed in the general population. The authors considered 6403 newborns of the general population who carried the HLA CD risk haplotypes HLA-DQ2.5 or HLA-DQ8. Specifically they selected children HLA-DQ2.5/HLA-DQ2.5, HLA-DQ2.5/DQ8, HLA-DQ8/DQ8 and HLA-DQ8/DQ4 (Liu et al., 2014). The mean follow-up time was 5 years. It was observed that 2% had a first-degree relative with CD, who showed higher frequency of the HLA-DQ2.5 haplotype (86% vs 62% among those with no affected family members). Of the total of screened children, 12% developed CD autoimmunity and 5% developed CD (25% before the age of 3 years). When looking at the HLA, the estimated cumulative risk for CD autoimmunity and for CD by the age of 5 years was higher for HLA-DQ2.5 homozygosity in both cases (26% and 11%, respectively), followed in a similar manner by HLA-DQ2.5/DQ8 (11% and 3%, respectively) and HLA-DQ8/DQ8 (9% and 3%, respectively), and finally by being HLA-DQ8/DQ4 (2% and $<1\%$, respectively). Homozygosity for HLA-DQ2.5 was also associated with the earliest onset.

Comparisons between both studies must take into account that Liu et al. do not consider low risk HLA categories except HLA-DQ8/DQ4. However, both studies observe that CD autoimmunity and overt CD appear at higher frequency in relatives of CD patients, especially overt CD. Also in both groups, the frequency increases with higher genetic risk.

Another point involves the distribution of HLA risk categories in families with several affected members. Our group observed increased HLA-DQ2.5 and HLA-DQ8 homozygosis in patients with a familial history of CD (Martinez-Ojinaga et al., 2019). Similar results had been obtained regarding HLA-DQ2.5 by other authors (Liu et al., 2014; Lopes et al., 2019), but further confirmation is needed for HLA-DQ8 since our study was based on a low number of patients. It is also notorious that the presence of the HLA-DQ2.5 heterodimer inherited in *trans* appears at a lower frequency in families with CD cases. It must be considered that *HLA-DQB1*02* and *HLA-DQA1*05* individually confer moderate-low risk, and that HLA-DQ2.5 in *trans* is inherited by 25% of the offspring having parents with those haplotypes. In contrast, when one of the progenitors carry HLA-DQ2.5 (*cis* configuration), 50% of the offspring will receive this haplotype, and this percentage rises to 100% with HLA-DQ2.5 homozygous progenitors.

6. Pathogenesis

The classical CD associated HLA-DQ variants are master pieces in CD pathogenesis. In fact, they are considered necessary, although not sufficient, for CD development (Kagnoff, 2007). HLA-DQ molecules are glycoproteins expressed on the surface of antigen presenting cells (APCs) and bind peptide fragments that are presented to $CD4^+$ T cells, mediating their activation. The range of peptides to be bound will depend on the specific HLA-DQ receptors and therefore, ultimately, on the specific HLA-DQ alleles. The peptide-binding grooves of HLA-DQ2 and HLA-DQ8 have high affinity for negatively charged peptides, such as those generated after deamidation of gluten peptides, which are able to be bound and elicit gluten-specific T-cell responses (Bodd et al., 2012a; Fallang et al., 2009). At this point, the role of the transglutaminase type 2 (TG2) enzyme must be highlighted. TG2 is an ubiquitous enzyme also present in the *lamina propria*, where higher expression can be observed after tissue injury (Siegel et al., 2008). Gluten presents a high content in proline and glutamine. Proline residues confer resistance to digestive proteases and allow gluten to be at high concentration levels in the gut epithelium and reach the *lamina propria* (Hausch et al., 2002). In this compartment, the enrichment in glutamine gains great importance, since those residues constitute the target for deamidation, which will be performed by the TG2 enzyme. Residues from glutamine change to negatively charged glutamate and these modified gluten-derived peptides bind to the groove of the HLA-DQ receptors associated to CD risk, being recognized by $CD4^+$ T cells of CD patients (Lundin et al., 1993), but not healthy subjects (Molberg et al., 1997). These gluten-reactive $CD4^+$ T cells have been detected in CD patients who specifically carry HLA-DQ2.5, HLA-DQ8 or HLA-DQ2.2 heterodimers (Lundin et al., 1994) and are the drivers of the CD pathophysiology.

The different HLA-DQ genetically determined receptors recognize diverse sets of gluten-derived peptides, with preference for negatively charged residues at different positions. The listing of the HLA-DQ restricted epitopes recognized by $CD4^+$ T cells of CD patients has been recently updated (Sollid et al., 2020). It contains the most important immunodominant epitopes, but as the authors state, most probably additional ones remain to be identified, since most of the studies have been focused on HLA-DQ2.5 patients. Studies including HLA-DQ7.5 have found unique repertoires of peptides also for this molecule (Bergseng et al., 2015).

Differences in the kinetic stability are crucial for the described differences to elicit T-cell responses and consequently for the differences in the risk to develop CD. The immunological reaction triggered relies on the formation of kinetically stable gluten peptides-HLA-DQ complexes. Its magnitude depends on the specific epitopes presented but also on the amount of those complexes HLA-DQ-peptides expressed at the surface of APCs (Gianfrani et al., 2018). This explains the different genetic risk depending on the specific HLA alleles. Patients carrying the HLA-DQ2.5 heterodimer can recognize the largest gluten peptide repertoire (Vader et al., 2003) and stable complexes between HLA-DQ and gluten T-cell epitopes are formed (Bodd et al., 2012a; Fallang et al., 2009), thus they show the highest risk to develop CD. However, the high complexity of the HLA region must be also considered, with non-additive effects also detected. In fact, HLA-DQ2.5 was found to interact with three other haplotypes: HLA-DQ2.2, *DQA1*01:01-HLA-DQB1*05:01* and HLA-DQ7.5, from the highest to the lowest intensity (Lenz et al., 2015). It is interesting to remark that an interaction was observed with a non-risk haplotype (*DQA1*01:01-HLA-DQB1*05:01*), but not with HLA-DQ8. That observation could explain why HLA-DQ2.5/HLA-DQ8 individuals do not show increased risk compared to other HLA-DQ2.5 combinations, as a priori expected.

To add further complexity to this issue, Pisapia et al. found that both alleles encoding HLA-DQ2.5, *HLA-DQA1*05* and *HLA-DQB1*02*, were preferentially expressed regarding non-CD associated alleles in APCs heterozygous for HLA-DQ2.5. This differential expression of CD risk alleles led to observe similar levels of *HLA-DQA1*05* and *HLA-DQB1*02* transcripts in HLA-DQ2.5 homozygous and heterozygous individuals and a similar amount of HLA-DQ2.5 receptors on the cell surface of their APCs. This could explain the similar magnitude of the specific T cells response when homozygous or heterozygous DR3-DQ2.5 cells were used as APCs observed by the authors (Pisapia et al., 2016). Epigenetic features could be involved.

It is also important to remember that TG2 expression can be increased by tissue injury and this injury can be propitiated by the native gluten itself, although other environmental factors, such as viral infections, can also contribute.

As a consequence of the triggered immunological response, an intestinal lesion in the small intestine is developed and autoantibodies appear. Those autoantibodies against TG2 are the most specific. Thus, TG2 is the major autoantigen in CD (Dieterich et al., 1997), and anti-TG2 antibodies are considered a hallmark of CD.

7. Role in diagnosis and clinical practice

HLA-DQ2 and HLA-DQ8 are present in the great majority of CD patients, but also in around 30–40% of the general population, at least in those of European descent (Husby et al., 2012). This confers HLA testing a very high negative predictive value, but it is not recommended as an initial screening test in the general population due to its poor positive predictive value. Since HLA genetics can be used for a lifelong exclusion of CD, it is very important the proper understanding by clinicians to enable a correct counsel to patients (Brown et al., 2019; Nunez et al., 2018).

7.1 Genetic test

The basis of the genetic test for CD is the genotyping of the *HLA-DQA1* and *HLA-DQB1* genes or, failing that, the identification of the specific alleles encoding HLA-DQ2.5 and HLA-DQ8, which means *HLA-DQA1*05*, *HLA-DQB1*02*, *HLA-DQA1*03* and *HLA-DQB1*03:02*. It must be considered that every subject shows two *HLA-DQA1* and two *HLA-DQB1* alleles and HLA-DQ receptors can be formed as combinations of alleles inherited in *cis* or in *trans* (this means, alleles inherited from the same or a different progenitor). This is the case, for example, of HLA-DQ2.5, that is encoded by the *HLA-DQA1*05* and *HLA-DQB1*02* alleles and can appear in presence of the HLA-DQ2.5 haplotype (*cis* configuration), as well as in presence of the HLA-DQ2.2 and HLA-DQ7.5 haplotypes (*trans* configuration). Consequently, every subject can carry up to four different HLA-DQ heterodimers.

As previously indicated in this chapter, CD risk depends on the specific HLA-DQ molecules present, which in turn is dictated by the *HLA-DQA1/-DBQ1* alleles. Therefore, the relevant information to be transmitted to clinicians is the presence of the HLA risk alleles and heterodimeric molecule/s. Specifically, it must be indicated whether the individual carry HLA-DQ2.5 or HLA-DQ8 receptors, but in absence of both of them, information about the presence of HLA-DQ2.2 or HLA-DQ7.5 must be included, since they also contribute to CD susceptibility.

In addition, it is necessary to add information about how to interpret those data, indicating whether the individual genetics is compatible or not compatible with CD development. As compatible genetics, the presence of both alleles encoding the complete HLA-DQ2.5 and/or HLA-DQ8 heterodimers must be considered, but also the presence of only one of the alleles necessary for HLA-DQ2.5. The high negative predictive value of the genetic study, almost 100%, is present in absence of the HLA-DQ8 haplotype and of both alleles encoding HLA-DQ2.5. It is also necessary to highlight that the classical HLA-DQ variants studied in the genetic report are considered necessary but not enough for CD development, and so it must be clarified to patients that compatible CD genetics does not mean that CD will develop at some point in their lives.

Note also that the predictive value of HLA testing has been only extensively studied in populations of European descent. Most of the studies including populations with different ethnicities are based on low numbers of patients and thus this issue needs further research. Therefore, we must be cautious before discarding CD in subjects of non-European descent based on HLA data, since lower negative predictive values may exist (Ozgenel et al., 2019).

7.2 Who must be tested

HLA testing was formally instituted in the diagnostic workup of CD in 2012, when the ESPGHAN published new recommendations for children and adolescents (Husby et al., 2012). A biopsy-avoiding approach was proposed for children with symptoms or signs suggesting CD and high anti-TG2 antibodies (higher than 10 times the upper limit of normality) when a second antibody determination showed anti-endomisium (EMA)-positive antibodies and, importantly, HLA-DQ2 (HLA-DQ2.5 or HLA-DQ2.2) or HLA-DQ8 heterodimers were present. Those recommendations also suggested HLA testing as the initial screening test for asymptomatic individuals belonging to risk groups, this means children with first-degree relatives with CD or with CD-related conditions (other immune-related conditions or some syndromes such as Down's or Turner's syndromes). In risk groups, the absence of the HLA predisposing alleles avoided further CD screening, while in individuals belonging to risk groups and disclosing HLA predisposing alleles, periodic screening for antibodies against IgA anti-TG2 should be considered to avoid a misdiagnosis of subclinical CD. This year, a new publication of the ESPGHAN modified those criteria and HLA genotyping is no longer considered a cost-effective practice for diagnosis of children with high anti-TG2 titers aimed to avoid the duodenal biopsy. It is considered that with such high antibody titers HLA-DQ2/8 will be always present (Husby et al., 2020). Currently, CD diagnosis in absence of biopsy is allowed only with high anti-TG2 antibodies and subsequent EMA-positivity. HLA testing continues being useful for initial screening at risk groups although excluding individuals with type 1 diabetes, disease that shows a strong HLA genetics background shared with CD and all subjects with type 1 diabetes are recommended to be tested for CD serology to reduce costs.

In adults, biopsy is still necessary for diagnosis in all cases (except medical contraindication), but sensitivity and specificity of serology is lower than in children, and HLA is considered useful in specific and complex cases such as seronegative disease, discrepancy serology-histology or low grade of enteropathy. It seems also useful when a gluten free diet is adopted before diagnosis and when a lack of response to the gluten free diet exists in order to avoid wrong diagnosis, being these options also applicable to children (Protocol, 2018) (Table 3).

It must be remembered that the result of the genetics test must be used only to rule out CD, but never as the only criterion for diagnosis. This is especially relevant when considering first-degree relatives of CD patients and potential CD. Considering that HLA-DQ2/8 is present in almost all CD patients, the presence of HLA risk genetics is very high in families with affected members, but that condition does not lead unequivocally to CD and specificity is reduced in this setting. Potential CD or coeliac autoimmunity are also developed in individuals at HLA genetic risk. Thus, HLA should not be used in those cases to confirm a diagnosis.

Table 3 Recommendations for HLA typing in clinical practice.

Groups of study
Asymptomatic children with increased CD risk (CD familiarity, immune-related diseases except type 1 diabetes, other CD-related conditions) as the initial screening test, previously to the serological study and, if needed, the duodenal biopsy
People with positive serology and normal histology in order to make periodically follow-up of those with compatible genetics
People with high clinical suspicion but seronegative for anti-TG2 antibodies in order to consider further study or alternative diagnosis
People with high clinical suspicion but a biopsy showing increased intestinal intraepithelial lymphocytes without the presence of atrophy in order to consider alternative diagnosis
People following a GFD without a previous diagnosis who reject gluten challenge in order to consider alternative diagnosis
People with a lack of response to the GFD in order to avoid a wrong diagnosis

CD: celiac disease; GFD: gluten free diet.

7.3 Future perspectives

Some studies suggest the utility of HLA testing in additional situations, but further research is needed.

The presence of different CD risk categories based on HLA genetics has not been translated into clinical practice. However, it has been suggested that it could be used for developing personalized programs in children belonging to risk groups. Specifically, it could be applied to establish the intervals of serological screening aimed to early detection of subclinical CD (Martinez-Ojinaga et al., 2018; Megiorni et al., 2009). In addition, the genetics of a recently diagnosed CD patient could be useful to know the chance of finding new CD cases in the same family, being highest in HLA-DQ2.5 homozygous subjects.

In first-degree relatives of CD patients younger than 10 years, some studies suggest to place HLA-DQ typing as the first-line screening to identify the HLA-DQ2/DQ8 positive cases and repeat periodically the serological study in order to identify most cases of CD. After 10 years old, it is considered more unlikely to develop CD and one serological testing in asymptomatic individuals could be enough (Lionetti et al., 2014; Wessels et al., 2018). This will be probably a good approach, since similar results have been even observed in general population at genetic risk (Fernandez-Fernandez et al., 2019; Hoffenberg et al., 2003; Liu et al., 2017).

According to all these studies, the recent ESPGHAN guidelines (Husby et al., 2020) state that future research should be focused on studying the utility of HLA testing in at-risk groups and its cost-effectiveness using health economic models. It could be also important to investigate the acceptability and family understanding of HLA testing.

8. Influence of HLA on phenotypic variation

Some works have investigated the influence of HLA on different features related to CD: age at onset, clinical outcomes, antibody levels and grade of histological lesion. The conclusions of those studies must be cautiously interpreted since most of them present limitations. Although in many cases they are properly pointed out by the authors, they are not usually considered for final remarks. The most common limitations are:

- *Sample size.* Most studies include a small number of patients, in many cases around only 100 subjects (Congia et al., 1994; Nenna et al., 2008; Vermeulen et al., 2009; Zubillaga et al., 2002). It must be considered that patients are stratified according to the HLA status. Therefore, some HLA categories will have few or even zero patients when the total number is low. This prevents a complete evaluation of HLA influence and clearly reduces the statistical power, which can lead to a lack of association although it is present. Some works find differences between the HLA groups, but no statistical significance is reached. The small sample sizes allow authors to detect significant associations only when strong differences between groups exist.
- *HLA risk stratification.* The at-risk HLA classification differs among studies and it makes difficult the comparison between works. Some studies compare HLA-DQ2.5 homozygous vs HLA-DQ2.5 heterozygous, others HLA-DQ2.5 with double dose of *HLA-DQB1*02* vs HLA-DQ2.5 with single dose, or vs lower risk categories, etc. Depending on the real differences, the type of analysis will or will not allow detecting them.
- *Confounding factors.* There are several variables than can be influencing one of more of the studied features. They can act as confounding variables when analyzing the influence of HLA factors on those features and need to be considered. A clear example is age at onset.
- *Population heterogeneity.* Slight differences in the HLA frequencies exist among populations, which can affect the associations present.

- *Family history of CD*. Different characteristics can be present regarding HLA or other features depending on having or not relatives with CD. Differences in the proportion of familial cases among studies may lead to different results.
- *Clinical criteria*. The clinical classification of the patients can differ among centers, and may be some subjective, specially the way to classify asymptomatic CD.

Further research is needed using larger sample sizes, adjusting at least by age of onset and stratifying according to the presence of a family history of CD. A prospective design should be preferred.

In a recent meta-analysis, studies evaluating the effect of double, single and zero dose of *HLA-DQB1*02* on several outcomes were included (Bajor et al., 2019a). A double dose of that allele was associated with classical clinical presentation and atrophic lesion, but no association with age at diagnosis was detected. Nevertheless, most of the concerns previously described were still present, even the low number of patients. It is notorious the work performed by our research group by studying the largest series up to this moment. We considered only children with no family history of CD and adjusted results by age of onset and sex, children carrying HLA-DQ2.5 with double dose of *HLA-DQB1*02* showed earlier onset, more frequent classical clinical presentation and severe histological lesions (Martinez-Ojinaga et al., 2019).

8.1 Age at onset

An earlier CD onset has been described in children showing HLA-DQ2.5 homozygosity (Zubillaga et al., 2002) and those with double dose of *HLA-DQB1*02* (Martinez-Ojinaga et al., 2019; Nenna et al., 2008). There are several studies not showing association (Greco et al., 1998; Pena-Quintana et al., 2003; Vermeulen et al., 2009), and it is notorious the study of Ploski et al., who described association with later onset (after 3 years old) in children homozygous for HLA-DQ2.5 with double dose of *HLA-DQB1*02* (Ploski et al., 1993).

In a prospective study investigating CD development in individuals at genetic risk, an earlier onset in subjects homozygous for HLA-DQ2.5 when compared to heterozygous HLA-DQ2.5/HLA-D8 or homozygous HLA-DQ8 was observed (Liu et al., 2014). In contrast, the only study including exclusively adult patients did not find an association between the number of copies of *HLA-DQB1*02* and age at onset, although only 86 subjects were studied (Murray et al., 2007). When children and adults were considered together, only the one with a majority of children (Nenna et al., 2008) showed earliest onset with double dose of *HLA-DQB1*02* (Bajor et al., 2019b; Cabrera et al., 2019; Jores et al., 2007). There are two studies including subjects with a family history of CD and both showed earlier age of diagnosis with double dose of *HLA-DQB1*02* (Karinen et al., 2006; Wessels et al., 2018).

8.2 Clinical presentation

The influence of HLA on clinical presentation has been studied more extensively. Two different approaches were followed. One of them considered sibling pairs and analyzed the concordance rate of the clinical outcome according to the HLA status. Doing this, a recent study including 66 sibling pairs affected with CD observed low concordance between relatives in the clinical presentation classified as malabsorption or anemia, gastrointestinal symptoms, extra-intestinal symptoms and asymptomatic. The HLA genetics did not seem to explain the observed clinical differences, since HLA was concordant in 70% of the studied sibling pairs and 74% of the studied pairs showed clinical discordance, and a higher proportion of similar genetics was even seen in those pairs with discordant clinical presentation (Kauma et al., 2019). Similarly, a major role of HLA underlying the development of classical clinical symptoms or dermatitis herpetiformis had been previously discarded after studying 100 sibling pairs belonging to 25 families, although the parallel case-control study developed showed a significantly increased percentage of HLA-DQ2.5 subjects with two copies of *HLA-DQA1*05* in the group with dermatitis herpetiformis (Karell et al., 2002). Considering 28 families with one sibling showing asymptomatic CD and the other sib pair showing classical CD, all the individuals carry the HLA-DQ2.5 haplotype. The authors also investigated the influence of the second haplotype looking at the proportion of each genotype in the two groups of siblings, and observed a similar percentage between groups even when comparing with the class of the highest risk (two copies of *HLA-DQB1*02* vs the others) (Mustalahti et al., 2002). Gudjónsdóttir et al. also did not find an association between HLA and clinical symptoms after studying 224 siblings belonging to 107 families. A different distribution of HLA risk categories (HLA-DQ2.5 with double dose, HLA-DQ2.5 with single dose and non-HLA-DQ2.5) was not detected in the different clinical groups: classical CD, milder typical symptoms and asymptomatic (Gudjonsdottir et al., 2009). Only Karinen et al. found more severe diarrhea more frequently in homozygous for *HLA-DQB1*02* when considering 156 CD patients from 54 families (Karinen et al., 2006).

Therefore, studies in siblings mainly discard the influence of the HLA status on clinical manifestations and point out environmental factors or non-HLA genetic factors as the possible factors underlying the clinical differences. It must be noted that there is not an only criterion to classify clinical symptoms: symptomatic vs asymptomatic, classical vs non-classical, classical vs dermatitis herpetiformis, or several subdivisions according to specific symptoms have been used depending on the authors.

Other kind of studies has compared HLA genetics between groups of independent CD patients classified according to the clinical outcome. In contrast, these studies offered several significant associations, but a more homogenous criteria was mostly used, by classifying patients according to classical or non-classical (including asymptomatic forms) symptoms. In 1994, Congia et al. reported the increased homozygosity for the HLA-DQ2.5 haplotype in children with classical CD compared to those with mono- or oligosymptomatic forms (Congia et al., 1994), later corroborated by Cakir et al. (2014) and by Liu et al. in a prospective study including individuals at genetic risk (Liu et al., 2014). A higher proportion of double dose of *HLA-DQB1*02* has been also described in children with classical clinical presentation (Martinez-Ojinaga et al., 2019; Zubillaga et al., 2002). Vermeulen et al. considered six different clinical symptoms and four HLA risk categories, and observed the genotypes HLA-HLA2.5/DQ7.5 and HLA-DQ2.2/DQ7.5 increased in children with abdominal distension and decreased in those with non-gastrointestinal symptoms (Vermeulen et al., 2009). Greco et al. did not find association between clinical outcome and HLA, but 9 risk categories were considered in 145 children, which highly decrease the statistical power of the study (Greco et al., 1998). In adults, Murray et al. did not observe a significant association with HLA when considering severity of symptomatic cases (Murray et al., 2007). Concordantly, only one work (Piccini et al., 2012) detected a significant association between HLA and clinical symptoms when combining children and adults (Bajor et al., 2019a; Cabrera et al., 2019; Jores et al., 2007).

Family-based approaches have the advantage of reducing the effect of the environmental and the genetic heterogeneity of the studied group. Population stratification, which can lead to false significant associations, will be avoided. However, most of the studies including independent patients were conducted by a single center in quite genetically homogenous populations. This also reduces differences in clinical practice. However, this may not be true for family studies, with sib pairs often diagnosed by different practitioners.

A different possibility involves the particularities of CD cases with a family history of CD. They show some differences in the proportion of HLA risk groups. Moreover, they probably show higher proportion of asymptomatic individuals and earlier diagnosis onset, since they are included in screening programs. Of note, our group developed the study including the highest number of patients (463 children) and observed 5% of subjects with affected relatives. Some differences in the HLA genetic background were observed between both groups and the effect of HLA on different features was analyzed independently in both groups. Gudjónsdóttir et al. described more severe symptoms and earlier diagnosis in the index cases of families with several affected members (Gudjonsdottir et al., 2009).

Another potential difference includes the classification of patients, which is more heterogeneous in family-based approaches. Differences in the clinical presentation of children and adults with CD are also well known. The described associations involved children, but further research in adults is necessary. This could be also a difference with studies in families, which, although usually in low proportion, include parents (and therefore adults) among the siblings.

8.3 Serology

Few works have investigated the presence of anti-TG2 or EMA antibodies, or their levels, in relation to HLA status. All of them suggest stronger HLA risk associated to increased probability of developing antibodies and showing higher antibody levels.

A higher frequency of EMA antibodies was described in adults carrying HLA-DQ2 compared to those carrying HLA-DQ8 by Thomas et al. (2009), who also observed increased proportion of EMA-positive patients at higher frequency of the *HLA-DQB1*02* allele. Our group found a concordant result in children, with HLA-DQ2.5 children with double dose of *HLA-DQB1*02* showing more frequently anti-TG2 and/or EMA antibodies, although only a nearly significant result was found (Martinez-Ojinaga et al., 2019).

Regarding anti-TG2 levels, higher titers with double dose of *HLA-DQB1*02* were found studying children and adults together (Cabrera et al., 2019; Nenna et al., 2008). In children, higher levels were also observed in subjects carrying HLA-DQ2.5 in *cis* with a single *DBQ1*02* allele compared to HLA-DQ2.2 subjects (Delgado et al., 2014).

There are other two works concordant with these results. Bajor et al. studied children and adults and observed that high levels of anti-TG2 antibodies (>10 times the upper limit of normality) were significantly more frequent in individuals with double dose of *HLA-DQB1*02* (Bajor et al., 2019a). Choung et al. described that adults with HLA-DQ2.2 or lower risk HLA never show high anti-TG2 levels (Choung et al., 2020).

A prospective study involving newborns at genetic risk followed during 15 years found that persistent anti-TG2 levels were associated to carry two copies of the HLA-DQ2.5 haplotype vs carrying one or zero (Agardh et al., 2015).

8.4 Histological damage

The association of HLA risk and the severity of mucosal damage has been scarcely investigated.

Vermeulen et al. studied this issue in a pediatric sample. Their data showed increased severe histology (Marsh 3b–3c) in HLA-DQ2.5 children with double or single dose of *HLA-DQB1*02* compared to HLA categories conferring lower risk (Vermeulen et al., 2009). Our group also observed increased Marsh 3b–3c in children with higher risk, although a significant difference emerged between carrying double vs single *HLA-DQB1*02* dose (Martinez-Ojinaga et al., 2019). This issue was not considered in works including exclusively adult patients, but when children and adults were studied together, higher frequency of total villous atrophy was described with double dose of *HLA-DQB1*02* (Jores et al., 2007; Nenna et al., 2008).

Karinen et al. performed a study including 156 probands from 54 families and also found more severe histology (Marsh 3b–3c compared to Marsh 3a) with double dose of *HLA-DQB1*02*.

Three studies did not find association between the grade of villous atrophy and HLA risk. One of them considered double vs single dose of *HLA-DQB1*02* in Marsh 3c vs Marsh 3a and Marsh 3b together (Cabrera et al., 2019). The other two works did not show enough data that allow comparisons (Bajor et al., 2019b; Delgado et al., 2014).

9. Microbiota and HLA

The gut microbiome, the genome of our gut microbiota, is considered as our second genome. It carries 150 times as many genes as those contained in our human genome and has been linked to many common complex diseases (Zmora et al., 2019). In addition, so far scientific evidence has supported a causal role for gut microbiota in regulating predisposition to many diseases. The gut microbiota is composed of around 800 different species, many of them very difficult to culture under laboratory conditions (Backhed et al., 2005). Profound colonization of the human intestine begins immediately after birth and by the age of 2 turns into a relatively stable community that is unique to each individual. The first bacteria to colonize the human gut are facultative anaerobes (*Enterobacteriaceae*, *Lactobacillus* and *Streptococcus*), but in the first weeks of life and due to the growing oxygen consumption of these first colonizers, anaerobic bacteria such as *Bifidobacterium*, *Bacteroides*, *Clostridium* and *Eubacterium* begin to succeed (Mackie et al., 1999). Many environmental factors linked to CD predisposition have been shown to confer an important role in colonization and maintenance of gut microbiota, being diet one of the main environmental drivers of gut microbiota composition and function (Rothschild et al., 2018). Besides environmental factors, different studies have shown that host genetics also influence the composition of the human gut microbiome. Studies in mice (Leamy et al., 2014; Org et al., 2015) and in human twins (Goodrich et al., 2016) have observed substantial heritability for some microbiota members. Most of the genes that have been described as genes associated with microbiota changes encode factors involved in bacterial sensing and immune reactions (Bonder et al., 2016). Remarkably, some genetic variants that could mediate CD pathogenesis through gut microbiota have been very recently identified (Garcia-Santisteban et al., 2020).

HLA class II molecules present bacterial antigens to T cells and activate the immune system. Thus, HLA plays a major role in the immune response to pathogens and in the tolerogenic response to symbiotic and commensal bacteria. Nowadays, it is well known that gut microbiota colonization of the newborn intestine contributes to the proper development of the host's immune function and seems to determine susceptibility to immune-mediated disorders such as CD among others (Cenit et al., 2014). In this context, in recent years alterations in the gut microbiota are being investigated as part of the modifiable factors possibly involved in CD development. In addition, different studies have been performed aiming to identify the effect of HLA CD risk variants on microbiota composition and to investigate whether HLA risk genotypes for CD could act in part via the microbiota modulating CD development. Several studies have suggested that infants with HLA associated to high and low CD risk have different intestinal microbiota from very early on. Those studies included infants with high (HLA-DQ2.5 in homozygosis or HLA-DQ2.5 *in trans* configuration), medium (HLA-DQ2.5 *in cis* configuration and in heterozygosis, or HLA-DQ8 in homozygosis) or low (other genotypes) HLA genetic risk to develop

CD and at least one first-degree relative affected of CD and considered both types of feeding (breast, formula or mixed feeding) and types of birth (vaginally delivered or C-section) as potential confounders. In some studies, genetic risk categories were reduced to high (presence of HLA-DQ2.5 and/or HLA-DQ8) and low risk (other genotypes). It should be noted that authors do not consider HLA-DQ2.2 or HLA-DQ7.5 by themselves as risk factors, and individuals with these moderate or low risk genotypes could have been classified in these studies within the low genetic risk group. First, a work focused on *Bacteroides* genus that classified genetic risk in high or low studied the effect of the type of milk-feeding and HLA-DQ genotype on intestinal colonization of *Bacteroides* species and described that both aspects influence the colonization process, probably modifying disease risk (Sanchez et al., 2011). The authors also described that infants with low risk HLA had a higher diversity of *Bacteroides* species and indicated a higher prevalence of *B. vulgatus* in infants with high genetic risk regardless of type of feeding, while a higher prevalence of *B. ovatus*, *B. plebeius* and *B. uniformis* in those infants with low genetic risk. Next, in 2012, the Proficel study, a prospective study in a cohort of 164 infants classified in three groups according to HLA risk (high, medium and low) showed also a specific decrease in *Bifidobacterium* spp. and *B. longum*, and an increase in *Staphylococcus* spp. and *Bacteroides fragilis* associated with higher HLA genetic risk of developing CD regardless of milk-feeding type (Palma et al., 2012). In 2014, a microbiome analysis performed using next generation sequencing on a cohort of 22 infants breastfed, vaginally delivered and classified in high or low risk according to HLA genotype confirmed that the HLA-DQ genotype itself influenced the gut microbiota composition (Olivares et al., 2015). In this study, an increased proportion of *Firmicutes* and *Proteobacteria* and a reduction in *Actinobacteria* (including the genus *Bifidobacterium*) in the infant group with high risk HLA were described.

A published nested case-control study in 2018 (Olivares et al., 2018b) carried out as part of a larger prospective cohort study, analyzed 10 cases of CD and the 10 best-matched controls who did not develop the disease after 5-year follow-up. In this study, individuals were classified in three risk groups according to HLA genotypes. The authors found an increase in bacterial diversity over time, characterized by an increase in *Firmicutes* families in infants who remained healthy, but not in those who developed CD. It was also described how this altered microbial trajectory coincided with immune changes, suggesting that alterations observed in the early trajectory of gut microbiota already in infants at CD risk who were only 1 month old could influence the immune maturation process and predispose to CD.

Additionally, another recent study (Olivares et al., 2018a), with individuals classified also in three risk groups according to HLA genetics, highlighted the higher prevalence, among breastfed infants, of enterotoxigenic *Escherichia coli* in infants with the highest HLA genetic risk, compared to either those with a low or intermediate HLA risk. Furthermore, different studies performed using different animal models have also shown the presence of certain MHC polymorphisms influencing fecal microbiota composition (Bolnick et al., 2014; Toivanen et al., 2001).

So far we have limited knowledge about the mechanisms by which HLA-DQ genotypes could modulate the composition of the gut microbiota. However, De Palma et al. described an increased abundance of *Staphylococcus* spp. in the group of infants with higher CD genetic risk (HLA-DQ2.5), and considering that *Staphylococcal* superantigens bind directly to HLA class II molecules and strongly activate T cells, it could lead to speculate that different degrees of T-cell activation, depending on the antigen presented to the T cells, may contribute to regulate the gut microbiota (Palma et al., 2012).

10. Evolutionary considerations

A positive correlation between wheat consumption and the frequency of either HLA-DQ2 or the sum of HLA-DQ2 and HLA-DQ8 was found (Lionetti and Catassi, 2014). This fact has been called the "evolutionary coeliac paradox." It has been hypothesized that it could be due to a positive selective pressure for HLA-DQ2/DQ8. HLA-DQ2 has been described to protect against dental caries, which had a strong negative impact in ancient populations. In addition, it could be also a consequence of the evolutionary adaptation to dietary changes imposed with the widespread of wheat consumption. Other consideration emerging from this observation is that the high current CD prevalence is probably resulting from a chain of events that may include a role of other environmental and genetic factors, intestinal microbiota and quantity and quality of ingested gluten (Lionetti and Catassi, 2014).

11. New therapies based on HLA

Although the only treatment for CD is a gluten free diet, substantial efforts have been made to develop alternative therapies. Some initiatives attempt to avoid immune activation by preventing HLA mediated antigen presentation. Two different approaches have been followed: blocking peptides and recombinant TCR ligands.

In our intestine, gluten is highly resistant to enzymatic digestion and proteolytically resistant gluten peptides are generated by physiological processes. These peptides can be efficiently presented to disease associated T cells in a HLA mediated process. These therapies try to transform these naturally occurring T-cell stimulatory agents into inhibitors of HLA mediated antigen presentation. They have been focused on HLA-DQ2.5, since it is the predominant receptor in CD patients and the immunodominant epitopes have been extensively studied (Anderson et al., 2000; Sollid et al., 2012).

Blocking peptides are small fragments of deamidated gliadin designed to bind the HLA-DQ2 groove, but that are modified to exert mild or null activation effect on gliadin-reactive T cells. Anderson et al. modified the dominant epitope in α-gliadin for T cells found in CD patients carrying the HLA-DQ2.5 heterodimer (deamidated α/β gliadin p57–73 QE65). Reactive T cells activated by this modified peptide were able to secrete the suppressive cytokine IL-10 but lacked the IFN-γ stimulatory activity, thus acting as antagonist of p57–73 QE65 (Anderson et al., 2006). Xia et al. focused on the immunogenic 33-mer peptide of gliadin and identified the most important residues for the high-affinity HLA-DQ2 binding, which were modified in order to get simple analogs that retain high HLA-DQ2 affinity but abrogate T-cell recognition (Xia et al., 2006). Kapoerchan et al. designed a series of gluten peptides in which proline residues were replaced by azidoprolines. Some peptides maintained good affinity for the HLA-DQ2 receptor and inhibit T-cell proliferation, but the ability to compete with natural peptides was not good enough (Kapoerchan et al., 2008). Cyclic and dimeric peptides have been also proposed as HLA blockers (Xia et al., 2007).

Recombinant TCR ligands (RTLs) are partial HLA molecules containing the α1 and β1 domains of the HLA-DQ molecule bound to specific antigenic peptides (Wang et al., 2003). In CD, Huan et al. tried this approach using HLA-DQ2.5-derived molecules bearing a gliadin derived peptide and a defective activation of gliadin-reactive T cells was observed (Huan et al., 2011).

This kind of therapy remains in the preclinical phase. It's in vivo efficacy and safety are unclear. One of the main concerns is that HLA is required in different immune responses, which could be affected by the suggested approaches. Moreover, these blockers have only shown moderate efficacy in inhibiting gluten-induced T-cell activation in vitro.

12. Common mistakes

The HLA region is very complex and a comprehensive understanding is often difficult. There are some important considerations that are not always taken into account when studying this region and can lead to misinterpretations. We will try to summarize the most common ones.

First, it must be remembered that the association between CD susceptibility and HLA lies on the encoded HLA-DQ heterodimeric receptors. *HLA-DQA1* and *HLA-DQB1* are encoding those heterodimers either in *cis* or in *trans*, which implies that up to four different receptors can be found in each individual (this will happen when four different alleles, considering both genes, are present). Therefore, information of both haplotypes (i.e. the genotype) must be shown; otherwise, it could lead to wrong conclusions.

Second, HLA-DQ2.5 is the predominant heterodimer in the totality of CD, being present in around 90–95% of the patients. Therefore, when less than 100 individuals are studied, very few non-HLA-DQ2.5 patients are expected and none could be observed. This can contribute to overestimate the frequency of the cited risk genotype. In the same manner, when we wanted to explore the association of specific HLA genotypes with different features (clinical presentation, histological damage, etc.) we must be sure of having enough representation of different HLA risk categories in order to be able to reject the presence of associations with enough statistical power. Representative samples of the studied populations must be used, which is not possible when including low numbers of patients.

Third, but very important, the extremely high representation of HLA-DQ2.5 in CD makes that the remaining haplotype/genotypes appear at low frequency. This should not be misinterpreted as a protective effect conferred by those haplotypes. Risk calculations when different risk factors are present require stratified analysis in which haplotypes with higher risk are removed from the sample before performing subsequent analyses. The totality of CD patients have a particular genetics, all of them, excepting the extremely rare included as non-HLA risk alleles, must be considered as permissive HLA genetics. Statistical analyses must take all these characteristics of CD genetics into consideration in their designs.

Acknowledgments

We appreciate MC Cénit for her expert scientific advice and critical reading of the HLA and microbiota section. C. Núñez receives a grant from "Fondo de Investigaciones Sanitarias, Instituto de Salud Carlos III-Fondo Europeo de Desarrollo Regional (FEDER)" (grant number PI18/00989).

References

Agardh, D., Lee, H.S., Kurppa, K., Simell, V., Aronsson, C.A., Jorneus, O., Hummel, M., Liu, E., Koletzko, S., TEDDY Study Group, 2015. Clinical features of celiac disease: a prospective birth cohort. Pediatrics 135, 627–634.

Ahmad, T., Neville, M., Marshall, S.E., Armuzzi, A., Mulcahy-Hawes, K., Crawshaw, J., Sato, H., Ling, K.L., Barnardo, M., Goldthorpe, S., Walton, R., Bunce, M., Jewell, D.P., Welsh, K.I., 2003. Haplotype-specific linkage disequilibrium patterns define the genetic topography of the human MHC. Hum. Mol. Genet. 12, 647–656.

Anderson, R.P., Degano, P., Godkin, A.J., Jewell, D.P., Hill, A.V., 2000. In vivo antigen challenge in celiac disease identifies a single transglutaminase-modified peptide as the dominant A-gliadin T-cell epitope. Nat. Med. 6, 337–342.

Anderson, R.P., van Heel, D.A., Tye-Din, J.A., Jewell, D.P., Hill, A.V., 2006. Antagonists and non-toxic variants of the dominant wheat gliadin T cell epitope in coeliac disease. Gut 55, 485–491.

Backhed, F., Ley, R.E., Sonnenburg, J.L., Peterson, D.A., Gordon, J.I., 2005. Host-bacterial mutualism in the human intestine. Science 307, 1915–1920.

Bajor, J., Szakacs, Z., Farkas, N., Hegyi, P., Illes, A., Solymar, M., Petervari, E., Balasko, M., Par, G., Sarlos, P., Szucs, A., Czimmer, J., Szemes, K., Huszar, O., Varju, P., Vincze, A., 2019a. Classical celiac disease is more frequent with a double dose of HLA-DQB1*02: a systematic review with meta-analysis. PLoS One 14, e0212329.

Bajor, J., Szakacs, Z., Juhasz, M., Papp, M., Kocsis, D., Szegedi, E., Foldi, I., Farkas, N., Hegyi, P., Vincze, A., 2019b. HLA-DQ2 homozygosis increases tTGA levels at diagnosis but does not influence the clinical phenotype of coeliac disease: a multicentre study. Int. J. Immunogenet. 46, 74–81.

Bergseng, E., Dorum, S., Arntzen, M.O., Nielsen, M., Nygard, S., Buus, S., de Souza, G.A., Sollid, L.M., 2015. Different binding motifs of the celiac disease-associated HLA molecules DQ2.5, DQ2.2, and DQ7.5 revealed by relative quantitative proteomics of endogenous peptide repertoires. Immunogenetics 67, 73–84.

Bodd, M., Kim, C.Y., Lundin, K.E., Sollid, L.M., 2012a. T-cell response to gluten in patients with HLA-DQ2.2 reveals requirement of peptide-MHC stability in celiac disease. Gastroenterology 142, 552–561.

Bodd, M., Tollefsen, S., Bergseng, E., Lundin, K.E., Sollid, L.M., 2012b. Evidence that HLA-DQ9 confers risk to celiac disease by presence of DQ9-restricted gluten-specific T cells. Hum. Immunol. 73, 376–381.

Bolnick, D.I., Snowberg, L.K., Caporaso, J.G., Lauber, C., Knight, R., Stutz, W.E., 2014. Major histocompatibility complex class IIb polymorphism influences gut microbiota composition and diversity. Mol. Ecol. 23, 4831–4845.

Bonder, M.J., Kurilshikov, A., Tigchelaar, E.F., Mujagic, Z., Imhann, F., Vila, A.V., Deelen, P., Vatanen, T., Schirmer, M., Smeekens, S.P., Zhernakova, D.V., Jankipersadsing, S.A., Jaeger, M., Oosting, M., Cenit, M.C., Masclee, A.A., Swertz, M.A., Li, Y., Kumar, V., Joosten, L., Harmsen, H., Weersma, R.K., Franke, L., Hofker, M.H., Xavier, R.J., Jonkers, D., Netea, M.G., Wijmenga, C., Fu, J., Zhernakova, A., 2016. The effect of host genetics on the gut microbiome. Nat. Genet. 48, 1407–1412.

Brown, N.K., Guandalini, S., Semrad, C., Kupfer, S.S., 2019. A clinician's guide to celiac disease HLA genetics. Am. J. Gastroenterol. 114, 1587–1592.

Cabrera, C.M., Sanchez-Godoy, L., Navas-Lopez, V.M., 2019. Is the double gene dose of DQ2.5 or DQ2.5/DQ2.2 an involved factor in the clinical features of celiac disease? Scand. J. Gastroenterol. 54, 960–964.

Cakir, M., Baran, M., Ucar, F., Akbulut, U.E., Kaklikkaya, N., Ersoz, S., 2014. Accuracy of HLA-DQ genotyping in combination with IgA anti-tissue transglutaminase serology and a "scoring system" for the diagnosis of celiac disease in Turkish children. Turk. J. Pediatr. 56, 347–353.

Cenit, M.C., Matzaraki, V., Tigchelaar, E.F., Zhernakova, A., 2014. Rapidly expanding knowledge on the role of the gut microbiome in health and disease. Biochim. Biophys. Acta 1842, 1981–1992.

Choung, R.S., Mills, J.R., Snyder, M.R., Murray, J.A., Gandhi, M.J., 2020. Celiac disease risk stratification based on HLA-DQ heterodimer (HLA-DQA1 ~ DQB1) typing in a large cohort of adults with suspected celiac disease. Hum. Immunol. 81, 59–64.

Congia, M., Cucca, F., Frau, F., Lampis, R., Melis, L., Clemente, M.G., Cao, A., de Virgiliis, S., 1994. A gene dosage effect of the DQA1*0501/DQB1*0201 allelic combination influences the clinical heterogeneity of celiac disease. Hum. Immunol. 40, 138–142.

Delgado, J.F., Amengual, M.J., Veraguas, A., Rodriguez, E., de los Santos, M. M. & Guallarte, M. P., 2014. Paediatric celiac patients carrying the HLA-DR7-DQ2 and HLA-DR3-DQ2 haplotypes display small clinical differences. Acta Paediatr. 103, e238–e242.

Demarchi, M., Borelli, I., Olivetti, E., Richiardi, P., Wright, P., Ansaldi, N., Barbera, C., Santini, B., 1979. Two HLA-D and DR alleles are associated with coeliac disease. Tissue Antigens 14, 309–316.

Demarchi, M., Carbonara, A., Ansaldi, N., Santini, B., Barbera, C., Borelli, I., Rossino, P., Rendine, S., 1983. HLA-DR3 and DR7 in coeliac disease: immunogenetic and clinical aspects. Gut 24, 706–712.

Dieterich, W., Ehnis, T., Bauer, M., Donner, P., Volta, U., Riecken, E.O., Schuppan, D., 1997. Identification of tissue transglutaminase as the autoantigen of celiac disease. Nat. Med. 3, 797–801.

Falchuk, Z.M., Rogentine, G.N., Strober, W., 1972. Predominance of histocompatibility antigen HL-A8 in patients with gluten-sensitive enteropathy. J. Clin. Invest. 51, 1602–1605.

Fallang, L.E., Bergseng, E., Hotta, K., Berg-Larsen, A., Kim, C.Y., Sollid, L.M., 2009. Differences in the risk of celiac disease associated with HLA-DQ2.5 or HLA-DQ2.2 are related to sustained gluten antigen presentation. Nat. Immunol. 10, 1096–1101.

Fernandez-Banares, F., Arau, B., Dieli-Crimi, R., Rosinach, M., Nunez, C., Esteve, M., 2017. Systematic review and meta-analysis show 3% of patients with celiac disease in Spain to be negative for HLA-DQ2.5 and HLA-DQ8. Clin. Gastroenterol. Hepatol. 15, 594–596.

Fernandez-Fernandez, S., Borrell, B., Cilleruelo, M.L., Tabares, A., Jimenez-Jimenez, J., Rayo, A.I., Perucho, T., Garcia-Garcia, M.L., 2019. Prevalence of celiac disease in a long-term study of a Spanish at-genetic-risk cohort from the general population. J. Pediatr. Gastroenterol. Nutr. 68, 364–370.

Garcia-Santisteban, I., Cilleros-Portet, A., Moyua-Ormazabal, E., Kurilshikov, A., Zhernakova, A., Garcia-Etxebarria, K., Fernandez-Jimenez, N., Bilbao, J.R., 2020. A two-sample Mendelian randomization analysis investigates associations between gut microbiota and Celiac disease. Nutrients 12, 1420–1429.

Gianfrani, C., Pisapia, L., Picascia, S., Strazzullo, M., del Pozzo, G., 2018. Expression level of risk genes of MHC class II is a susceptibility factor for autoimmunity: new insights. J. Autoimmun. 89, 1–10.

Goodrich, J.K., Davenport, E.R., Beaumont, M., Jackson, M.A., Knight, R., Ober, C., Spector, T.D., Bell, J.T., Clark, A.G., Ley, R.E., 2016. Genetic determinants of the gut microbiome in UK twins. Cell Host Microbe 19, 731–743.

Greco, L., Percopo, S., Clot, F., Bouguerra, F., Babron, M.C., Eliaou, J.F., Franzese, C., Troncone, R., Clerget-Darpoux, F., 1998. Lack of correlation between genotype and phenotype in celiac disease. J. Pediatr. Gastroenterol. Nutr. 26, 286–290.

Greco, L., Romino, R., Coto, I., di Cosmo, N., Percopo, S., Maglio, M., Paparo, F., Gasperi, V., Limongelli, M.G., Cotichini, R., D'agate, C., Tinto, N., Sacchetti, L., Tosi, R., Stazi, M.A., 2002. The first large population based twin study of coeliac disease. Gut 50, 624–628.

Gudjonsdottir, A.H., Nilsson, S., Naluai, A.T., Ek, J., Amundsen, S.S., Wahlstrom, J., Ascher, H., 2009. Association between genotypes and phenotypes in coeliac disease. J. Pediatr. Gastroenterol. Nutr. 49, 165–169.

Gutierrez-Achury, J., Zhernakova, A., Pulit, S.L., Trynka, G., Hunt, K.A., Romanos, J., Raychaudhuri, S., van Heel, D.A., Wijmenga, C., de Bakker, P.I., 2015. Fine mapping in the MHC region accounts for 18% additional genetic risk for celiac disease. Nat. Genet. 47, 577–578.

Hausch, F., Shan, L., Santiago, N.A., Gray, G.M., Khosla, C., 2002. Intestinal digestive resistance of immunodominant gliadin peptides. Am. J. Physiol. Gastrointest. Liver Physiol. 283, G996–G1003.

Hoffenberg, E.J., Mackenzie, T., Barriga, K.J., Eisenbarth, G.S., Bao, F., Haas, J.E., Erlich, H., Bugawan, T.L., Sokol, R.J., Taki, I., Norris, J.M., Rewers, M., 2003. A prospective study of the incidence of childhood celiac disease. J. Pediatr. 143, 308–314.

Huan, J., Meza-Romero, R., Mooney, J.L., Vandenbark, A.A., Offner, H., Burrows, G.G., 2011. Single-chain recombinant HLA-DQ2.5/peptide molecules block alpha2-gliadin-specific pathogenic CD4+ T-cell proliferation and attenuate production of inflammatory cytokines: a potential therapy for celiac disease. Mucosal Immunol. 4, 112–120.

Husby, S., Koletzko, S., Korponay-Szabo, I.R., Mearin, M.L., Phillips, A., Shamir, R., Troncone, R., Giersiepen, K., Branski, D., Catassi, C., Lelgeman, M., Maki, M., Ribes-Koninckx, C., Ventura, A., Zimmer, K.P., ESPGHAN Working Group on Coeliac Disease Diagnosis, ESPGHAN Gastroenterology Committee, European Society for Pediatric Gastroenterology, Hepatology, and Nutrition, 2012. European society for pediatric gastroenterology, hepatology, and nutrition guidelines for the diagnosis of coeliac disease. J. Pediatr. Gastroenterol. Nutr. 54, 136–160.

Husby, S., Koletzko, S., Korponay-Szabo, I., Kurppa, K., Mearin, M.L., Ribes-Koninckx, C., Shamir, R., Troncone, R., Auricchio, R., Castillejo, G., Christensen, R., Dolinsek, J., Gillett, P., Hrobjartsson, A., Koltai, T., Maki, M., Nielsen, S.M., Popp, A., Stordal, K., Werkstetter, K., Wessels, M., 2020. European society paediatric gastroenterology, hepatology and nutrition guidelines for diagnosing coeliac disease 2020. J. Pediatr. Gastroenterol. Nutr. 70, 141–156.

Jores, R.D., Frau, F., Cucca, F., Grazia Clemente, M., Orru, S., Rais, M., de Virgiliis, S., Congia, M., 2007. HLA-DQB1*0201 homozygosis predisposes to severe intestinal damage in celiac disease. Scand. J. Gastroenterol. 42, 48–53.

Kagnoff, M.F., 2007. Celiac disease: pathogenesis of a model immunogenetic disease. J. Clin. Invest. 117, 41–49.

Kapoerchan, V.V., Wiesner, M., Overhand, M., van der Marel, G.A., Koning, F., Overkleeft, H.S., 2008. Design of azidoproline containing gluten peptides to suppress CD4+ T-cell responses associated with celiac disease. Bioorg. Med. Chem. 16, 2053–2062.

Karell, K., Korponay-Szabo, I., Szalai, Z., Holopainen, P., Mustalahti, K., Collin, P., Maki, M., Partanen, J., 2002. Genetic dissection between coeliac disease and dermatitis herpetiformis in sib pairs. Ann. Hum. Genet. 66, 387–392.

Karell, K., Louka, A.S., Moodie, S.J., Ascher, H., Clot, F., Greco, L., Ciclitira, P.J., Sollid, L.M., Partanen, J., European Genetics Cluster on Celiac Disease, 2003. HLA types in celiac disease patients not carrying the DQA1*05-DQB1*02 (DQ2) heterodimer: results from the European genetics cluster on celiac disease. Hum. Immunol. 64, 469–477.

Karinen, H., Karkkainen, P., Pihlajamaki, J., Janatuinen, E., Heikkinen, M., Julkunen, R., Kosma, V.M., Naukkarinen, A., Laakso, M., 2006. Gene dose effect of the DQB1*0201 allele contributes to severity of coeliac disease. Scand. J. Gastroenterol. 41, 191–199.

Karnes, J.H., Bastarache, L., Shaffer, C.M., Gaudieri, S., Xu, Y., Glazer, A.M., Mosley, J.D., Zhao, S., Raychaudhuri, S., Mallal, S., Ye, Z., Mayer, J.G., Brilliant, M.H., Hebbring, S.J., Roden, D.M., Phillips, E.J., Denny, J.C., 2017. Phenome-wide scanning identifies multiple diseases and disease severity phenotypes associated with HLA variants. Sci. Transl. Med. 9, 389–416.

Kauma, S., Kaukinen, K., Huhtala, H., Kivela, L., Pekki, H., Salmi, T., Saavalainen, P., Lindfors, K., Kurppa, K., 2019. The phenotype of celiac disease has low concordance between siblings, despite a similar distribution of HLA haplotypes. Nutrients 11, 479–488.

Keuning, J.J., Pena, A.S., van Leeuwen, A., van Hooff, J.P., va rood, J. J., 1976. HLA-DW3 associated with coeliac disease. Lancet 1, 506–508.

Kuja-Halkola, R., Lebwohl, B., Halfvarson, J., Wijmenga, C., Magnusson, P.K., Ludvigsson, J.F., 2016. Heritability of non-HLA genetics in coeliac disease: a population-based study in 107 000 twins. Gut 65, 1793–1798.

Leamy, L.J., Kelly, S.A., Nietfeldt, J., Legge, R.M., Ma, F., Hua, K., Sinha, R., Peterson, D.A., Walter, J., Benson, A.K., Pomp, D., 2014. Host genetics and diet, but not immunoglobulin A expression, converge to shape compositional features of the gut microbiome in an advanced intercross population of mice. Genome Biol. 15, 552.

Lenz, T.L., Deutsch, A.J., Han, B., Hu, X., Okada, Y., Eyre, S., Knapp, M., Zhernakova, A., Huizinga, T.W., Abecasis, G., Becker, J., Boeckxstaens, G.E., Chen, W.M., Franke, A., Gladman, D.D., Gockel, I., Gutierrez-Achury, J., Martin, J., Nair, R.P., Nothen, M.M., Onengut-Gumuscu, S., Rahman, P., Rantapaa-Dahlqvist, S., Stuart, P.E., Tsoi, L.C., van Heel, D.A., Worthington, J., Wouters, M.M., Klareskog, L., Elder, J.T., Gregersen, P.K., Schumacher, J., Rich, S.S., Wijmenga, C., Sunyaev, S.R., de Bakker, P.I., Raychaudhuri, S., 2015. Widespread non-additive and interaction effects within HLA loci modulate the risk of autoimmune diseases. Nat. Genet. 47, 1085–1090.

Lionetti, E., Catassi, C., 2014. Co-localization of gluten consumption and HLA-DQ2 and -DQ8 genotypes, a clue to the history of celiac disease. Dig. Liver Dis. 46, 1057–1063.

Lionetti, E., Castellaneta, S., Francavilla, R., Pulvirenti, A., Tonutti, E., Amarri, S., Barbato, M., Barbera, C., Barera, G., Bellantoni, A., Castellano, E., Guariso, G., Limongelli, M.G., Pellegrino, S., Polloni, C., Ughi, C., Zuin, G., Fasano, A., Catassi, C., Weaning, S.W.G.O., Risk, C.D., 2014. Introduction of gluten, HLA status, and the risk of celiac disease in children. N. Engl. J. Med. 371, 1295–1303.

Liu, J., Juo, S.H., Holopainen, P., Terwilliger, J., Tong, X., Grunn, A., Brito, M., Green, P., Mustalahti, K., Maki, M., Gilliam, T.C., Partanen, J., 2002. Genomewide linkage analysis of celiac disease in Finnish families. Am. J. Hum. Genet. 70, 51–59.

Liu, E., Lee, H.S., Aronsson, C.A., Hagopian, W.A., Koletzko, S., Rewers, M.J., Eisenbarth, G.S., Bingley, P.J., Bonifacio, E., Simell, V., Agardh, D., TEDDY Study Group, 2014. Risk of pediatric celiac disease according to HLA haplotype and country. N. Engl. J. Med. 371, 42–49.

Liu, E., Dong, F., Baron, A.E., Taki, I., Norris, J.M., Frohnert, B.I., Hoffenberg, E.J., Rewers, M., 2017. High incidence of celiac disease in a long-term study of adolescents with susceptibility genotypes. Gastroenterology 152 (1329–1336), e1.

Lopes, L.H.C., Muniz, J.G., Oliveira, R.P., Sdepanian, V.L., 2019. Celiac disease in Brazilian first-degree relatives: the odds are five times greater for HLA DQ2 homozygous. J. Pediatr. Gastroenterol. Nutr. 68, e77–e80.

Louka, A.S., Nilsson, S., Olsson, M., Talseth, B., Lie, B.A., Ek, J., Gudjonsdottir, A.H., Ascher, H., Sollid, L.M., 2002. HLA in coeliac disease families: a novel test of risk modification by the 'other' haplotype when at least one DQA1*05-DQB1*02 haplotype is carried. Tissue Antigens 60, 147–154.

Lundin, K.E., Scott, H., Hansen, T., Paulsen, G., Halstensen, T.S., Fausa, O., Thorsby, E., Sollid, L.M., 1993. Gliadin-specific, HLA-DQ(alpha 1*0501,beta 1*0201) restricted T cells isolated from the small intestinal mucosa of celiac disease patients. J. Exp. Med. 178, 187–196.

Lundin, K.E., Scott, H., Fausa, O., Thorsby, E., Sollid, L.M., 1994. T cells from the small intestinal mucosa of a DR4, DQ7/DR4, DQ8 celiac disease patient preferentially recognize gliadin when presented by DQ8. Hum. Immunol. 41, 285–291.

Mackie, R.I., Sghir, A., Gaskins, H.R., 1999. Developmental microbial ecology of the neonatal gastrointestinal tract. Am. J. Clin. Nutr. 69, 1035S–1045S.

Margaritte-Jeannin, P., Babron, M.C., Bourgey, M., Louka, A.S., Clot, F., Percopo, S., Coto, I., Hugot, J.P., Ascher, H., Sollid, L.M., Greco, L., Clerget-Darpoux, F., 2004. HLA-DQ relative risks for coeliac disease in European populations: a study of the European genetics cluster on coeliac disease. Tissue Antigens 63, 562–567.

Martinez-Ojinaga, E., Molina, M., Polanco, I., Urcelay, E., Nunez, C., 2018. HLA-DQ distribution and risk assessment of celiac disease in a Spanish center. Rev. Esp. Enferm. Dig. 110, 421–426.

Martinez-Ojinaga, E., Fernandez-Prieto, M., Molina, M., Polanco, I., Urcelay, E., Nunez, C., 2019. Influence of HLA on clinical and analytical features of pediatric celiac disease. BMC Gastroenterol. 19, 91.

Mearin, M.L., Biemond, I., Pena, A.S., Polanco, I., Vazquez, C., Schreuder, G.T., de Vries, R.R., van Rood, J.J., 1983. HLA-DR phenotypes in Spanish coeliac children: their contribution to the understanding of the genetics of the disease. Gut 24, 532–537.

Medrano, L.M., Dema, B., Lopez-Larios, A., Maluenda, C., Bodas, A., Lopez-Palacios, N., Figueredo, M.A., Fernandez-Arquero, M., Nunez, C., 2012. HLA and celiac disease susceptibility: new genetic factors bring open questions about the HLA influence and gene-dosage effects. PLoS One 7, e48403.

Megiorni, F., Mora, B., Bonamico, M., Barbato, M., Nenna, R., Maiella, G., Lulli, P., Mazzilli, M.C., 2009. HLA-DQ and risk gradient for celiac disease. Hum. Immunol. 70, 55–59.

Molberg, O., Kett, K., Scott, H., Thorsby, E., Sollid, L.M., Lundin, K.E., 1997. Gliadin specific, HLA DQ2-restricted T cells are commonly found in small intestinal biopsies from coeliac disease patients, but not from controls. Scand. J. Immunol. 46, 103–109.

Murray, J.A., Moore, S.B., van Dyke, C.T., Lahr, B.D., Dierkhising, R.A., Zinsmeister, A.R., Melton 3rd, L.J., Kroning, C.M., El-Youssef, M., Czaja, A.J., 2007. HLA DQ gene dosage and risk and severity of celiac disease. Clin. Gastroenterol. Hepatol. 5, 1406–1412.

Mustalahti, K., Holopainen, P., Karell, K., Maki, M., Partanen, J., 2002. Genetic dissection between silent and clinically diagnosed symptomatic forms of coeliac disease in multiplex families. Dig. Liver Dis. 34, 842–845.

Nenna, R., Mora, B., Megiorni, F., Mazzilli, M.C., Magliocca, F.M., Tiberti, C., Bonamico, M., 2008. HLA-DQB1*02 dose effect on RIA anti-tissue transglutaminase autoantibody levels and clinicopathological expressivity of celiac disease. J. Pediatr. Gastroenterol. Nutr. 47, 288–292.

Nistico, L., Fagnani, C., Coto, I., Percopo, S., Cotichini, R., Limongelli, M.G., Paparo, F., D'alfonso, S., Giordano, M., Sferlazzas, C., Magazzu, G., Momigliano-Richiardi, P., Greco, L., Stazi, M.A., 2006. Concordance, disease progression, and heritability of coeliac disease in Italian twins. Gut 55, 803–808.

Nunez, C., Garrote, J.A., Arranz, E., Bilbao, J.R., Fernandez Banares, F., Jimenez, J., Perucho, T., Ruiz Casares, E., Sanchez-Valverde, F., Serrano, J.I., 2018. Recommendations to report and interpret HLA genetic findings in coeliac disease. Rev. Esp. Enferm. Dig. 110, 458–461.

Olivares, M., Neef, A., Castillejo, G., Palma, G.D., Varea, V., Capilla, A., Palau, F., Nova, E., Marcos, A., Polanco, I., Ribes-Koninckx, C., Ortigosa, L., Izquierdo, L., Sanz, Y., 2015. The HLA-DQ2 genotype selects for early intestinal microbiota composition in infants at high risk of developing coeliac disease. Gut 64, 406–417.

Olivares, M., Benitez-Paez, A., de Palma, G., Capilla, A., Nova, E., Castillejo, G., Varea, V., Marcos, A., Garrote, J.A., Polanco, I., Donat, E., Ribes-Koninckx, C., Calvo, C., Ortigosa, L., Palau, F., Sanz, Y., 2018a. Increased prevalence of pathogenic bacteria in the gut microbiota of infants at risk of developing celiac disease: the PROFICEL study. Gut Microbes 9, 551–558.

Olivares, M., Walker, A.W., Capilla, A., Benitez-Paez, A., Palau, F., Parkhill, J., Castillejo, G., Sanz, Y., 2018b. Gut microbiota trajectory in early life may predict development of celiac disease. Microbiome 6, 36.

Org, E., Parks, B.W., Joo, J.W., Emert, B., Schwartzman, W., Kang, E.Y., Mehrabian, M., Pan, C., Knight, R., Gunsalus, R., Drake, T.A., Eskin, E., Lusis, A.J., 2015. Genetic and environmental control of host-gut microbiota interactions. Genome Res. 25, 1558–1569.

Ozgenel, S.M., Temel, T., Uskudar Teke, H., Yildiz, P., Korkmaz, H., Ozakyol, A., 2019. HLA-DQ2/DQ8 frequency in adult patients with celiac disease, their first-degree relatives, and normal population in Turkey. Turk. J. Gastroenterol. 30, 321–325.

Palma, G.D., Capilla, A., Nova, E., Castillejo, G., Varea, V., Pozo, T., Garrote, J.A., Polanco, I., Lopez, A., Ribes-Koninckx, C., Marcos, A., Garcia-Novo, M.D., Calvo, C., Ortigosa, L., Pena-Quintana, L., Palau, F., Sanz, Y., 2012. Influence of milk-feeding type and genetic risk of developing coeliac disease on intestinal microbiota of infants: the PROFICEL study. PLoS One 7, e30791.

Pena-Quintana, L., Torres-Galvan, M.J., Deniz-Naranjo, M.C., Ortigosa-Castillo, L., Ramos-Varela, J.C., Calvo-Hernandez, F., Fiuza-Perez, M.D., Rodriguez-Gallego, J.C., Sanchez-Garcia, F., 2003. Assessment of the DQ heterodimer test in the diagnosis of celiac disease in the Canary Islands (Spain). J. Pediatr. Gastroenterol. Nutr. 37, 604–608.

Piacantelli, D., Ben El Barhdadi, I., Oumhani, K., Sebastiani, P., Colanardi, A., Essaid, A., 2017. HLA typing and celiac disease in moroccans. Med Sci (Basel) 5, 2–12.

Piccini, B., Vascotto, M., Serracca, L., Luddi, A., Margollicci, M.A., Balestri, P., Vindigni, C., Bassotti, G., Villanacci, V., 2012. HLA-DQ typing in the diagnostic algorithm of celiac disease. Rev. Esp. Enferm. Dig. 104, 248–254.

Pisapia, L., Camarca, A., Picascia, S., Bassi, V., Barba, P., del Pozzo, G., Gianfrani, C., 2016. HLA-DQ2.5 genes associated with celiac disease risk are preferentially expressed with respect to non-predisposing HLA genes: implication for anti-gluten T cell response. J. Autoimmun. 70, 63–72.

Ploski, R., Ek, J., Thorsby, E., Sollid, L.M., 1993. On the HLA-DQ(alpha 1*0501, beta 1*0201)-associated susceptibility in celiac disease: a possible gene dosage effect of DQB1*0201. Tissue Antigens 41, 173–177.

Protocol, 2018. Early Diagnosis of Celiac Disease Protocol. Working group for Early Diagnosis of Celiac Disease protocol, Ministerio de Sanidad, Servicios Sociales e Igualdad. Servicio de Evaluación del Servicio Canario de la Salud (SESCS).

Rothschild, D., Weissbrod, O., Barkan, E., Kurilshikov, A., Korem, T., Zeevi, D., Costea, P.I., Godneva, A., Kalka, I.N., Bar, N., Shilo, S., Lador, D., Vila, A.V., Zmora, N., Pevsner-Fischer, M., Israeli, D., Kosower, N., Malka, G., Wolf, B.C., Avnit-Sagi, T., Lotan-Pompan, M., Weinberger, A., Halpern, Z., Carmi, S., Fu, J., Wijmenga, C., Zhernakova, A., Elinav, E., Segal, E., 2018. Environment dominates over host genetics in shaping human gut microbiota. Nature 555, 210–215.

Ruiz-Ortiz, E., Montraveta, M., Cabre, E., Herrero-Mata, M.J., Pujol-Borrell, R., Palou, E., Faner, R., 2014. HLA-DQ2/DQ8 and HLA-DQB1*02 homozygosity typing by real-time polymerase chain reaction for the assessment of celiac disease genetic risk: evaluation of a Spanish celiac population. Tissue Antigens 84, 545–553.

Sanchez, E., de Palma, G., Capilla, A., Nova, E., Pozo, T., Castillejo, G., Varea, V., Marcos, A., Garrote, J.A., Polanco, I., Lopez, A., Ribes-Koninckx, C., Garcia-Novo, M.D., Calvo, C., Ortigosa, L., Palau, F., Sanz, Y., 2011. Influence of environmental and genetic factors linked to celiac disease risk on infant gut colonization by Bacteroides species. Appl. Environ. Microbiol. 77, 5316–5323.

Siegel, M., Strnad, P., Watts, R.E., Choi, K., Jabri, B., Omary, M.B., Khosla, C., 2008. Extracellular transglutaminase 2 is catalytically inactive, but is transiently activated upon tissue injury. PLoS One 3, e1861.

Singh, P., Arora, S., Lal, S., Strand, T.A., Makharia, G.K., 2015. Risk of celiac disease in the first- and second-degree relatives of patients with celiac disease: a systematic review and meta-analysis. Am. J. Gastroenterol. 110, 1539–1548.

Sollid, L.M., Markussen, G., Ek, J., Gjerde, H., Vartdal, F., Thorsby, E., 1989. Evidence for a primary association of celiac disease to a particular HLA-DQ alpha/beta heterodimer. J. Exp. Med. 169, 345–350.

Sollid, L.M., Qiao, S.W., Anderson, R.P., Gianfrani, C., Koning, F., 2012. Nomenclature and listing of celiac disease relevant gluten T-cell epitopes restricted by HLA-DQ molecules. Immunogenetics 64, 455–460.

Sollid, L.M., Tye-Din, J.A., Qiao, S.W., Anderson, R.P., Gianfrani, C., KONING, F., 2020. Update 2020: nomenclature and listing of celiac disease-relevant gluten epitopes recognized by CD4(+) T cells. Immunogenetics 72, 85–88.

Spurkland, A., Sollid, L.M., Polanco, I., Vartdal, F., Thorsby, E., 1992. HLA-DR and -DQ genotypes of celiac disease patients serologically typed to be non-DR3 or non-DR5/7. Hum. Immunol. 35, 188–192.

Thomas, H.J., Ahmad, T., Rajaguru, C., Barnardo, M., Warren, B.F., Jewell, D.P., 2009. Contribution of histological, serological, and genetic factors to the clinical heterogeneity of adult-onset coeliac disease. Scand. J. Gastroenterol. 44, 1076–1083.

Tinto, N., Cola, A., Piscopo, C., Capuano, M., Galatola, M., Greco, L., Sacchetti, L., 2015. High frequency of haplotype HLA-DQ7 in celiac disease patients from South Italy: retrospective evaluation of 5,535 subjects at risk of celiac disease. PLoS One 10, e0138324.

Toivanen, P., Vaahtovuo, J., Eerola, E., 2001. Influence of major histocompatibility complex on bacterial composition of fecal flora. Infect. Immun. 69, 2372–2377.

Vader, W., Stepniak, D., Kooy, Y., Mearin, L., Thompson, A., van Rood, J.J., Spaenij, L., Koning, F., 2003. The HLA-DQ2 gene dose effect in celiac disease is directly related to the magnitude and breadth of gluten-specific T cell responses. Proc. Natl. Acad. Sci. U. S. A. 100, 12390–12395.

van Belzen, M.J., Koeleman, B.P., Crusius, J.B., Meijer, J.W., Bardoel, A.F., Pearson, P.L., Sandkuijl, L.A., Houwen, R.H., Wijmenga, C., 2004. Defining the contribution of the HLA region to cis DQ2-positive coeliac disease patients. Genes Immun. 5, 215–220.

Vermeulen, B.A., Hogen Esch, C.E., Yuksel, Z., Koning, F., Verduijn, W., Doxiadis, I.I., Schreuder, G.M., Mearin, M.L., 2009. Phenotypic variance in childhood coeliac disease and the HLA-DQ/DR dose effect. Scand. J. Gastroenterol. 44, 40–45.
Wang, C., Mooney, J.L., Meza-Romero, R., Chou, Y.K., Huan, J., Vandenbark, A.A., Offner, H., Burrows, G.G., 2003. Recombinant TCR ligand induces early TCR signaling and a unique pattern of downstream activation. J. Immunol. 171, 1934–1940.
Wang, H., Zhou, G., Luo, L., Crusius, J.B., Yuan, A., Kou, J., Yang, G., Wang, M., Wu, J., von Blomberg, B.M., Morre, S.A., Pena, A.S., Xia, B., 2015. Serological screening for celiac disease in adult Chinese patients with diarrhea predominant irritable bowel syndrome. Medicine (Baltimore) 94, e1779.
Wessels, M.M.S., de Rooij, N., Roovers, L., Verhage, J., de Vries, W., Mearin, M.L., 2018. Towards an individual screening strategy for first-degree relatives of celiac patients. Eur. J. Pediatr. 177, 1585–1592.
Xia, J., Siegel, M., Bergseng, E., Sollid, L.M., Khosla, C., 2006. Inhibition of HLA-DQ2-mediated antigen presentation by analogues of a high affinity 33-residue peptide from alpha2-gliadin. J. Am. Chem. Soc. 128, 1859–1867.
Xia, J., Bergseng, E., Fleckenstein, B., Siegel, M., KIM, C.Y., Khosla, C., Sollid, L.M., 2007. Cyclic and dimeric gluten peptide analogues inhibiting DQ2-mediated antigen presentation in celiac disease. Bioorg. Med. Chem. 15, 6565–6573.
Zmora, N., Suez, J., Elinav, E., 2019. You are what you eat: diet, health and the gut microbiota. Nat. Rev. Gastroenterol. Hepatol. 16, 35–56.
Zubillaga, P., Vidales, M.C., Zubillaga, I., Ormaechea, V., Garcia-Urkia, N., Vitoria, J.C., 2002. HLA-DQA1 and HLA-DQB1 genetic markers and clinical presentation in celiac disease. J. Pediatr. Gastroenterol. Nutr. 34, 548–554.

CHAPTER THREE

Human intestinal dendritic cell and macrophage subsets in coeliac disease

Eduardo Arranz[a], Ángel De Prado[a], Aida Fiz-López[a], Elisa Arribas[a], José A. Garrote[a,b], and David Bernardo[a,*]

[a]Mucosal Immunology Lab. Instituto de Biología y Genética Molecular (IBGM), Universidad de Valladolid-CSIC, Valladolid, Spain

[b]Servicio de Análisis Clínicos, Hospital Universitario Río Hortega, Valladolid, Spain

[*]Corresponding author: e-mail address: d.bernardo.ordiz@gmail.com

Contents

Abstract

Dendritic cells (DC) and macrophages (Mϕ) constitute the most abundant antigen presenting cells in the human intestinal mucosa. In resting conditions, they are essential to maintain the mechanisms of immune tolerance toward food antigens and commensals, at the time that they keep the capacity to initiate and maintain antigen-specific proinflammatory immune responses toward invading pathogens. Nevertheless, this delicate equilibrium between immunity and tolerance is not perfect, like in coeliac disease (CD), where DC and Mϕ drive the development of antigen-specific immune responses toward dietary gluten peptides. In this review, we provide therefore a comprehensive discussion about CD pathogenesis, the human intestinal immune system and the biology of intestinal DC and Mϕ both in resting conditions and in CD. Last, but not least, we discuss about all the remaining issues pending to be studied regarding DC and Mϕ

International Review of Cell and Molecular Biology, Volume 358
ISSN 1937-6448
https://doi.org/10.1016/bs.ircmb.2020.09.006

contribution toward CD pathogenesis. This may allow the identification of unique and specific factors which may be useful in the clinical practice, as well as identify new therapeutic targets in order to reestablish the loss intestinal homeostasis in CD.

1. Coeliac disease pathogenesis

Coeliac disease (CD) is an inflammatory disorder of the small intestine caused by dietary gluten from wheat and other cereals. The etiology is multifactorial, with environmental factors and both MHC and non-MHC genes contributing to its pathogenesis. More than 90% of patients carry the HLA-DQA1*05 and HLA-DQB1*02 alleles, encoding for the HLA-DQ2.5 molecule; and most of the remainder carry DQA1*03 and DQB1*03:02 (HLA-DQ8) (Sollid and Lie, 2005), but only few individuals with these haplotypes develop the disease.

Specific $CD4^+$ T-cells have a central role in CD pathogenesis by recognizing gluten peptides following post-transcriptional modification induced by the enzyme tissue transglutaminase type 2 (TG2), which selects T-cell epitopes with the highest affinity for HLA-DQ molecules. TG2 is also the target for disease-specific autoantibodies. The coeliac lesion is manifested by several degrees of mucosal remodeling, with villous blunting, crypt cell hyperplasia, intraepithelial lymphocytosis, and selective destruction of epithelial cells.

1.1 The adaptive immune response to gluten

Gluten contains several T-cell epitopes with different immunogenic capacity (Sollid et al., 2012). Some of them are immunodominant and activate $CD4^+$ T-cells from most HLA-$DQ2.5^+$ patients, such as the 6 overlapping immunogenic epitopes of the 33mer peptide from α-gliadin (Shan et al., 2002; Tye-Din et al., 2010). The generation of T-cell epitopes relies on the resistance of gluten to digestive enzymes due to its high proline and glutamine content, the deamination by TG2, and its binding affinity for HLA molecules (Du Pré and Sollid, 2015). TG2 converts glutamine residues into glutamic acid (negative charge) in QXP sequences (Willemijn Vader et al., 2002), hence increasing its binding affinity for HLA-DQ molecules and stabilizing, therefore, the HLA-peptide complex (Arentz-Hansen et al., 2000; Molberg et al., 1998). The binding ability of HLA molecules for different peptides depends on the positions of negatively charged residues in the peptide binding groove, HLA-DQ2.5 has preference for P4, P6 and P7; whereas HLA-DQ8 hast it for P1 and/or P9 (Tollefsen et al., 2006).

Gluten-specific CD4$^+$ T-cells can been isolated from the intestinal lamina propria from CD patients, where gluten peptides gain access after passing through the epithelium by the transcellular or paracellular routes (Abadie et al., 2011; Lundin et al., 1993). Most of these cells preferentially recognize deaminated T-cell epitopes with the highest binding affinity for the HLA-DQ2.5 and HLA-DQ8 molecules expressed by antigen presenting cells (APC) (MOLBERG et al., 1997). The presentation of a particular repertoire of gluten T-cell epitopes may determine, as well, the selection of effector CD4$^+$ T-cells with high affinity T-cell receptors (TCR) (Qiao et al., 2011). This T-cell response to gluten is mediated by pro-inflammatory cytokines, such as interferon (IFN)-γ and IL-21, but not IL-17, and a lower expression of the immunorregulatory cytokines like IL-10 and transforming growth factor (TGF)-β (Bodd et al., 2010; Nilsen et al., 1995). This cytokine profile may serve as a link between adaptive and innate immunity (Sarra et al., 2013), enhancing as well the activation of CD8$^+$ T-cell cytotoxicity.

T and B-cell cooperation is crucial for the generation of the immune response to gluten. Anti-TG2 antibody production occurs in HLA-DQ2 and HLA-DQ8 individuals and depends on dietary gluten, suggesting the involvement of gluten-reactive CD4$^+$ T-cells. These cells may provide help to gluten-specific and TG2-specific B-cells, leading to the differentiation of antibody-secreting plasma cells directed toward deaminated gliadin peptides and TG2 (Stamnaes and Sollid, 2015). Although TG2-specific T-cells have not been found, the production of anti-TG2 antibodies could be explained if TG2-specific B-cells act as APC of covalent TG2-gluten complexes, via HLA-DQ molecules, to gluten-reactive T cells (Sollid et al., 1997). Intestinal deposits of IgA anti-TG2 antibodies can be found in CD patients underlying the epithelial layer and surrounding blood vessels, even before antibody production is detected in serum (Korponay-Szabó et al., 2004).

Both gluten-specific and TG2-specific B-cells may have a role too as APC for gluten-reactive CD4$^+$ T-cells, leading to the amplification of the immune response to gluten and the inflammatory environment requested for TG2 activation (Du Pré and Sollid, 2015). A link has been suggested between gluten peptide deamination, antigen presentation by TG2-reactive B-cells and the activation of gluten-specific T cells, which in turn may activate gluten-reactive B cells. Nevertheless, it is not yet known whether antigen presentation by B-cells is similar to the role performed by pro-inflammatory DC expressing HLA-DQ molecules (Abadie et al., 2011; Jabri and Sollid, 2017).

1.2 The innate immune response to gluten

Gluten induces epithelial stress and innate immune activation, though it is not yet known how interacts with the epithelium, or which signaling pathways are activated. Several peptides, not recognized by $CD4^+$ T-cells, have been reported to elicit stimulatory effects on epithelial cells and DC where they would induce the production of IL-15, IL-18 and type I interferons (Abadie et al., 2011; Malamut et al., 2010). The dysregulated IL-15 production plays indeed a central role in CD development by licensing IELs to epithelial cell damage and by inhibiting regulatory $CD4^+$ T-cells (Mention et al., 2003). The increased expression of the IL-15Rα receptor in the intestine of CD patients may also confer a lower IL-15 response threshold in these patients (Bernardo et al., 2008). Other environmental factors, such as α-amylase-trypsin inhibitor (ATI) from wheat, may stimulate innate cells acting through Toll-like receptor 4 (TLR4), as well as intestinal changes related to dysbiosis, or viral infections (Bouziat et al., 2017; Zevallos et al., 2017).

The coeliac lesion is characterized by increased numbers of $CD8^+ CD4^- TCR\alpha\beta^+$ and $CD8^- CD4^- TCR\gamma\delta^+$ IELs. These cells are not gluten-specific, but under the effects of both activated gluten-specific $CD4^+$ T-cells, and stressed epithelial cells expressing high levels of IL-15, $CD8^+ CD4^- TCR\alpha\beta +$ IELs undergo cytotoxic transformation by increasing the expression of the activating receptor NKG2D, while losing the inhibitory CD94/NKG2A receptor (Meresse et al., 2004). Stressed epithelial cells also express the unconventional HLA-class I molecule MICA, main ligand of NKG2D, which results is the destruction of the epithelium and the loss of the intestinal villi (Hüe et al., 2004).

2. Introduction to the human gut immune system

The intestinal mucosa is a vulnerable surface continuously exposed to external antigens, hence maintaining the delicate balance between tolerance toward commensals and dietary compounds, and immunity against invading pathogens. To maintain this balance, the intestine contains the largest component of the immune system of the body, the gut-associated lymphoid tissue (GALT), with organized structures including Peyer's patches, lymphoid follicles and the appendix, and draining mesenteric lymph nodes; and numerous intestinal immune cells with memory and effector functions scattered throughout the lamina propria and the overlying epithelium (Brandtzaeg, 2009; Mowat and Agace, 2014).

The induction of adaptive immune responses occurs in the organized lymphoid tissue after the uptake and transport of particulate antigens by epithelium-associated M cells from the lumen to the subepithelial dome region, where naive T- and B-cells are primed by APC. Peyer's patches are the main source of small intestine-homing IgA plasmablasts, after specialized T-cells induce antigen-activated IgM^+ B-cells to switch isotype to IgA (Masahata et al., 2014; Mora et al., 2006). Activated T-cells migrate from mesenteric lymph nodes to both the lamina propria and the epithelium. Intestinal homeostasis depends on two main adaptative immune effector mechanisms, immune exclusion mediated by secretory IgA antibodies and induction of active tolerogenic T and B-cell responses. Other specific and non-specific mechanisms might have also a role depending to its distribution in different segments of the small and large intestine (Mowat and Agace, 2014).

The epithelium and the lamina propria are the main effector sites, and contain the largest population of T-cells, plasma cells and other immune cells of the body. Lymphocytes are found in the organized lymphoid tissue, but also in the epithelial cell layer (as intraepithelial lymphocytes, IEL), and scattered within the lamina propria, with other lymphoid and non-lymphoid immune cells (Mowat and Agace, 2014). In the small intestine, luminal antigens may reach the lamina propria by different mechanisms other than through lymphoid structures, and these include the transfer of protein to lamina propria dendritic cells (DC) via channels formed by neighboring goblet cells, paracellular transport across the epithelium and uptake by lamina propria DC, or antigen carriage in lymph to DC located in the mesenteric lymph nodes (McDole et al., 2012).

2.1 The intestinal epithelium

The intestine is covered by a single layer of epithelial cells separating the lumen from the underlying connective tissue. These cells contribute to the first line of defense against microorganisms by forming a physical and biochemical barrier with a rapid turnover via intercellular tight junctions, mucus and antimicrobial peptides; and have also immunoregulatory functions. The gut epithelium includes several differentiated and specialized cell types, all derived from stem cells at the bottom of the crypts, including absorptive epithelial cells (enterocytes) covered by a brush border in which digestive enzymes are embedded; Paneth cells containing granules with anti-microbial peptides; mucus-secreting goblet cells controlled by inflammatory mediators; enteroendocrine cells; chemosensory tuft cells with a role in detecting helminth parasites; and M cells, involved in the uptake and transport of luminal antigens to the underlying lymphoid structures (Allaire et al., 2018; Peterson and Artis, 2014).

2.2 Intestinal B cells

From Peyer's patches, IgA plasmablasts migrate via lymphatics and mesenteric lymph nodes to the thoracic duct and hence to the circulation, and then return preferentially to the intestinal lamina propria, undergoing maturation into IgA-producing cells. These cells produce monomeric IgA, converted into a dimer by the J chain, a small peptide produced by plasma cells, which binds to the polymeric immunoglobulin receptor synthesized by epithelial cells, forming a complex which is endocytosed and transported to the luminal surface where polymeric Ig receptor is proteolyzed. The IgA dimer is then released attached to a fragment of the receptor called secretory component, which stabilizes the secretory IgA molecule (SIgA) and protects from proteolysis. As part of the epithelial barrier, SIgA neutralizes viruses, bacteria and toxins, and prevents the adherence of pathogens to the epithelium by a mechanism of immune exclussion (Brandtzaeg, 2009; Mantis and Forbes, 2010).

2.3 Intraepithelial lymphocytes

First identified by Ferguson at a frequency of 10–15 per 100 epithelial cells (Ferguson, 1977), IELs express both non-rearranged innate immune receptors and arranged adaptive immune receptors. Unlike conventional T-cells, they are found in the gut before birth. These cells also express the human mucosal lymphocyte antigen 1 (HML-1, CD103), an adhesion molecule of the β7 integrin family, whose ligand (E-cadherin) is expressed by epithelial cells. In the adult human jejunum, IEL form an heterogeneous population consisting of $\alpha\beta TCR^{+}CD8\alpha\beta^{+}$ cells, the main subset of tissue-resident effector memory $CD8^{+}$ T-cells (Jabri and Ebert, 2007); $\alpha\beta TCR^{+}CD4^{-}CD8^{-}$ cells, more relevant in ileum and colon; and a minor subtype of γδTCR cells, with a limited TCR diversity, but a relatively higher frequency in the human epithelium, compared with peripheral lymphocytes. IELs have regulatory and effector activities without antigen-specific priming, functioning as cytotoxic or natural killer cells, and having a role in controlling the epithelial cell barrier function (Cheroutre et al., 2011; Olivares-Villagómez and Van Kaer, 2018).

2.4 Lamina propria T cells

The lamina propria contains both $CD4^+$ and $CD8^+$ T-cells (proportion 2:1), being most of them effector memory cells. Upon priming within GALT, activated T-cells express at least two homing molecules, the α4β7 integrin, whose ligand is the mucosal vascular addressin cell-adhesion molecule (MAdCAM), and the gut-homing chemokine receptor 9 (CCR9), which binds to the chemokine CCL25, both expressed in the small bowel lamina propria(Mora et al., 2003; Mowat and Agace, 2014). Among $CD4^+$ T-cells, there are different subsets, IL-17 producing cells (Th17), IFNγ-producing (Th1) cells, and two types of regulatory T-cells, IL-10 producing cells expressing forkhead box P3 ($Foxp3^+$) named regulatory T-cells (Tregs), and T regulatory type 1 (Tr1) cells not expressing Foxp3 ($Foxp3^-$) (Sathaliyawala et al., 2013). Regulatory T-cells are involved in immunosuppression by several mechanisms, such as production of inhibitory cytokines (IL-10), expansion of innate lymphoid cells (ILC)1-like NK cells and ILC2, or even DC function modulation (Vignali et al., 2008). On the other hand, the generation of Th1 and Th17 is associated with chronic inflammation. Although Th17 cells contribute to maintain intestinal homeostasis via IL-22(Martínez-López et al., 2019), they can also drive pro-inflammatory responses (Gagliani et al., 2015; Huber et al., 2012). Lamina propria Th1 and Th2 cells do not seem to differ through the length of the human gastrointestinal tract, although the proportion of IL-17 producing $CD4^+$ T-cells is higher in colon and ileum, compared to jejunum (Sathaliyawala et al., 2013).

2.5 Tolerance in the intestinal mucosa

The gut immune system is characterized by its ability to develop tolerance to harmless foreign antigens (such as food proteins), which is relevant to prevent damaging T-cell mediated (DTH) reactions and IgE dependent responses in the small intestine; whereas active immune responses against pathogens are maintained. Tolerance can be defined as the local and systemic hyporesponsiveness to subsequent challenge when exogenous antigens are administered by the oral route (Mowat, 2018). The induction of oral tolerance may depend on the generation of antigen-specific Tregs (Du Pré et al., 2011). In the colon, these cells are also responsible of inhibiting effector T-cells recognizing commensal microorganisms by producing IL-10 or TGFβ (Asseman et al., 1999; Li et al., 2007). However, unlike food antigens, specific immune responses to the microbiota (Treg cells and secretory IgA antibodies) are confined to the intestine.

Local conditioning of lamina propria monocytes, macrophages (Mϕ) and dendritic cells (DC) determine a central role of these cells in the induction and maintenance of oral tolerance (see below), by taking up soluble antigen (food proteins) directly from the lumen (Niess et al., 2005) or by other specific pathways including via goblet cells (McDole et al., 2012), and migrating to the draining mesenteric lymph nodes (NLM), which contains naive CD4 and CD8 T-cells. Primed $CD4^{+}Foxp3^{+}$ Tregs return to the lamina propria mucosa where they prevent the activation of effector T cells by direct contact (via CTLA-4) or by producing inhibitory cytokines. The entry route of commensals is mainly by transcitosis via microfold (M) cells in Peyer patches (PP) and isolated lymph follicles (ILF), which contains DC and naive $CD4^{+}$ T-cells. However, the mechanisms controlling the migration of microbiota-specific Tregs to the colon are still unknown (Mowat, 2018).

3. Dendritic cells in the healthy gut

DC, the most potent APC, determine the outcome (pro-inflammatory or tolerogenic) of antigen-specific adaptive immune responses (Banchereau and Steinman, 1998; Knight et al., 1983). DC are divided into two major subsets: myeloid, conventional or classical DC (cCD), and plasmocytoid DC (pDC). cDC can be further divided into subsets based on their ontogeny and function named type 1 (cDC1), specialized in cross-presentation and type 2 (cDC2), specialized in classical presentation (Guilliams et al., 2014; Guilliams and van de Laar, 2015; Sichien et al., 2017). cDC1 require transcription factors IRF8 and Baft3. In mice, they express surface marker CD8α, although in human they are characterized by the expression of CD141. Nevertheless, both mice and human cDC1 express XCR1 and CD103. cDC2, on the contrary, require transcription factors IRF4 and Notch2. While murine cDC2 express surface marker CD11b, in human they are characterized by the expression of CD1c, while both species express SIRPα. Nevertheless, novel transcriptomic approaches have provided better understanding on how cDC can be altered during inflammation with cDC2 acquiring a hybrid phenotype and function

(Bosteels et al., 2020) and $CD14^{+}$ cDC being redefined as pro-inflammatory cDC3 (Bourdely et al., 2020).

In the intestine, cDC are essential to maintain the balance between tolerance toward food antigens/commensals and immunity against pathogens (Chirdo et al., 2005; Mowat, 2018; Pabst and Mowat, 2012; Stagg et al., 2003). Intestinal cDC can be found all over the GALT, including PP, ILF and diffusely scattered across LP mucosa. In addition, cDC are also present in MLN where they elicit their function (Bekiaris et al., 2014). Indeed, intestinal cDC display a regulatory phenotype given their capacity to secrete IL-10 (Mowat, 2018) and induce IL-22 (Martínez-López et al., 2019). As opposed to other tissues, where cDC can be divided into cDC1 and cDC2, the gut harbors a third and unique subset of $CD103^{+}$ cDC2 (Bernardo et al., 2016; Mann et al., 2016; Watchmaker et al., 2014). This gut-specific subset mediates most of the mechanisms of intestinal tolerance as they capture luminal antigens and carry them to the MLN in a CCR7 dependent manner. In the MLN, in non-inflammatory conditions, $CD103^{+}$ cDC2 mediate most of the mechanisms of immune tolerance given their unique capacity to generate gut-homing Tregs and IgA-producing B-cells, although they can also drive Th17 responses (Kinnebrew et al., 2012; Persson et al., 2013; Scott et al., 2011). These properties of cDC, which are essential to maintain the mechanisms of tissue homeostasis, are dependent on TFGβ and on the enzyme RALDH2 (which mediates the conversion of dietary vitamin A into retinoic acid -RA-) (Stagg, 2018). Moreover, tolerance toward oral antigens may be transferred without previous contact by the infusion of LP-isolated $CD11c^{+}MHCII^{+}$ cells (Bernardo et al., 2008).

This regulatory activity displayed by intestinal cDC is dependent on the local microenvironment (Bernardo et al., 2012; Iliev et al., 2009a; Iliev et al., 2009b; Mann et al., 2012) including the presence of regulatory cytokines like IL-10 and TGFβ, but also by epithelial secreted TSLP, PGE2, VIP, biliary salts and dietary vitamin A derived RA. Indeed, regulatory $CD103^{+}$ cDC2 are derived from newly arrived $CD103^{-}$ cDC2 which are further conditioned by the local microenvironment in a TFGβ-dependent manner (Bain et al., 2017). Given that cDC phenotype and function is dependent on the tissue microenvironment, changes in the mucosal like the induction of an innate immune response can modulate their properties hence initiating pro-inflammatory immune responses. Indeed, intestinal cDC from patients with inflammatory bowel disease have a reduced capacity to generate suppressor Treg coupled with an enhanced production of pro-inflammatory

Th1 and Th17 effector T-cells, which is thought to be associated with lower numbers of $CD103^+$ cDC (Mann et al., 2014; Matsuno et al., 2017; Ogino et al., 2013).

4. Macrophages in the healthy gut

As opposed to cDC, which can initiate antigen-specific adaptive immune responses by performing antigen presentation to naïve T-cells, Mϕ are APC highly adapted to the tissue that they inhabit (Lavin et al., 2014). Indeed, in the intestine, they are highly specialized to avoid overt immunity in response to the gut microbiota (Cerovic et al., 2014; Mowat et al., 2014; Persson et al., 2013).

Most current knowledge about Mϕ biology has been obtained from murine models, revealing that as opposed to Mϕ from other tissues, in the gut they are constantly replenished from circulating $Ly6C^{high}$ monocytes entering the mucosa in a CCR2-dependent manner (Bain et al., 2013, 2014). Once there, they are subsequently conditioned by the tissue microenvironment through several intermediates via the "monocyte waterfall" (Bain et al., 2013, 2014; Cerovic et al., 2014; Joeris et al., 2017; Mowat et al., 2014; Zigmond and Jung, 2013) which results in the differentiation of tissue resident tolerogenic Mϕ ($Ly6C^- MHC\text{-}II^{high}CX3CR1^{high}CCR2^-$). Mature intestinal Mϕ maintain local homeostasis via their hypo-responsiveness to microbial stimulation, but also via removal of apoptotic/dead cells and microbes, mediation of tissue and epithelial remodeling and secretion of regulatory cytokines, including IL-10, which maintains the survival of local regulatory Tregs. However, this process is disrupted during intestinal inflammation (Cerovic et al., 2014; Joeris et al., 2017; Mowat et al., 2014; Mowat and Bain, 2011; Rivollier et al., 2012; Tamoutounour et al., 2012; Zigmond et al., 2012).

It has been recently described that a similar mechanism occurs in the human intestinal mucosa. Hence, monocytes newly arrived in the human intestinal mucosa can be distinguished from mature Mϕ based on the expression levels of the CD11c integrin present in newly arrived pro-inflammatory monocytes and absent in tolerant or regulatory Mϕ (Bujko et al., 2018). Indeed, $CD14^+$ monocytes infiltrate the intestinal mucosa as pro-inflammatory $CD11c^+CX3CR1^+CCR2^+$ cells through a CCR2-dependent mechanism. Once in the tissue, they are subsequently conditioned by the local microenvironment to differentiate toward IL-10 producing $CD11c^-CX3CR1^-CCR2^-$ tolerogenic Mϕ

(Bernardo et al., 2018). This process, nevertheless, is abrogated in the presence of an ongoing intestinal inflammation where there is an increased migration of monocytes to the intestinal mucosa where, however, they do not differentiate into tolerant Mϕ hence accumulating as pro-inflammatory cells that exacerbate the immune response (Bain and Schridde, 2018; Na et al., 2019).

5. Intestinal dendritic cells and macrophages in coeliac disease

The study of human intestinal DC and Mϕ can be controversial given their many times overlapping phenotype and function. Indeed, some authors discuss the presence of pro-inflammatory $CD14^+$ DC, recently renamed as cDC3 (Bourdely et al., 2020), while cDC2 can also display a hybrid phenotype under inflammatory conditions (Bosteels et al., 2020).

Although cDC are central to trigger CD pathogenesis as they mediate the differentiation of gut-homing pro-inflammatory T-cells in a DQ-restricted manner against different gliadin peptides (Di Sabatino et al., 2011; Escudero-Hernández et al., 2020; Nilsen et al., 1995), little is known about their specific implication in vivo. Hence, most studies have traditionally focused on the study of monocyte-derived DC (MoDC) where gliadin activates then in a NK-κB dependent manner, resulting in an increase capacity of the cells to migrate toward the lymph nodes, where they will process and present their peptides to T-cells irrespectively of CD or genetic predisposition (Chladkova et al., 2011; Palová-Jelínková et al., 2005; Rakhimova et al., 2009). Nevertheless, MoDC do not truly resemble the properties of real cDC (Alcántara-Hernández et al., 2017; Benlahrech et al., 2015) reason why it is essential to get a deeper insight into the study of tissue *bona-fide* cDC. Indeed, the CD mucosa carries higher numbers of cDC (Rajaguru et al., 2013; Vorobjova et al., 2015), likely because of an increased gut-homing migration pattern which favors their enrichment in the intestinal mucosa (Comino et al., 2014; Escudero-Hernández et al., 2020) and would explain their reduced numbers in the blood of these patients (Ciccocioppo et al., 2007). Once in the mucosa, the local tissue microenvironment, mainly mediated by IL-15 (Bernardo et al., 2007, 2008), can reshape cDC function where, in combination with retinoic acid, induces a pro-inflammatory phenotype which abrogates cDC tolerance toward food antigens (Depaolo et al., 2011). Indeed, it may also be possible that cDC fail to establish immune tolerance toward gluten peptides due to a miss-match reaction toward viral and/or microbial antigens (Bouziat et al., 2017; Petersen et al., 2020). On the other hand, it has also been suggested that pDC may provide a source of IFNα in the CD mucosa (Di Sabatino et al., 2007), although other authors have not been able to validate these findings (Ráki et al., 2013) suggesting that pDC numbers are scarce in the human intestinal mucosa (Bernardo et al., 2016) and therefore may not have a direct role in CD pathogenesis.

Mϕ also have a direct role toward CD pathogenesis. Hence, it has been described how ex-vivo stimulation with gliadin, or its derived peptides, induces pro-inflammatory activation not only of intestinal Mϕ (Halstensen et al., 1993; Tucková et al., 2002; Barilli et al., 2018), but also of circulating monocytes from which they are derived (Barilli et al., 2014, 2015; Cinova et al., 2007; Jelínková et al., 2004). In fact, the oligomerization of the 33-mer peptide (main peptide involved in the pathogenesis of CD) is recognized by monocytes and Mϕ in a TLR2 and TLR4 dependent manner, where it induces the activation of the NF-κB signaling pathway and the activation of the NLRP3 inflammasome. This suggests that monocytes and Mϕ may be the first immune cell types to respond to gluten in patients with CD and, therefore, the initial triggers of the pathology (Herrera et al., 2018; Palová-Jelínková et al., 2013). Indeed, Mϕ are expanded in the inflamed tissue from CD patients where they, in turn, contribute to the proliferation of antigen-specific T-cells together with cDC (Beitnes et al., 2011; Ráki et al., 2006). In fact, in vivo dietary gluten stimulation rapidly increases the presence of Mϕ in the duodenal mucosa, even prior to the presence of intestinal lesions, due to an increased migration of monocytes into the tissue in a CCR2-dependent manner (Beitnes et al., 2012). Hence, these findings suggest that, together with lamina effector T-cells, Mϕ contribute as an important source of pro-inflammatory cytokines in CD, thus contributing to its pathogenesis (Sollid and Jabri, 2013).

Last, but not least, recent evidence has revealed that plasma cells are actually the most abundant APC present in the inflamed intestinal mucosal tissue from CD patients, suggesting that their role may be not just restricted to generate IgA antibodies, but that they may also have a direct role triggering the immune response (Høydahl et al., 2019).

6. Concluding remarks and future perspectives

Although the involvement of cDC and monocyte/Mϕ in CD pathogenesis seems clear, their specific phenotype and function in the CD intestinal mucosa remains elusive in the light of their new classification into different subsets (Bosteels et al., 2020; Bourdely et al., 2020; Guilliams et al., 2014; Guilliams and van de Laar, 2015; Sichien et al., 2017). Hence, the specific contribution of cDC, monocytes and Mϕ in CD pathogenesis needs to be readdressed in the light of their new classification into different subsets (Bosteels et al., 2020; Bourdely et al., 2020; Guilliams et al., 2014; Guilliams and van de Laar, 2015; Sichien et al., 2017). Indeed, all these cell types express the HLA-risk alleles, reason why they can initiate and perpetuate antigen-specific immune responses against gluten antigen peptides. Nevertheless, not all carriers of the HLA-risk alleles develop CD. Therefore, the relevance of each cDC and monocyte/Mϕ cell type in the establishment of oral tolerance or the development of an antigen-specific immune response remains elusive not just in CD and healthy controls, but also in non-CD HLA-DQ2^{+} individuals and first degree relatives of CD patients.

In a similar manner, the relevance of the recruitment and differentiation mechanisms governing cDC and monocyte/Mϕ biology remain unknown. Hence, it is necessary to identify the specific chemoattractants that mediate the recruitment of monocytes and cDC toward the mucosa in non-inflammatory conditions, as well as to determine whether the mechanisms are the same (but increased) in patients with active CD, or whether, on the contrary, there are differences in the mechanisms mediating such recruitment. Likewise, it is essential to identify the factors present in the intestinal mucosa that mediate the conditioning of newly arrived CD103^{-} cDC2 and monocytes toward CD103^{+} cDC2 and tolerogenic Mϕ, respectively, in order to determine, subsequently, whether these factors are absent or masked in the inflamed tissue of patients with CD. Unraveling therefore the factors mediating cell migration to the gut and their further conditioning may allow us to (i) prevent the infiltration of pro-inflammatory cells; (ii) reprogram cDC and/or Mϕ differentiation toward a tolerogenic phenotype and function; and/or (iii) target the pro-inflammatory cell subsets. This would translate into the development of novel therapies, alternative or complementary to the gluten-free diet (GFD), something of relevance not just to CD patients, but also to those refractory to the GFD (Vaquero et al., 2019).

In summary, the identification of the factors that mediate the recruitment of cDC and monocytes to the intestinal mucosa, as well as their subsequent conditioning in healthy mucosa or inflamed tissue, will contribute to increase our current knowledge about the biology of these cell populations. Furthermore, the identification of the mechanisms specifically altered in patients with CD may allow the identification of unique and specific factors which may be useful in the clinical practice, as well as identify new therapeutic targets in order to reestablish the loss intestinal homeostasis in CD.

References

Abadie, V., et al., 2011. Integration of genetic and immunological insights into a model of celiac disease pathogenesis. Annu. Rev. Immunol. https://doi.org/10.1146/annurev-immunol-040210-092915.

Alcántara-Hernández, M., et al., 2017. High-dimensional phenotypic mapping of human dendritic cells reveals Interindividual variation and tissue specialization. Immunity. https://doi.org/10.1016/j.immuni.2017.11.001.

Allaire, J.M., et al., 2018. The intestinal epithelium: central coordinator of mucosal immunity. Trends Immunol. https://doi.org/10.1016/j.it.2018.04.002.

Arentz-Hansen, H., et al., 2000. The intestinal T cell response to α-gliadin in adult celiac disease is focused on a single deamidated glutamine targeted by tissue transglutaminase. J. Exp. Med. https://doi.org/10.1084/jem.191.4.603.

Asseman, C., et al., 1999. An essential role for interleukin 10 in the function of regulatory T cells that inhibit intestinal inflammation. J. Exp. Med. https://doi.org/10.1084/jem.190.7.995.

Bain, C.C., Schridde, A., 2018. Origin, differentiation, and function of intestinal macrophages. Front. Immunol. https://doi.org/10.3389/fimmu.2018.02733.

Bain, C.C., et al., 2013. Resident and pro-inflammatory macrophages in the colon represent alternative context-dependent fates of the same Ly6C hi monocyte precursors. Mucosal Immunol. https://doi.org/10.1038/mi.2012.89.

Bain, C.C., et al., 2014. Constant replenishment from circulating monocytes maintains the macrophage pool in the intestine of adult mice. Nat. Immunol. https://doi.org/10.1038/ni.2967.

Bain, C.C., et al., 2017. TGFβR signalling controls CD103+CD11b+ dendritic cell development in the intestine. Nat. Commun. https://doi.org/10.1038/s41467-017-00658-6.

Banchereau, J., Steinman, R.M., 1998. Dendritic cells and the control of immunity. Nature 392 (6673), 245–252. https://doi.org/10.1038/32588.

Barilli, A., et al., 2014. Gliadin activates arginase pathway in RAW264.7 cells and in human monocytes. Biochim. Biophys. Acta, Mol. Basis Dis. https://doi.org/10.1016/j.bbadis.2014.04.021.

Barilli, A., et al., 2015. Gliadin-mediated production of polyamines by RAW264.7 macrophages modulates intestinal epithelial permeability in vitro. Biochim. Biophys. Acta, Mol. Basis Dis. https://doi.org/10.1016/j.bbadis.2015.06.001.

Barilli, A., et al., 2018. Gluten peptides drive healthy and celiac monocytes toward an M2-like polarization. J. Nutr. Biochem. https://doi.org/10.1016/j.jnutbio.2017.10.017.

Beitnes, A.C.R., et al., 2011. Density of CD163 +CD11c + dendritic cells increases and CD103 + dendritic cells decreases in the coeliac lesion. Scand. J. Immunol. https://doi.org/10.1111/j.1365-3083.2011.02549.x.

Beitnes, A.C.R., et al., 2012. Rapid accumulation of CD14 +CD11c + dendritic cells in gut mucosa of celiac disease after in vivo gluten challenge. PLoS One. https://doi.org/10.1371/journal.pone.0033556.

Bekiaris, V., Persson, E.K., Agace, W.W., 2014. Intestinal dendritic cells in the regulation of mucosal immunity. Immunol. Rev. https://doi.org/10.1111/imr.12194.

Benlahrech, A., et al., 2015. Human blood CD1c dendritic cells stimulate IL-12-independent IFN- γ responses and have a strikingly low inflammatory profile. J. Leukoc. Biol. https://doi.org/10.1189/jlb.1a0114-058rr.
Bernardo, D., et al., 2007. Is gliadin really safe for non-coeliac individuals? Production of interleukin 15 in biopsy culture from non-coeliac individuals challenged with gliadin peptides [8]. Gut. https://doi.org/10.1136/gut.2006.118265.
Bernardo, D., et al., 2008. Higher constitutive IL15Rα expression and lower IL-15 response threshold in coeliac disease patients. Clin. Exp. Immunol. https://doi.org/10.1111/j.1365-2249.2008.03743.x.
Bernardo, D., et al., 2012. IL-6 promotes immune responses in human ulcerative colitis and induces a skin-homing phenotype in the dendritic cells and Tcells they stimulate. Eur. J. Immunol. https://doi.org/10.1002/eji.201142327.
Bernardo, D., et al., 2016. Chemokine (C-C motif) receptor 2 mediates dendritic cell recruitment to the human Colon but is not responsible for differences observed in dendritic cell subsets, phenotype, and function between the proximal and distal Colon. CMGH. https://doi.org/10.1016/j.jcmgh.2015.08.006.
Bernardo, D., et al., 2018. Human intestinal pro-inflammatory CD11chighCCR2+ CX3CR1+ macrophages, but not their tolerogenic CD11c-CCR2-CX3CR1- counterparts, are expanded in inflammatory bowel disease article. Mucosal Immunol. https://doi.org/10.1038/s41385-018-0030-7.
Bodd, M., et al., 2010. HLA-DQ2-restricted gluten-reactive T cells produce IL-21 but not IL-17 or IL-22. Mucosal Immunol. https://doi.org/10.1038/mi.2010.36.
Bosteels, C., et al., 2020. Inflammatory type 2 cDCs acquire features of cDC1s and macrophages to orchestrate immunity to respiratory virus infection. Immunity. https://doi.org/10.1016/j.immuni.2020.04.005.
Bourdely, P., et al., 2020. Transcriptional and functional analysis of CD1c+ human dendritic cells identifies a CD163+ subset priming CD8+CD103+ T cells. Immunity. https://doi.org/10.1016/j.immuni.2020.06.002.
Bouziat, R., et al., 2017. Reovirus infection triggers inflammatory responses to dietary antigens and development of celiac disease. Science. https://doi.org/10.1126/science.aah5298.
Brandtzaeg, P., 2009. Mucosal immunity: induction, dissemination, and effector functions. Scand. J. Immunol. https://doi.org/10.1111/j.1365-3083.2009.02319.x.
Bujko, A., et al., 2018. Transcriptional and functional profiling defines human small intestinal macrophage subsets. J. Exp. Med. https://doi.org/10.1084/jem.20170057.
Cerovic, V., et al., 2014. Intestinal macrophages and dendritic cells: What's the difference? Trends Immunol. https://doi.org/10.1016/j.it.2014.04.003.
Cheroutre, H., Lambolez, F., Mucida, D., 2011. The light and dark sides of intestinal intraepithelial lymphocytes. Nat. Rev. Immunol. https://doi.org/10.1038/nri3007.
Chirdo, F.G., et al., 2005. Immunomodulatory dendritic cells in intestinal lamina propria. Eur. J. Immunol. https://doi.org/10.1002/eji.200425882.
Chladkova, B., et al., 2011. Gliadin fragments promote migration of dendritic cells. J. Cell. Mol. Med. https://doi.org/10.1111/j.1582-4934.2010.01066.x.
Ciccocioppo, R., et al., 2007. Reduced number and function of peripheral dendritic cells in coeliac disease. Clin. Exp. Immunol. https://doi.org/10.1111/j.1365-2249.2007.03431.x.
Cinova, J., et al., 2007. Gliadin peptides activate blood monocytes from patients with celiac disease. J. Clin. Immunol. https://doi.org/10.1007/s10875-006-9061-z.
Comino, I., et al., 2014. Constitutive gut-homing capacity on circulating myeloid dendritic cells in coeliac disease. Rev. Esp. Enferm. Dig. https://doi.org/10.4321/s1130-01082014000100013.

Depaolo, R.W., et al., 2011. Co-adjuvant effects of retinoic acid and IL-15 induce inflammatory immunity to dietary antigens. Nature. https://doi.org/10.1038/nature09849.
Di Sabatino, A., et al., 2007. Evidence for the role of interferon-alfa production by dendritic cells in the Th1 response in celiac disease. Gastroenterology. https://doi.org/10.1053/j.gastro.2007.08.018.
Di Sabatino, A., et al., 2011. Involvement of CD40-CD40 ligand in uncomplicated and refractory celiac disease. Am. J. Gastroenterol. https://doi.org/10.1038/ajg.2010.450.
Du Pré, M.F., Sollid, L.M., 2015. T-cell and B-cell immunity in celiac disease. Best Pract. Res. Clin. Gastroenterol. https://doi.org/10.1016/j.bpg.2015.04.001.
Du Pré, M.F., et al., 2011. Tolerance to ingested deamidated gliadin in mice is maintained by splenic, type 1 regulatory T cells. Gastroenterology. https://doi.org/10.1053/j.gastro.2011.04.048.
Escudero-Hernández, C., et al., 2020. Circulating dendritic cells from celiac disease patients display a gut-homing profile and are differentially modulated by different gliadin-derived peptides. Mol. Nutr. Food Res. https://doi.org/10.1002/mnfr.201900989.
Ferguson, A., 1977. Intraepithelial lymphocytes of the small intestine. Gut. https://doi.org/10.1136/gut.18.11.921.
Gagliani, N., et al., 2015. TH17 cells transdifferentiate into regulatory T cells uring resolution of inflammation. Nature. https://doi.org/10.1038/nature14452.
Guilliams, M., van de Laar, L., 2015. A hitchhiker's guide to myeloid cell subsets: practical implementation of a novel mononuclear phagocyte classification system. Front. Immunol. https://doi.org/10.3389/fimmu.2015.00406.
Guilliams, M., et al., 2014. Dendritic cells, monocytes and macrophages: a unified nomenclature based on ontogeny. Nat. Rev. Immunol. https://doi.org/10.1038/nri3712.
Halstensen, T.S., et al., 1993. Gluten stimulation of coeliac mucosa in vitro induces activation (CD25) of Lamina Propria CD4H T cells and macrophages but no crypt-cell hyperplasia. Scand. J. Immunol. https://doi.org/10.1111/j.1365-3083.1993.tb03245.x.
Herrera, M.G., et al., 2018. Large supramolecular structures of 33-mer gliadin peptide activate toll-like receptors in macrophages. Nanomed. Nanotechnol. Biol. Med. https://doi.org/10.1016/j.nano.2018.04.014.
Høydahl, L.S., et al., 2019. Plasma cells are the Most abundant gluten peptide MHC-expressing cells in inflamed intestinal tissues from patients with celiac disease. Gastroenterology. https://doi.org/10.1053/j.gastro.2018.12.013.
Huber, S., Gagliani, N., Flavell, R.A., 2012. Life, death, and miracles: Th17 cells in the intestine. Eur. J. Immunol. https://doi.org/10.1002/eji.201242619.
Hüe, S., et al., 2004. A direct role for NKG2D/MICA interaction in villous atrophy during celiac disease. Immunity. https://doi.org/10.1016/j.immuni.2004.06.018.
Iliev, I.D., Mileti, E., et al., 2009a. Intestinal epithelial cells promote colitis-protective regulatory T-cell differentiation through dendritic cell conditioning. Mucosal Immunol. https://doi.org/10.1038/mi.2009.13.
Iliev, I.D., Spadoni, I., et al., 2009b. Human intestinal epithelial cells promote the differentiation of tolerogenic dendritic cells. Gut. https://doi.org/10.1136/gut.2008.175166.
Jabri, B., Ebert, E., 2007. Human CD8+ intraepithelial lymphocytes: a unique model to study the regulation of effector cytotoxic T lymphocytes in tissue. Immunol. Rev. https://doi.org/10.1111/j.1600-065X.2006.00481.x.
Jabri, B., Sollid, L.M., 2017. T cells in celiac disease. J. Immunol. https://doi.org/10.4049/jimmunol.1601693.
Jelínková, L., et al., 2004. Gliadin stimulates human monocytes to production of IL-8 and TNF-α through a mechanism involving NF-κB. FEBS Lett. https://doi.org/10.1016/j.febslet.2004.06.057.
Joeris, T., et al., 2017. Diversity and functions of intestinal mononuclear phagocytes. Mucosal Immunol. https://doi.org/10.1038/mi.2017.22.

Kinnebrew, M.A., et al., 2012. Interleukin 23 production by intestinal CD103 +CD11b + dendritic cells in response to bacterial Flagellin enhances mucosal innate immune Defense. Immunity. https://doi.org/10.1016/j.immuni.2011.12.011.
Knight, S.C., et al., 1983. Induction of immune responses in vivo with small numbers of veiled (dendritic) cells. Proc. Natl. Acad. Sci. U. S. A. https://doi.org/10.1073/pnas.80.19.6032.
Korponay-Szabó, I.R., et al., 2004. In vivo targeting of intestinal and extraintestinal transglutaminase 2 by coeliac autoantibodies. Gut. https://doi.org/10.1136/gut.2003.024836.
Lavin, Y., et al., 2014. Tissue-resident macrophage enhancer landscapes are shaped by the local microenvironment. Cell. https://doi.org/10.1016/j.cell.2014.11.018.
Li, M.O., Wan, Y.Y., Flavell, R.A., 2007. T cell-produced transforming growth factor-β1 controls T cell tolerance and regulates Th1- and Th17-cell differentiation. Immunity. https://doi.org/10.1016/j.immuni.2007.03.014.
Lundin, K.E.A., et al., 1993. Gliadin-specific, HLA-DQ(α1*0501, β1*0201) restricted t cells isolated from the small intestinal mucosa of celiac disease patients. J. Exp. Med. https://doi.org/10.1084/jem.178.1.187.
Malamut, G., et al., 2010. IL-15 triggers an antiapoptotic pathway in human intraepithelial lymphocytes that is a potential new target in celiac disease-associated inflammation and lymphomagenesis. J. Clin. Investig. https://doi.org/10.1172/JCI41344.
Mann, E.R., et al., 2012. Human gut-specific homeostatic dendritic cells are generated from blood precursors by the gut microenvironment. Inflamm. Bowel Dis. https://doi.org/10.1002/ibd.21893.
Mann, E.R., et al., 2014. Human gut dendritic cells drive aberrant gut-specific T-cell responses in ulcerative colitis, characterized by increased IL-4 production and loss of IL-22 and IFNγ. Inflamm. Bowel Dis. https://doi.org/10.1097/MIB.0000000000000223.
Mann, E.R., et al., 2016. Compartment-specific immunity in the human gut: properties and functions of dendritic cells in the colon versus the ileum. Gut. https://doi.org/10.1136/gutjnl-2014-307916.
Mantis, N.J., Forbes, S.J., 2010. Secretory IgA: arresting microbial pathogens at epithelial borders. Immunol. Invest. https://doi.org/10.3109/08820131003622635.
Martínez-López, M., et al., 2019. Microbiota sensing by Mincle-Syk Axis in dendritic cells regulates Interleukin-17 and -22 production and promotes intestinal barrier integrity. Immunity 50 (2), 446–461.e9. https://doi.org/10.1016/j.immuni.2018.12.020. Cell Press.
Masahata, K., et al., 2014. Generation of colonic IgA-secreting cells in the caecal patch. Nat. Commun. https://doi.org/10.1038/ncomms4704.
Matsuno, H., et al., 2017. CD103+ dendritic cell function is altered in the colons of patients with ulcerative colitis. Inflamm. Bowel Dis. https://doi.org/10.1097/MIB.0000000000001204.
McDole, J.R., et al., 2012. Goblet cells deliver luminal antigen to CD103 + dendritic cells in the small intestine. Nature. https://doi.org/10.1038/nature10863.
Mention, J.J., et al., 2003. Interleukin 15: a key to disrupted intraepithelial lymphocyte homeostasis and lymphomagenesis in celiac disease. Gastroenterology. https://doi.org/10.1016/S0016-5085(03)01047-3.
Meresse, B., et al., 2004. Coordinated induction by IL15 of a TCR-independent NKG2D signaling pathway converts CTL into lymphokine-activated killer cells in celiac disease. Immunity. https://doi.org/10.1016/j.immuni.2004.06.020.
Molberg, O., et al., 1997. Gliadin specific, HLA DQ2-restricted T cells are commonly found in small intestinal biopsies from coeliac disease patients, but not from controls. Scand. J. Immunol. https://doi.org/10.1046/j.1365-3083.1997.d01-93.x.
Molberg, Ø., et al., 1998. Tissue transglutaminase selectively modifies gliadin peptides that are recognized by gut-derived T cells in celiac disease. Nat. Med. https://doi.org/10.1038/nm0698-713.

Mora, J.R., et al., 2003. Selective imprinting of gut-homing T cells by Peyer's patch dendritic cells. Nature. https://doi.org/10.1038/nature01726.

Mora, J.R., et al., 2006. Generation of gut-homing IgA-secreting B cells by intestinal dendritic cells. Science. https://doi.org/10.1126/science.1132742.

Mowat, A.M.I., 2018. To respond or not to respond—a personal perspective of intestinal tolerance. Nat. Rev. Immunol. https://doi.org/10.1038/s41577-018-0002-x.

Mowat, A.M., Agace, W.W., 2014. Regional specialization within the intestinal immune system. Nat. Rev. Immunol. https://doi.org/10.1038/nri3738.

Mowat, A.M., Bain, C.C., 2011. Mucosal macrophages in intestinal homeostasis and inflammation. J. Innate Immun. https://doi.org/10.1159/000329099.

Mowat, A.M., Bain, C.C., Mowat, A.M., 2014. Macrophages in intestinal homeostasis and inflammation. Immunol. Rev.

Na, Y.R., et al., 2019. Macrophages in intestinal inflammation and resolution: a potential therapeutic target in IBD. Nat. Rev. Gastroenterol. Hepatol. https://doi.org/10.1038/s41575-019-0172-4.

Niess, J.H., et al., 2005. CX3CR1-mediated dendritic cell access to the intestinal lumen and bacterial clearance. Science. https://doi.org/10.1126/science.1102901.

Nilsen, E.M., et al., 1995. Gluten specific, HLA-DQ restricted T cells from coeliac mucosa produce cytokines with Th1 or Th0 profile dominated by interferon γ. Gut. https://doi.org/10.1136/gut.37.6.766.

Ogino, T., et al., 2013. Increased Th17-inducing activity of CD14+ CD163low myeloid cells in intestinal lamina propria of patients with Crohn's disease. Gastroenterology. https://doi.org/10.1053/j.gastro.2013.08.049.

Olivares-Villagómez, D., Van Kaer, L., 2018. Intestinal intraepithelial lymphocytes: sentinels of the mucosal barrier. Trends Immunol. https://doi.org/10.1016/j.it.2017.11.003.

Pabst, O., Mowat, A.M., 2012. Oral tolerance to food protein. Mucosal Immunol. https://doi.org/10.1038/mi.2012.4.

Palová-Jelínková, L., et al., 2005. Gliadin fragments induce phenotypic and functional maturation of human dendritic cells. J. Immunol. https://doi.org/10.4049/jimmunol.175.10.7038.

Palová-Jelínková, L., et al., 2013. Pepsin digest of wheat gliadin fraction increases production of IL-1β via TLR4/MyD88/TRIF/MAPK/NF-κB Signaling pathway and an NLRP3 Inflammasome activation. PLoS One. https://doi.org/10.1371/journal.pone.0062426.

Persson, E.K., et al., 2013. Dendritic cell subsets in the intestinal lamina propria: ontogeny and function. Eur. J. Immunol. https://doi.org/10.1002/eji.201343740.

Petersen, J., et al., 2020. T cell receptor cross-reactivity between gliadin and bacterial peptides in celiac disease. Nat. Struct. Mol. Biol. https://doi.org/10.1038/s41594-019-0353-4.

Peterson, L.W., Artis, D., 2014. Intestinal epithelial cells: regulators of barrier function and immune homeostasis. Nat. Rev. Immunol. https://doi.org/10.1038/nri3608.

Qiao, S.-W., et al., 2011. Posttranslational modification of gluten shapes TCR usage in celiac disease. J. Immunol. https://doi.org/10.4049/jimmunol.1101526.

Rajaguru, P., et al., 2013. Increased accumulation of dendritic cells in celiac disease associates with increased expression of autophagy protein LC3. Indian J. Pathol. Microbiol. https://doi.org/10.4103/0377-4929.125282.

Rakhimova, M., et al., 2009. In vitro differentiation of human monocytes into dendritic cells by peptic-tryptic digest of gliadin is independent of genetic predisposition and the presence of celiac disease. J. Clin. Immunol. https://doi.org/10.1007/s10875-008-9228-x.

Ráki, M., et al., 2006. A unique dendritic cell subset accumulates in the celiac lesion and efficiently activates gluten-reactive T cells. Gastroenterology. https://doi.org/10.1053/j.gastro.2006.06.002.

Ráki, M., et al., 2013. Plasmacytoid dendritic cells are scarcely represented in the human gut mucosa and are not recruited to the celiac lesion. Mucosal Immunol. https://doi.org/10.1038/mi.2012.136.

Rivollier, A., et al., 2012. Inflammation switches the differentiation program of Ly6chi monocytes from antiinflammatory macrophages to inflammatory dendritic cells in the colon. J. Exp. Med. https://doi.org/10.1084/jem.20101387.

Sarra, M., et al., 2013. IL-15 positively regulates IL-21 production in celiac disease mucosa. Mucosal Immunol. https://doi.org/10.1038/mi.2012.65.

Sathaliyawala, T., et al., 2013. Distribution and compartmentalization of human circulating and tissue-resident memory T cell subsets. Immunity. https://doi.org/10.1016/j.immuni.2012.09.020.

Scott, C.L., Aumeunier, A.M., Mowat, A.M., 2011. Intestinal CD103 + dendritic cells: master regulators of tolerance? Trends Immunol. https://doi.org/10.1016/j.it.2011.06.003.

Shan, L., et al., 2002. Structural basis for gluten intolerance in Celiac Sprue. Science. https://doi.org/10.1126/science.1074129.

Sichien, D., et al., 2017. Development of conventional dendritic cells: from common bone marrow progenitors to multiple subsets in peripheral tissues. Mucosal Immunol. https://doi.org/10.1038/mi.2017.8.

Sollid, L.M., Jabri, B., 2013. Triggers and drivers of autoimmunity: lessons from coeliac disease. Nat. Rev. Immunol. https://doi.org/10.1038/nri3407.

Sollid, L.M., Lie, B.A., 2005. Celiac disease genetics: current concepts and practical applications. Clin. Gastroenterol. Hepatol. https://doi.org/10.1016/S1542-3565(05)00532-X.

Sollid, L.M., et al., 1997. Autoantibodies in coeliac disease: tissue transglutaminase guilt by association? Gut. https://doi.org/10.1136/gut.41.6.851.

Sollid, L.M., et al., 2012. Nomenclature and listing of celiac disease relevant gluten T-cell epitopes restricted by HLA-DQ molecules. Immunogenetics. https://doi.org/10.1007/s00251-012-0599-z.

Stagg, A.J., 2018. Intestinal dendritic cells in health and gut inflammation. Front. Immunol. https://doi.org/10.3389/fimmu.2018.02883.

Stagg, A.J., et al., 2003. The dendritic cell: its role in intestinal inflammation and relationship with gut bacteria. Gut. https://doi.org/10.1136/gut.52.10.1522.

Stamnaes, J., Sollid, L.M., 2015. Celiac disease: autoimmunity in response to food antigen. Semin. Immunol. https://doi.org/10.1016/j.smim.2015.11.001.

Tamoutounour, S., et al., 2012. CD64 distinguishes macrophages from dendritic cells in the gut and reveals the Th1-inducing role of mesenteric lymph node macrophages during colitis. Eur. J. Immunol. https://doi.org/10.1002/eji.201242847.

Tollefsen, S., et al., 2006. HLA-DQ2 and -DQ8 signatures of gluten T cell epitopes in celiac disease. J. Clin. Investig. https://doi.org/10.1172/JCI27620.

Tucková, L., et al., 2002. Activation of macrophages by gliadin fragments: isolation and characterization of active peptide. J. Leukoc. Biol. https://doi.org/10.1189/jlb.71.4.625.

Tye-Din, J.A., et al., 2010. Comprehensive, quantitative mapping of T cell epitopes in gluten in celiac disease. Sci. Transl. Med. https://doi.org/10.1126/scitranslmed.3001012.

Vaquero, L., et al., 2019. Challenges to drug discovery for celiac disease and approaches to overcome them. Expert Opin. Drug Discovery. https://doi.org/10.1080/17460441.2019.1642321.

Vignali, D.A.A., Collison, L.W., Workman, C.J., 2008. How regulatory T cells work. Nat. Rev. Immunol. https://doi.org/10.1038/nri2343.

Vorobjova, T., et al., 2015. Increased density of tolerogenic dendritic cells in the small bowel mucosa of celiac patients. World J. Gastroenterol. https://doi.org/10.3748/wjg.v21.i2.439.

Watchmaker, P.B., et al., 2014. Comparative transcriptional and functional profiling defines conserved programs of intestinal DC differentiation in humans and mice. Nat. Immunol. https://doi.org/10.1038/ni.2768.
Willemijn Vader, L., et al., 2002. Specificity of tissue transglutaminase explains cereal toxicity in celiac disease. J. Exp. Med. https://doi.org/10.1084/jem.20012028.
Zevallos, V.F., et al., 2017. Nutritional wheat amylase-trypsin inhibitors promote intestinal inflammation via activation of myeloid cells. Gastroenterology. https://doi.org/10.1053/j.gastro.2016.12.006.
Zigmond, E., Jung, S., 2013. Intestinal macrophages: well educated exceptions from the rule. Trends Immunol. https://doi.org/10.1016/j.it.2013.02.001.
Zigmond, E., et al., 2012. Ly6C hi monocytes in the inflamed Colon give rise to Proinflammatory effector cells and migratory antigen-presenting cells. Immunity. https://doi.org/10.1016/j.immuni.2012.08.026.

CHAPTER FOUR

The tight junction and the epithelial barrier in coeliac disease

Amaia Jauregi-Miguel*
Wallenberg Centre for Molecular Medicine, Linköping University, Linköping, Sweden
Department of Biomedical and Clinical Sciences, Faculty of Health Science, Linköping University, Linköping, Sweden
*Corresponding author: e-mail address: amaia.jauregimiguel@gmail.com

Contents

Abstract

Epithelial barriers are essential to maintain multicellular organisms well compartmentalized and protected from external environment. In the intestine, the epithelial layer orchestrates a dynamic balance between nutrient absorption and prevention of microorganisms, and antigen intrusion. Intestinal barrier function has been shown to be altered in coeliac disease but whether it contributes to the pathogenesis development or if it is merely a phenomenon secondary to the aberrant immune response is still unknown. The tight junction complexes are multiprotein cell-cell adhesions that seal the epithelial intercellular space and regulate the paracellular permeability of ions and solutes. These structures have a fundamental role in epithelial barrier integrity as well as in signaling mechanisms that control epithelial-cell polarization, the formation of apical domains and cellular processes such as cell proliferation, migration,

International Review of Cell and Molecular Biology, Volume 358
ISSN 1937-6448
https://doi.org/10.1016/bs.ircmb.2020.09.010

differentiation, and survival. In coeliac disease, the molecular structures and function of tight junctions appear disrupted and are not completely recovered after treatment with gluten-free diet. Moreover, zonulin, the only known physiological regulator of the tight junction permeability, appears augmented in autoimmune conditions associated with TJ dysfunction, including coeliac disease. This chapter will examine recent discoveries about the molecular architecture of tight junctions and their functions. We will discuss how different factors contribute to tight junction disruption and intestinal barrier impairment in coeliac disease. To conclude, new insights into zonulin-driven disruption of tight junction structures and barrier integrity in coeliac disease are presented together with the advancements in novel therapy to treat the barrier defect seen in pathogenesis.

1. Intestinal epithelial barrier: Definition and function

The intestinal epithelium is the single layer of cells that define the gastrointestinal wall. It is composed of several elements; (i) the luminal mucus layer with the commensal gut microbiota and immune regulators; (ii) the central epithelial monolayer with specialized intestinal epithelial cells; and (iii) the inner lamina propria which is home to both innate and adaptive immune cells, such as T cells, B cells, macrophages and dendritic cells. A fundamental function of the intestinal mucosa is to act as a physical and immunological defence barrier against external molecules. While the mucosal component provides the very first line of chemical protection against pathogenic and commensal microbes, the underlying epithelium is by far the strongest determinant of the physical intestinal barrier. To maintain intestinal homeostasis, the epithelial layer will function in close relation with the immune system underneath, which has a crucial role in creating healthy communication between the microbiome and the immune cells (Eri and Chieppa, 2013).

The small intestinal epithelium is highly dynamic monolayer that undergoes constant turnover self-renewing along the crypt-villus axis every couple of days. A pool of pluripotent stem cells residing at the crypts gives rise to five distinct cell types; (1) enterocytes: absorptive cells that form most of the gut epithelium; (2) goblet cells: mucosa-secreting columnar cells; (3) enteroendocrine cells: hormones-secreting cells; (4) paneth cells: α-defensins secreting cells located in small intestinal crypts; and (5) microfold cells: cells specialized for antigen sampling (Chia and Kuo, 2010). These cells together form a continuous and polarized monolayer in which the apical domain faces the lumen while the basolateral domain contacts the basement membrane and the neighboring cells (for more information, see the chapter of Epithelial cell dysfunction in coeliac disease). Establishing a selective permeability to transmucosal water or solute flux will be crucial for the maintaining of homeostasis of the intestinal mucosa (Van Itallie and Anderson, 2006). The intestinal epithelium mediates selective permeability via two major routes: transcellular and paracellular pathways (Tsukita et al., 2001). The transcellular permeability consists of a selective transport of solutes through the epithelial cells. The transportation across the cell membranes will be done either by diffusion and endocytosis of lipophilic and large molecules, or by specific transporters in case of hydrophilic solutes (Ferraris and Diamond, 1997). In contrast, the paracellular permeability is associated with transport in the space between epithelial cells, and it is regulated through the presence of junctional complexes that mechanically link adjacent cells and seal the paracellular space (Fig. 1). These structures play a key role in intestinal epithelial barrier and require a tight regulation to avoid any imbalance in homeostasis (Williams et al., 2015). Defects in intestinal permeability can flare up into an uncontrollable immune reaction or an imbalanced growth of microbiota and thus, it has been associated with a

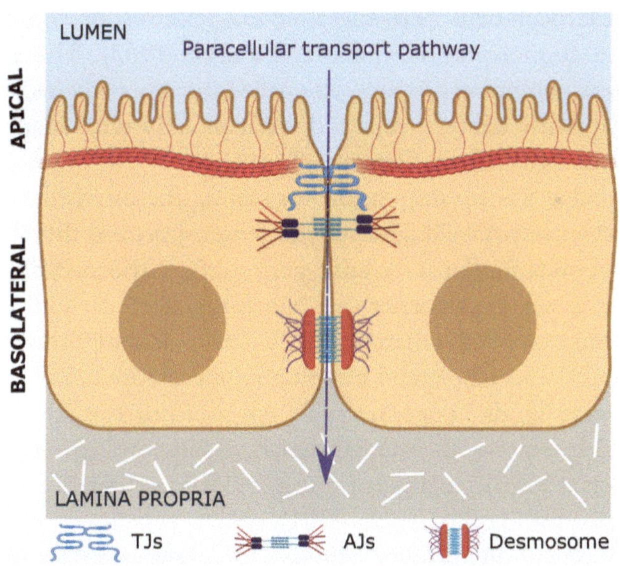

Fig. 1 Cell-cell adhesion complexes in intestinal epithelium. The paracellular space between epithelial cells is sealed and regulated by adhesion structures that join adjacent cells. These complexes are, from apical to basal, tight junctions (TJs), adherens junctions (AJs), and desmosomes. The apical side of the cell is encircled by a dense network of actin and myosin microtubules which helps in epithelial barrier regulation and supports the dense microvillus brush border. *Image created with BioRender.com.*

broad range of diseases, including intestinal inflammatory disorders such as food allergy, coeliac disease (CD), inflammatory bowel disease (IBD) and colon carcinoma; extra-intestinal autoimmune disorders such as chronic liver disease, rheumatoid arthritis and multiple sclerosis; and metabolic disorders including type 1 diabetes and obesity. For all these diseases, it is commonly hypothesized that dysfunction of the intestinal barrier lead to excessive entrance of dietary antigens such as gliadin or microbe-derived macromolecules. Enhanced permeability challenges the immune system of susceptible individuals and initiates an inflammatory reaction in the gut or more distant organ systems (De Mey and Freund, 2013; Meddings, 2008; Yu, 2012).

2. The tight junction and the paracellular permeability

2.1 Apical junctional complex

The epithelium maintains its selective barrier function through the formation of cell-cell junctions that comprise the apical junctional complex (APC). From apical-to-basal, the intercellular junctions are the tight junction or zonula occludens (TJ), the adherens junction or zonula adherens (AJ), and the desmosome (Farquhar and Palade, 1963). These complexes consist of transmembrane proteins, also called integral proteins, that interact extracellularly with adjacent cells and intracellularly with adaptor proteins that link to the cytoskeletal filaments (Fig. 1).

The epithelial TJs are responsible for sealing the intercellular space and regulating selective paracellular ionic solute transport and therefore, are the primary determinant of paracellular permeability (Förster, 2008). These structures act as selective barrier that control paracellular diffusion of ions and solutes and support the maintenance of cell polarity by restricting intermixing of apical and basolateral transmembrane components, such as proteins membrane lipids. Together with AJ conform the apical junctional complex and interact with internal signaling pathways to regulate diverse processes such as apical-basal polarization, cellular proliferation, polarization and differentiation (Tsukita et al., 2001).

AJs and desmosomes provide adhesive forces necessary for maintenance of cell-cell interactions (Schneeberger and Lynch, 1992). Desmosomes are spot-like adhesions on the lateral sides of plasma membranes and are crucial to tissues that experience mechanical stress, such as cardiac muscle, bladder, gastrointestinal mucosa, and skin (Garrod and Chidgey, 2008). The AJ is composed mainly of single spanning transmembrane proteins cadherins.

The extracellular portion of cadherins interact homotypically with cadherins on adjacent cells while on the cytoplasmic side interacts directly with p120-catenin and β-catenin, which in turn interact with α-catenin. The latter regulates perijunctional actin assembly which provides further strength to these structures (Buckley et al., 2014; Ireton et al., 2002; Yap et al., 1998). The AJ structures are necessary for efficient TJ and desmosome assembly (Capaldo and Macara, 2007; Maiers et al., 2013; Perez-Moreno et al., 2003).

2.2 Tight junction composition

The TJ consists of a series of anastomosing strands that form a continuous belt-like ring which seals the border between the apical and basolateral regions in epithelial cells. Although composed of many constituents, TJ proteins can be broadly divided into transmembrane proteins and a cytosolic plaque formed by scaffolding and regulatory proteins (Fig. 2).

The transmembrane proteins are divided in tetraspan proteins formed by the large family of claudins and the family of tight junction-associated Marvel proteins (TAMPs); and single transmembrane proteins as junctional adhesion molecules (JAM). TAMPs itself are divided in integral components predominantly localizing in bicellular TJs (occludin and marvelD3) or tricellular TJ (tricellulin) (Campbell et al., 2017).

The family of 26 claudins constitute the major components of TJ strand. These integral proteins form size- and charge-specific pores to regulate TJ ion selectivity, playing a crucial role in the paracellular ion barrier and in the polarity of epithelial cells (Günzel and Yu, 2013). The diversity in composition and interactions explains the tissue-specific heterogeneity found in TJ formations and permeability (Garcia-Hernandez et al., 2017). From their primary function as "gate keeper," intestinal claudins are grouped according to their channel- and barrier-forming properties into those that support cation-selectivity (e.g., claudin 2, claudin 10b and claudin 15), anion-selectivity (e.g., claudin 10a and claudin 17) or sealing (e.g., Claudin-1, -3, -4, -5, -7, and -19) (Tsukita et al., 2001; Zihni et al., 2016). Additionally, claudins also have functions in cytoskeleton organization, several signaling pathways and "fence" function that prevents intermixing of apical and basolateral membrane proteins (Krause et al., 2008; van Meer and Simons, 1986). Due to their key role in permeability, alterations in the expression of claudins are related to disturbance in gut homeostasis and contribute to several diseases including CD (Kominsky et al., 2003; Zeissig et al., 2007; Zhu et al., 2006).

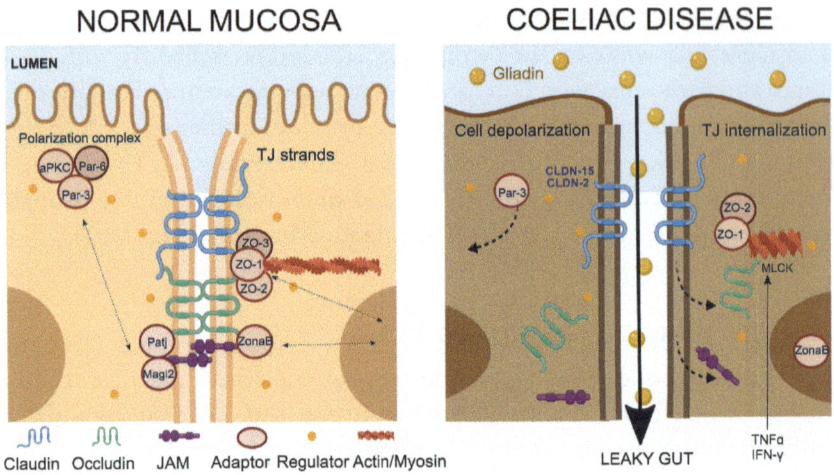

Fig. 2 Tight Junction in normal mucosa and in coeliac disease. In physiological conditions, paracellular diffusion of molecules is restricted by the tight junctions, a network of transmembrane proteins (mainly claudin, occludin and JAM) that form pores and connect to the actomyosin ring through a cytoplasmique plaque consisting of adaptor proteins and signaling regulators. Apart from the barrier function, tight junctions also regulate the polarization of the cells through the polarization complex, which will keep the apicobasal asymmetry by restricting intermixing of apical and basolateral transmembrane components. In disease and upon gluten intake, the aberrant inflammatory response will produce high levels of cytokines such as tumor necrosis factor-α (TNFα) and interferon gamma (IFNγ), which increase myosin light chain kinase (MLCK) production. MLCK in turn, causes reduction in ZO-1 protein and internalization of both ZO-1 and occludin proteins. In parallel, pore-forming Claudin-2 and Claudin 15 expression is activated while barrier-forming claudins are silenced. The expression of polarization complex is also reduced, allowing Par-3 to spread out to the lateral membrane instead of staying at the apical junction, resulting in the depolarization of the cell. CLDN, Claudin; MLCK; myosin light chain kinase; TGF-α, transforming growth factor alpha; IFNγ, interferon gamma. *Image created with BioRender.com.*

The rest of integral proteins are not directly regulating the paracellular permeability. However, they have an important role in the barrier formation (Ebnet, 2017) and recruitment of signaling complex proteins to TJs (Raleigh et al., 2010) in order to regulate processes such as wound healing (Chandhoke and Mooseker, 2012) or cell adhesion (Van Itallie and Anderson, 1997).

Intracellularly, the transmembrane strand interacts with a complex network that underlies the junctional membrane also known as the "cytoplasmic plaque." This plaque consists of scaffolding and signaling molecules that form the interface between the integral membrane proteins and the filamentous cytoskeleton (Matter and Balda, 2003). The most studied scaffolding components, are zonula occludens family ZO-1, -2, and -3; also known as TJ protein-1, -2, and -3. These adaptor proteins contain multiple protein-protein interaction motifs to facilitate recruiting of other proteins for structural organization and signaling (Cordenonsi et al., 1999; Van Itallie and Anderson, 2014). As example, ZO-1 has three PDZ domains that bind claudins and JAMs, a GUK domain that binds occludin, a SH3 domain that binds the transcription factor ZONAB, and a carboxy-terminus that interacts with cytoskeletal F-actin (Rodgers et al., 2013; Stevenson et al., 1986). Other studied adaptor proteins are the membrane-associated guanylate kinase inverted (MAGI) proteins, or the multi-PDZ domain proteins MUPP1 and PALS1-associated TJ (PATJ; also known as INADL) protein as well as PAR3-PAR6-aPKC complex, an evolutionarily conserved signaling module that drives apical polarization (Dobrosotskaya et al., 1997; Ide, 1999; Lemmers, 2002; Roh, 2002; Poliak et al., 2002; Hamazaki et al., 2002; Yu et al., 2015).

Together with scaffolding proteins, several molecules involved in the signal transduction processes, including protein kinases, myosin light chain (MLC) kinase (MLCK), phosphatases, small GTP-binding proteins and transcriptional regulators, are recruited to junctions, indicating that the TJ proteins are regulated by multiple signaling proteins and signaling molecules (Quiros and Nusrat, 2014; Stuart and Nigam, 1995; Zihni et al., 2014). There are also components of the plaque such as ZONAB and ZO-1 with dual localization that are able to translocate to the nucleus to promote proliferation. In addition, several miRNA have been linked to TJ proteins, affecting intestinal TJ permeability (Barisani, 2018).

Interestingly, spatiotemporal patterns of the TJ proteins varies significantly over both longitudinal (i.e., proximal small intestine to distal colon), and vertical (i.e., crypt to villus) axes, suggesting that TJ contributes to the homeostasis of epithelial layer (Holmes et al., 2006). In addition, recent study in intestinal crypts-derived organoids revealed differential TJ composition and epithelial permeability among different intestinal cell types, pointing to a cell-type specific paracellular pathway regulation (Pearce et al., 2018). The vast diversity of proteins and molecules, the redundant motifs and overlapping function as well as the context specific regulation makes difficult to understand the functional relevance of the intramolecular interactions happening in the cytoplasmic plaque. However, it is known that the major mechanism of TJ regulation is a complex network of cell signaling together with post-translational phosphorylation, which drastically alters their membrane distribution and turnover. Even the exact function of the individual TJ proteins remains elusive, it is well-accepted that TJ integrity is essential for homeostasis. Hence, alterations in expression, phosphorylation and localization patterns of the TJs are common in pathological conditions such as chronic inflammatory diseases that generates intestinal barrier dysfunction. It may lead to increased paracellular transport of solutes and water that causes leak-flux diarrhea (Sandle, 2005) and/or increased permeability to large molecules, such as gliadin, that induce an immune response (Al-Sadi et al., 2011; Buschmann et al., 2013). In the next section, we summarize the most recent data on components of TJs and focus on the implication of TJ dysfunction in CD.

3. The role of the intestinal barrier in coeliac disease

Compromised intestinal barrier integrity is observed in both intestinal and systemic diseases, including CD. In active CD, paracellular permeability is increased and TJ morphology is substantially altered. The molecular mechanisms underlying the structural and functional modifications of TJs include the reduction and internalization of the TJ proteins as well as the activation of myosin light chain kinase (MLCK) phosphorylation to promote cytoskeletal contraction (Matsuda et al., 2004; Utech et al., 2005; Zolotarevsky et al., 2002). Despite the indisputable role of intestinal permeability in pathogenesis, still is not determined whether the loss of barrier integrity is the cause or consequence of diseases such as CD. It is also hard to know whether the gliadin peptides are able to freely diffuse along the disrupted epithelial junctions or whether an increased transcytosis drives the antigen actively toward the lamina propria. Pathophysiological, environmental, and genetic factors seem to be crucial for the development of an increased barrier permeability.

3.1 Tight junction dysfunction in coeliac disease

The earliest evidence of defective epithelial barrier in CD comes from experiments in 1970s. Coeliac patients with villous atrophy showed an enhanced intestinal permeability to different sugars, which is shown to increase further when the patients are given a diet containing gluten (Cobden et al., 1978; Hamilton et al., 1982; Menzies et al., 1979; Pearson et al., 1982). Following this, one-path impedance spectroscopy was done to determine the transepithelial electrical resistance of the coeliac intestinal mucosa in an ex-vivo system. A reduction of 50% of epithelial barrier function was reported in active CD patients, which was reversed partially after GFD treatment (Fromm et al., 2009; Schulzke et al., 1995; Schumann et al., 2012, 2017). Ultrastructurally, freeze-fracture electron microscopy demonstrated a thin a reduced TJ strands as well as strand discontinuities in active CD. Again, a GFD was shown to reverse the defects only partially presenting TJ strands in a smaller number than non-coeliac controls (Schulzke et al., 1998). Overall, these studies showed that even barrier function and TJ integrity are mainly compromised in the active disease, the disruption persists in asymptomatic patients who were on a gluten-free diet, introducing the potential of an intrinsic coeliac barrier defect in CD patients (Schulzke et al., 1998).

At the molecular lever, TJ strand defect in CD reflects a complex molecular alteration in TJ proteins content and localization (Fig. 2). The main affected TJ proteins are the claudin family, that has a delicate balance of spatiotemporal and crypt-luminal axis expression in homeostasis and show a complex dysregulation of their expression and function in inflammatory disorders. In active CD, claudins with a possible role in the villus atrophy-related malabsorptive mechanism has been seen altered. Specifically, an upregulation of the channel-formers claudin-2 and claudin-15 and downregulation of the barrier-formers for ions a larger solutes such as claudin-3, claudin-5, and claudin-7 was found (Nagy Szakál et al., 2010; Schumann et al., 2012). Claudin-2 has been already suggested to have a prominent role in CD (Nagy Szakál et al., 2010). It predominantly localizes to the crypts of CD patients and its expression is induced by TNFα, also upregulated in CD (Zeissig et al., 2007). Claudin-2 overexpression has been also linked to increased epithelial cell proliferation in adenocarcinoma by promoting translocation of ZONAB into the nucleus (Dhawan et al., 2011; Ikari et al., 2014). Remarkably, refractory CD patients, who present impaired barrier function similar to active patients, claudin-2 protein content is even higher (Schumann et al., 2012). Regarding to Claudin-3, apart from expression reduction, it is not localizing to TJ at all in CD, suggesting a complete loose of its function at disease (Schumann et al., 2012). In contrast, expression and localization of claudin-1 and claudin-4, more related to the role of TJ changes for the diarrhea, is unchanged (Schumann et al., 2008a). Together with claudins, occludin and TJ-associated molecule ZO-1 proteins expression are reduced and interaction among them defective, resulting in membrane-displaced ZO-1 in CD (Ciccocioppo et al., 2006; Schumann et al., 2012). ZO-1 pattern is associated with F-actin redistribution and resembles the distribution in undifferentiated cells pointing to enterocyte immaturity in CD (Montalto et al., 2002; Pizzuti et al., 2004). ZO-1 expression alteration is reversible as gluten withdrawal allows a normalization of it in treated CD patients (Montalto et al., 2002; Pizzuti et al., 2004). These data suggest that gluten-induced intestinal inflammation leads to a redistribution of a phosphorylated ZO-1 that creates a weakening of the epithelial barrier. This dysregulation might be an early event that could precede to TJ structural disorganization. As a proof of that, reintroduction of ZO-1 in ZO-1/ZO-2-deficient epithelial cells determines the correct localization of TJ complex to the membrane (Umeda et al., 2006). In line to ZO-1 displacement, the barrier-forming claudin-5 and pore forming claudin-15 are also found intracellular evidencing a relevant protein shift from the TJ into the cytoplasm (Tamura et al., 2011). Concordantly, the AJ component β-catenin is extensively phosphorylated in active CD patients and might be correlated with a decrease of E-cadherin localization to the membrane (Ciccocioppo et al., 2006).

Although much progress has been made in the knowledge of TJ barrier integrity disruption in CD, the regulatory mechanisms that control TJ function and cellular distribution is still unknown. Improved understanding of TJ regulation in homeostasis and pathogenesis will provide strategies to strengthen the epithelial barrier function and permeability in gut disorders.

3.2 Impairment of tight junction-dependent epithelial polarization in coeliac disease

In addition to a barrier impairment, the epithelial polarity is affected at molecular level in coeliac intestinal mucosa (Fig. 2). This includes reduction in polarization proteins partition defective-3 (Par-3) and its opponent protein phosphatase-1 (PP-1) (Schumann et al., 2017). PP-1 mediates phosphorylation of Par-3 and may contribute in its reduction at the TJ (McCaffrey and Macara, 2012; Traweger et al., 2008). Dysfunction of Par-3 is known to cause lateral exclusion of Par-3 and depolarization of epithelial cells (Benton and St Johnston, 2003; Suzuki and Ohno, 2006). All these results suggest that in CD Par-3 could be localized to the basolateral space rather than at the apical junction, affecting TJ formation. Moreover, a genetic association with CD of Par-3 and Membrane-associated guanylate kinase, WW and PDZ domain-containing protein 2 (Magi2) was reported, suggesting that this two regulators of epithelial polarity play a role in the CD barrier defect (Wapenaar et al., 2008). In addition, TGF-β, secreted by T cells in active CD, also regulates the Par complex involved in epithelial-to mesenchymal transformation (EMT) (Thiery and Sleeman, 2006). It is still not known if epithelial depolarization is a primary or secondary event in CD because even if genetic association points to be a dysregulation that occurs at preliminary stage, the fact that it is induced by TFG-β, a cytokine released secondarily after immune activation, supports the hypothesis of depolarization being a secondary barrier defect with a role in an amplification loop of the inflammatory response.

3.3 The role of genetics in intestinal barrier defect in coeliac disease

As mentioned above, altered intestinal permeability persists in asymptomatic CD patients treated with a gluten-free diet pointing an existences of increased intestinal permeability prior to disease onset (Bjarnason and Peters, 1984; Schulzke et al., 1998). Consistent with these observations, Irish Setter dogs, which spontaneously develop a gluten dependent enteropathy that resembles CD, have increased gut permeability prior to gluten exposure and disease onset (Hall and Batt, 1992).

In line with functional evidence, healthy first-degree relatives of patients with active CD are much more likely to have increased gut permeability than the normal population, which may indicate a genetical predisposition to the development of CD (Van Elburg et al., 1993). Several genome-wide association and follow-up studies revealed that 39 loci, besides the extensively studied HLA-DQ2 and HLA-DQ8 variants, are associated with CD (Dubois et al., 2010; Trynka et al., 2011; Van Heel et al., 2007). Based on coexpression algorithm, pathway analysis on prioritized CD loci showed that four genes, *LPP*, *C1orf106*, *ARHGAP31* and *PTPRK*, are involved in intestinal barrier function, especially in the actin-cytoskeleton rearrangement and cell-cell adhesion pathways (Almeida et al., 2014; Kumar et al., 2015). Posterior functional data corroborate that the LPP protein colocalizes with E-cadherin from AJs and has a role in cell adhesion of epithelial cells (Van Itallie et al., 2014). Mice studies showed that *C1orf106* is involved in maintaining adherens junction stability as its knockdown reduces intestinal barrier integrity (Mohanan et al., 2018). *ARHGAP31* is a novel regulator of cell-cell contact maintenance and *PTPRK* is a receptor present at epithelial cell-cell junctions that promotes junction integrity by dephosphorylating regulators of cell-cell adhesion (Fearnley et al., 2019; McCormack et al., 2017).

In addition to the previous evidences, another studies genetically associated *MYO9B*, *PARD3*, and MAGI2 genes involved in TJ formation and cell polarity, with CD (Monsuur et al., 2005; Wapenaar et al., 2008).

However, this finding was not replicated in larger GWASs and follow-up studies (Dubois et al., 2010; Jauregi-Miguel et al., 2019; Trynka et al., 2011). Moreover, an independent transcriptomic analysis in a celiac family also revealed cell-cell junctions as one of the pathways shutdown in treated patients (Ciccocioppo et al., 2018).

All these studies concluded that mechanisms determining the intestinal barrier function in CD contribute to disease development rather than being secondary to it. However, the function of these CD-associated genes in the regulation of paracellular permeability is still unknown.

3.4 Gliadin and immune system induced tight junction barrier defect in coeliac disease

Although TJ dysfunction that occurs in CD can induce molecular leakage, later studies indicate that the immune response to ingested gluten also influence the intestinal barrier integrity. In this case, epithelial barrier disruption would be a secondary event that has a role as an amplification loop of immune reaction. As evidence of that, in vitro exposure of intestinal cells to TNFα and IFNγ proinflammatory cytokines affects intestinal barrier integrity through ZO-1, occludin and JAM internalization as well as AMPK-dependent ZO-1 and occludin expression reduction (Ciccocioppo et al., 2006; Marchiando et al., 2010; Ozaki et al., 1999; Utech et al., 2005). These modifications were shown to be dependent on micropinocytosis-like processes and Rho/ROCK signaling. A synergistic effect of both cytokines in inducing barrier damage has been observed and may induce the activity of the myosin light chain kinase (MLCK), which will disrupt the interaction between the TJ proteins and the cytoskeleton, subsequently damaging the TJ integrity (Capaldo and Nusrat, 2009; Shen et al., 2006). IL-15 has also been linked to epithelial barrier impairment, as IL-15 exerts important barrier effects by inducing enterocyte cytolysis in CD (Maiuri et al., 2003).

Gliadin exposure alone also alters the barrier properties of intestinal epithelial cell as shown in electrical experiments done in intestinal cell lines, where an increase in permeability was associated with reduced expression of occludin, ZO-1, and claudins-3 and -4 (Sander et al., 2005). Whether barrier disturbance produced by the gliadin antigen reflects upregulated epithelial transcytosis or paracellular leakage is poorly defined. While in healthy individuals only small molecules diffuse through TJs, in active CD, TJ integrity disruption leads to paracellular flux of higher molecules. Yet, the gliadin peptides such as p31–49 and 33 mer are 10 times bigger than markers used to evaluate the paracellular permeability and hence, it has been hypothesized that gliadin could be transported to the lamina propria via the transcellular pathway. Recent studies done in CD showed that SIgA, the protective luminal immunoglobulin that in homeostasis protects intestinal epithelia from antigens, in active disease allows the transcytosis of gliadin peptides (Matysiak-Budnik et al., 2008). SIgA creates a protective complex with gliadin and promotes the entry of intact peptides into the lamina propria via the transferrin receptor CD71 at the apical surface. This receptor in homeostasis is confined to the basolateral membrane but shows an ectopic expression at the apical membrane of enterocytes in CD. Taking together previous evidences of a transcellular transport of gliadin peptides in intestinal epithelial cell lines, it is hypothesized that in active CD, the retrotransport of IgA-gliadin immune complexes most likely triggers exacerbated adaptive and innate immune responses that result in mucosal lesions (Barone et al., 2010; Schumann et al., 2008b; Zimmer et al., 2010). Accordingly, it has been shown that the proinflammatory cytokine TNFα, secreted in CD, increases endosomal uptake of antigens, reinforcing the importance of the transcellular pathway in the barrier dysfunction of CD (Rodriguez et al., 1995; Schmitz et al., 1999). However, the increased permeability to intact gliadin peptides observed in active CD (which support the theory of transcytosis transportation of the gliadin peptides) is reversed in most of treated patients on GFD, even if intestinal permeability is not completely normalized (Matysiak-Budnik et al., 2008). Hence, a primary defect in intestinal permeability that promotes a paracellular flux of gliadin peptides in early stage of the disease cannot be excluded.

In summary, it seems that in CD intestine the gliadin uptake is driven by both paracellular and transcellular transport pathways which contribute to activation of the intestinal immune system. Even if the mechanism is not fully understood, the inflammatory reaction creates a vicious cycle in which luminal antigens reach the lamina propria and promotes a further increment of epithelial barrier dysfunction.

3.5 Additional external factors: Infections

How the environmental factors influence the development of CD is complex and poorly understood. It is well known that the genetic susceptibility and the ingested gluten are not sufficient to explain the onset of disease. As proof of that, CD can be diagnosed at any age from infancy well into senior adulthood (Rampertab et al., 2006).

A potential factor related to CD etiology are infection-related alterations of the intestinal barrier. In the late 1980s, CD incidence in Sweden increased by threefold within few years and turned back to the initial incidence level in the late 1990s (Ivarsson et al., 2000). Among the possible hypotheses, repeated infectious episodes that disturb the epithelial barrier was proposed as potential risk factor. This assumption was based on a previous study that identified sequence similarities between peptide sequences of gluten and adenovirus peptides (Kagnoff et al., 1987). Other recent studies has associated rotavirus infection, campylobacter and high incidence of infections in the first few years of life with increased CD risk (Mårild et al., 2015; Riddle et al., 2013; Stene et al., 2006). The mechanism through which infections affect the onset of disease are yet obscure. Recent studies point to a mechanism resembling molecular mimicry or by alteration of the intestinal barrier, being preferred the second one (Schumann et al., 2008b).

4. Zonulin, a reversible regulator of the epithelial barrier in coeliac disease

The only known physiological regulators of the intestinal TJ was brought forward by Alessio Fasano's group and it has been intensely discussed mechanism (Wang et al., 2000). Fasano identified zonulin, a family of paracrine proteins that reversibly modulate the permeability of TJs upon gliadin exposure. In fact, zonulin was described as an endogenous homologue to the *Vibrio cholerae* derived ZO toxin (Zot), an enterotoxin that modulates the intestinal TJs leading to a "leaky gut" and the consequent watery diarrhea (Tripathi et al., 2009). Through proteomic analysis of human sera, zonulin was identified to be the precursor protein to the haptoglobin glycoprotein-2 (pre-HP2) (Tripathi et al., 2009). There are two human genetic variants for these protein and only those who possess the HP2 allele, not HP1, produce active zonulin. Remarkably, an increased frequency of HP2 genotype is seen in CD as well as in inflammatory bowel disease, pointing to a potential genetic risk factor for intestinal barrier disruption in inflammatory diseases (Márquez et al., 2012; Papp et al., 2008).

The mechanism by which gliadin leads to zonulin release from intestinal epithelial cells and modulates the opening of the intestinal TJs is well-characterized (Sturgeon and Fasano, 2016) (Fig. 3). Luminal zonulin release is stimulated upon gliadin exposure, the toxic component of gluten, or microbiome imbalance (El Asmar et al., 2002). On enterocytes and monocytes, gliadin binds to the CXCR3 chemokine receptor and initiate a

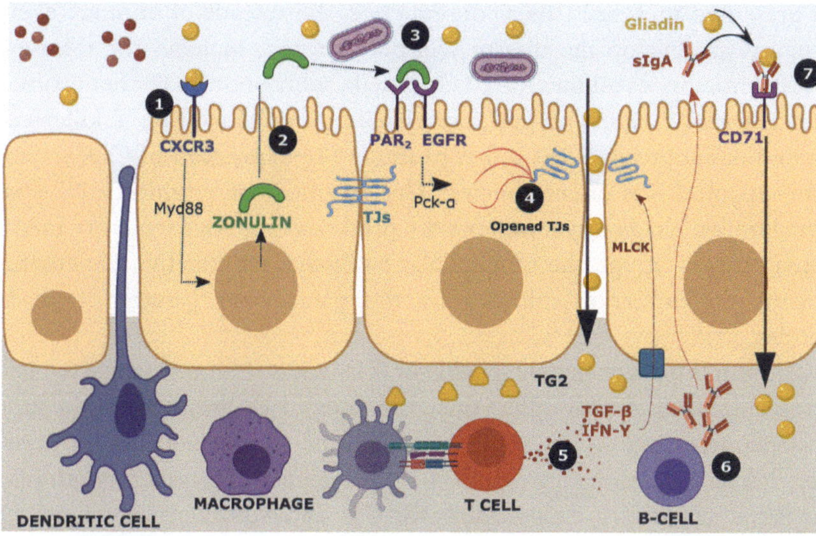

Fig. 3 Mechanism of gliadin-induced zonulin release and subsequent increase in intestinal permeability. Gliadin specific peptides or bacteria (1) cause a CXCR-3-mediated, MyD88-dependent zonulin release (2). Zonulin transactivates EGFR through PAR2 leading to PCK-α dependent tight junction disassembly (3). Increased intestinal permeability leads to paracellular flux of antigens (4) into the lamina propria where they are able to interact with the immune system. The immune reaction leads to TNFβ and IFNγ production, which alter the MLCK activity leading to higher TJ disruption and increased permeability (5). At the same time, activated B cells will differentiate into IgA plasma cells, including secretory IgA antibodies (sIgA) that are secreted to the luminal area (6). In pathogenic conditions, sIgA creates a complex with gliadin and permits the transcellular entry of intact peptides into the lamina propria via binding to the transferrin receptor CD71 at the apical surface (7), promoting an enhanced loop activation of the immune system. Pck-α, *protein kinase C alpha; sIgA, secretory* immunoglobulin A; MLCK, myosin light-chain kinase; TJ, tight junction; TG2, transglutaminase 2; TGF-β, transforming growth factor beta; IFNγ, interferon gamma. *Image created with BioRender.com.*

MyD88-dependent luminal release of zonulin (Lammers et al., 2008). Zonulin binds epidermal growth factor receptor (EGFR) in IECs through direct binding and/or through protease activated receptor 2 (PAR_2) transactivation. The protein then activates membrane-associated PKC-α that catalyses the phosphorylation of target proteins, including ZO-1 and myosin 1C, as well as actin polymerization, causing the rearrangement of the cytoskeleton and the subsequent displacement of proteins (including ZO-1 and occludin) from the junctional complex. As a result, the intestinal TJ becomes looser leading to increased paracellular passage of luminal antigens in the lamina propria where they are processed by mucosal antigen presenting cells and presented to T cells. As a consequence, the cascade of immune events necessary to develop the disease will be activated, includes the release of proinflammatory cytokines by T helper cells, which in turn further promote TJs disassembly increasing gut permeability. Simultaneously, T killer cells initiate the apoptosis of IECs that promote the expression of CD71 transporter in apical side of enterocytes. This receptor will permit the transport of the luminal gliadin (in complex with protective mucosal secretory IgA) in lamina propria trough the transcellular pathway and thereby, perpetuating intestinal inflammation initiated by the paracellular passage of gliadin peptides.

Zonulin is not a molecule specific of CD. It has also been associated with bacterial gastrointestinal infections, rheumatoid arthritis, systemic lupus erythematosus, and type 1 diabetes mellitus (Fasano, 2011; Vanuytsel et al., 2013). Therefore, it might be part of a general inflammatory gut reaction rather than specific mechanism for CD. Intestinal permeability studies in zonulin transgenic mouse model showed that the expression of zonulin causes barrier impairment before the onset of chronic inflammatory diseases (Miranda-Ribera et al., 2019; Sturgeon et al., 2017). However, despite major progress in our knowledge on the mechanism by which zonulin modulates TJ barrier, its role in CD pathogenesis remains controversial. Up to date, there is no zonulin inhibition approach that completely blocks the barrier defect in CD patients. In addition, the high zonulin concentrations affecting barrier function as well as the fact of zonulin and pre-HP2 being identical create controversy among scientist (Tripathi et al., 2009). Recently, several studies questioned the use of serum zonulin as serological marker of mucosal barrier integrity. They have demonstrated that actual commercial assays that use monoclonal antibodies to detect pre-haptoglobin 2, are indeed not detecting pre-haptoglobin2, neither the Complement-associated (C3), which shares similar homology. Instead these kits are detecting several structural and maybe functional analog proteins (Ajamian et al., 2019; Scheffler et al., 2018). Despite the controversy, a zonulin receptor antagonist called larazotide acetate has been designed to modulate tight junctions. This drug is in clinical trials and is being tested to use as a supplement to a gluten-free diet (see novel therapies below).

5. Correlation between barrier dysfunction in coeliac disease and liver injury

CD is a chronic immune-mediated multisystem disorder and one of common extraintestinal manifestations is coeliac hepatitis. Coeliac hepatitis is defined by the presence of liver injury such as histological changes in CD patients that is reversed after GFD treatment (Bardella et al., 1995; Kaukinen et al., 2002). The mechanisms by which CD cause liver injury are unknown. However, the correlation between intestinal barrier defect and the development of a transaminitis (high levels of certain liver enzymes) suggest that liver dysfunction is a consequence of the coeliac "leaky gut" (Rubio-Tapia and Murray, 2007). Intestinal inflammation and barrier impairment may facilitate the circulation of toxins, antigens, cytokines and/or other mediators of liver injury through the vascular system to the liver (Bardella et al., 1995). It is also believed that proinflammatory cytokine secretion by hepatocytes in the liver induce an amplifying loop that consist on an intestinal barrier defect that allows the passage of bacterial products which in turn aggravates the already existing liver damage (Hagander et al., 1977; Wiest et al., 2014). Although it seems that the epithelial barrier has a role in the onset of coeliac hepatitis, liver injury is not common in other intestinal disorders associated with inflammation and increased intestinal permeability and therefore, other unknown genetic or CD-specific factors may contribute in the pathogenesis.

6. Novel therapies targeting barrier defects in coeliac disease

The only treatment for CD established to date is lifelong adherence to a gluten-free diet as it reduces symptoms and heals the mucosal damage within weeks to months. However, evidence indicate that observed barrier disruption in disease remains affected, being only partially restored. Recent advances and insights in the pathogenesis of CD together with the challenge in establishing a lifelong adherence to gluten-free diet has opened the gates for development of newer targeted drugs. Increased intestinal permeability could play a major role in the pathogenesis of CD and therefore, alternative therapeutic strategies are being directed toward preventing the passage of gliadin through the intestinal cell layer by reducing paracellular permeability (Fasano, 2009; Lammers et al., 2008).

Recent studies identified the AT-1001 peptide as the inhibitor molecule of zonulin system. Larazotide acetate (formerly named AT-1001, developed by Alba Therapeutics, USA) is an octapeptide derived from a cholera ZOT toxin, and contains amino acid sequences shared by the receptor binding-motif of zonulin. This peptide binds to the zonulin receptor present on the apical membrane of enterocytes in a competitive manner and, therefore, prevents the paracellular transport caused by gliadin and the consequent pathological inflammatory response. Since the ZOT/zonulin receptor is expressed in the small intestine, the effect of AT-1001 will be restricted to that area. Several clinical trials have been completed with larazotide acetate. During the Phase 1 trials, in vivo and in vitro experiments tested that larazotide acetate reduces gluten-induced intestinal barrier dysfunction (Gopalakrishnan et al., 2012; Paterson et al., 2007). Moreover, the safety profile of larazotide was comparable to that of the placebo. In the phase II, placebo-controlled trials failed in significantly reducing intestinal permeability but symptoms triggered by gluten challenge were significantly ameliorated by larazotide acetate (Kelly et al., 2013; Leffler et al., 2012). The last study published on larazotide in CD suggested that patients suffering from persistent inflammatory symptoms despite adherence to a GFD might benefit from GFD plus larazotide (Leffler et al., 2015). Recently, first ever phase 3 clinical trial for CD has started to be planned and first patient has been dosed to evaluate the effect of larazotide acetate.

Another promising approach to reduce the epithelial barrier impairment in CD is the use of probiotic bacteria to fight pathogens and reconstitute a protective microflora. It is known that some bacterial pathogens are able to modulate the TJ barrier function, via the MLCK activation in enteropathogenic *E. coli* infection (Yuhan et al., 1997), via direct interactions with claudins by *C. perfringens* enterotoxin (Saitoh et al., 2015) or by rho inhibition by *C. difficile* toxins (Just et al., 1995). Therefore, it is thought that commensal bacteria and probiotic bacterial strains could combat pathogenic effect and may exert beneficial effect in the preservation of epithelial barrier function. In addition, it has been hypothesized that probiotics modulate inflammatory processes and sustain peptide hydrolysis at the intestinal level. Indeed, animal model studies showed that the administration of *B. longum* CECT 7347 reduces the production of inflammatory cytokines and the CD4+ T-cell mediated immune response (Laparra et al., 2012). In addition, *Lactobacillus rhamnosus* prevents the reduction of TJ proteins and thus, protects the intestinal mucosa from PT-gliadin-induced damage (Orlando et al., 2018). Recently, a study assessed in intestinal cell line the capacity of probiotic mixture to hydrolyse gluten peptides and counteract its toxic effect. Digestion of the probiotic strains reversed the toxic effects of gliadin peptides in TJ barrier integrity, including IL-6 production and zonulin and occludin redistribution and modulation (Giorgi et al., 2020).

Lastly, a number of next generation ROCK inhibitors are under active evaluation for other indications such as asthma and multiple sclerosis, and some of these drugs may hold promise as therapeutic agent for CD as increase in intestinal permeability is also dependent on Rho kinase

(ROCK) activity (Sollid and Khosla, 2011). However, the side effect of inhibiting this pathway are unknown and could be it inappropriate for chronic, long-term use in CD patients with CD. More studies are needed to establish whether ROCK inhibition can reverse gluten dependent increased in intestinal permeability in patients with CD.

7. Concluding remarks

All evidences confirm the involvement of tight junction genes related to permeability, polarity, and cell proliferation in the epithelial destruction observed in CD. Barrier loss as a consequence of reduced TJ protein expression and altered distribution is a common feature in CD. However, gut permeability and its role in the pathogenesis of CD remains controversial. It is not yet clear the aberrant immune system reaction causes barrier defect or inversely. While the association between CD and TJ-related genetic variations point to a primary effect, the partial recovery of intestinal permeability in treated CD patients suggest a potential inflammatory cause behind. Although our knowledge of TJs has improved over the recent years, little is known about the regulation of the multiple functions and signals pathways involved in the TJ structures. In addition, we have seen that TJ function and composition is dynamic and specific to epithelial cell-type or tissue location. Due to its complexity, the specific role of each TJ proteins for the onset and perpetuation of the coeliac inflammatory response is still poorly understood. Future studies should be conducted to establish how different proteins cooperate to regulate such functions and integrate the complex network of signaling pathways.

Up to date, zonulin is the only known physiological modulator of paracellular permeability that reversibly opens the TJ structures permitting uncontrolled influx of dietary and microbial antigens. The mechanisms are well described, and its action has been linked to the development of several chronic inflammatory disorders, suggesting a potential role in the pathogenesis of these disease. Therefore, a novel therapy targeting its activity is promising for the field. The zonulin antagonist larazotide, by directly targeting the physiological regulator of TJ opening, is nowadays the only promising novel treatment for CD with long-term benefits. Larazotide acetate inhibits the paracellular flux of gliadin peptides to the lamina propria, but it is not clear if it will affect the transcellular pathway. However, even if the blocking of the intestinal epithelial impairment alone could be a limited approach, in combination with other treatments could be beneficial for the patient, improving the quality of life. Considering the high prevalence of resistance and recurrence happening in CD, even a slight reduction in symptoms could have positive impact in the health of patients. In the future, a fundamental understanding of the mechanisms triggering the barrier defect might open the door to the identification of yet unknown additional triggers that contribute to CD pathogenesis and the development of new, barrier-effective therapeutics that reduce long-term side effects and increases maintenance of remission.

References

Ajamian, M., et al., 2019. Serum zonulin as a marker of intestinal mucosal barrier function: may not be what it seems. PLoS ONE 14 (1), e0210728. https://doi.org/10.1371/journal.pone.0210728.

Almeida, R., et al., 2014. Fine mapping of the celiac disease-associated LPP locus reveals a potential functional variant. Hum. Mol. Genet. 23 (9), 2481–2489. https://doi.org/10.1093/hmg/ddt619.

Al-Sadi, R., et al., 2011. Occludin regulates macromolecule flux across the intestinal epithelial tight junction barrier. Am. J. Physiol. Gastrointest. Liver Physiol. 300 (6), G1054–G1064. https://doi.org/10.1152/ajpgi.00055.2011.

Bardella, M.T., et al., 1995. Prevalence of hypertransaminasemia in adult celiac patients and effect of gluten-free diet. Hepatology 22 (3), 833–836, (no longer published by Elsevier). https://doi.org/10.1016/0270-9139(95)90305-4.

Barisani, D., 2018. miRNAs and their role in the pathogenesis of celiac disease: a review. In: Patel, V., Preedy, V. (Eds.), Handbook of Nutrition, Diet, and Epigenetics. Springer International Publishing, pp. 1–21, https://doi.org/10.1007/978-3-319-31143-2_124-1.

Barone, M.V., et al., 2010. Gliadin peptide P31-43 localises to endocytic vesicles and interferes with their maturation. PLoS One 5 (8), e12246. https://doi.org/10.1371/journal.pone.0012246.

Benton, R., St Johnston, D., 2003. Drosophila PAR-1 and 14-3-3 inhibit Bazooka/PAR-3 to establish complementary cortical domains in polarized cells. Cell 115 (6), 691–704, (Cell Press). https://doi.org/10.1016/S0092-8674(03)00938-3.

Bjarnason, I., Peters, T.J., 1984. In vitro determination of small intestinal permeability: demonstration of a persistent defect in patients with coeliac disease. Gut 25 (2), 145–150. https://doi.org/10.1136/gut.25.2.145.

Buckley, C.D., et al., 2014. The minimal cadherin-catenin complex binds to actin filaments under force. Science 346 (6209), 1254211. https://doi.org/10.1126/science.1254211.

Buschmann, M.M., et al., 2013. Occludin OCEL-domain interactions are required for maintenance and regulation of the tight junction barrier to macromolecular flux. Mol. Biol. Cell 24 (19), 3056–3068. https://doi.org/10.1091/mbc.E12-09-0688.

Campbell, H.K., Maiers, J.L., DeMali, K.A., 2017. Interplay between tight junctions & adherens junctions. Exp. Cell Res. 358 (1), 39–44. https://doi.org/10.1016/j.yexcr.2017.03.061.

Capaldo, C.T., Macara, I.G., 2007. Depletion of E-cadherin disrupts establishment but not maintenance of cell junctions in Madin-Darby canine kidney epithelial cells. Mol. Biol. Cell 18 (1), 189–200. https://doi.org/10.1091/mbc.E06-05-0471.

Capaldo, C.T., Nusrat, A., 2009. Cytokine regulation of tight junctions. Biochim. Biophys. Acta 1788 (4), 864–871. https://doi.org/10.1016/j.bbamem.2008.08.027.

Chandhoke, S.K., Mooseker, M.S., 2012. A role for myosin IXb, a motor-RhoGAP chimera, in epithelial wound healing and tight junction regulation. Mol. Biol. Cell 23 (13), 2468–2480. https://doi.org/10.1091/mbc.E11-09-0803.

Chia, L.A., Kuo, C.J., 2010. The intestinal stem cell. Prog. Mol. Biol. Transl. Sci. 96, 157–173. https://doi.org/10.1016/B978-0-12-381280-3.00007-5.

Ciccocioppo, R., et al., 2006. Altered expression, localization, and phosphorylation of epithelial junctional proteins in celiac disease. Am. J. Clin. Pathol. 125 (4), 502–511, (Oxford University Press, OUP). https://doi.org/10.1309/dtyr-a91g-8r0k-tm8m.

Ciccocioppo, R., et al., 2018. The transcriptomic analysis of circulating immune cells in a celiac family unveils further insights into disease pathogenesis. Front. Med. 5, 182. https://doi.org/10.3389/fmed.2018.00182.

Cobden, I., et al., 1978. Intestinal permeability assessed by excretion ratios of two molecules: results in coeliac disease. Br. Med. J. 2 (6144), 1060 (BMJ Publishing Group) https://doi.org/10.1136/bmj.2.6144.1060.

Cordenonsi, M., et al., 1999. Cingulin contains globular and coiled-coil domains and interacts with ZO-1, ZO-2, ZO-3, and myosin. J. Cell Biol. 147 (7), 1569–1581, (The Rockefeller University Press). https://doi.org/10.1083/jcb.147.7.1569.

De Mey, J.R., Freund, J.-N., 2013. Understanding epithelial homeostasis in the intestine. Tissue Barriers 1 (2), e24965, (Informa UK Limited). https://doi.org/10.4161/tisb.24965.

Dhawan, P., et al., 2011. Claudin-2 expression increases tumorigenicity of colon cancer cells: role of epidermal growth factor receptor activation. Oncogene 30 (29), 3234–3247, (NIH Public Access). https://doi.org/10.1038/onc.2011.43.

Dobrosotskaya, I., Guy, R.K., James, G.L., 1997. MAGI-1, a membrane-associated guanylate kinase with a unique arrangement of protein-protein interaction domains. J. Biol. Chem. 272 (50), 31589–31597. https://doi.org/10.1074/jbc.272.50.31589.

Dubois, P.C.A., et al., 2010. Multiple common variants for celiac disease influencing immune gene expression. Nat. Genet. 42 (4), 295–302, (Nature Publishing Group). https://doi.org/10.1038/ng.543.

Ebnet, K., 2017. Junctional adhesion molecules (JAMs): cell adhesion receptors with pleiotropic functions in cell physiology and development. Physiol. Rev. 97 (4), 1529–1554, (American Physiological Society). https://doi.org/10.1152/physrev.00004.2017.

El Asmar, R., et al., 2002. Host-dependent zonulin secretion causes the impairment of the small intestine barrier function after bacterial exposure. Gastroenterology 123 (5), 1607–1615. https://doi.org/10.1053/gast.2002.36578.

Eri, R., Chieppa, M., 2013. Messages from the inside. The dynamic environment that favors intestinal homeostasis. Front. Immunol. 4, 323. https://doi.org/10.3389/fimmu.2013.00323.

Farquhar, M.G., Palade, G.E., 1963. Junctional complexes in various epithelia. J. Cell Biol. 17 (2), 375–412. https://doi.org/10.1083/jcb.17.2.375.

Fasano, A., 2009. Surprise from celiac disease. Sci. Am. 301 (2), 54–61. https://doi.org/10.1038/scientificamerican0809-54.

Fasano, A., 2011. Zonulin and its regulation of intestinal barrier function: the biological door to inflammation, autoimmunity, and cancer. Physiol. Rev. 151–175. https://doi.org/10.1152/physrev.00003.2008.

Fearnley, G.W., et al., 2019. The homophilic receptor PTPRK selectively dephosphorylates multiple junctional regulators to promote cell-cell adhesion. Elife 8, 1–41, (eLife Sciences Publications Ltd). https://doi.org/10.7554/eLife.44597.

Ferraris, R.P., Diamond, J., 1997. Regulation of intestinal sugar transport. Physiol. Rev. 77 (1), 257–302, (American Physiological Society). https://doi.org/10.1152/physrev.1997.77.1.257.

Förster, C., 2008. Tight junctions and the modulation of barrier function in disease. In: Histochemistry and Cell Biology. Springer, pp. 55–70, https://doi.org/10.1007/s00418-008-0424-9.

Fromm, M., et al., 2009. High-resolution analysis of barrier function. Ann. N. Y. Acad. Sci. 1165, 74–81, (Blackwell Publishing Inc). https://doi.org/10.1111/j.1749-6632.2009.04047.x.

Garcia-Hernandez, V., Quiros, M., Nusrat, A., 2017. Intestinal epithelial claudins: expression and regulation in homeostasis and inflammation. Ann. N. Y. Acad. Sci. 1397 (1), 66–79. https://doi.org/10.1111/nyas.13360.

Garrod, D., Chidgey, M., 2008. Desmosome structure, composition and function. Biochim. Biophys. Acta 1778 (3), 572–587, (Elsevier). https://doi.org/10.1016/j.bbamem.2007.07.014.

Giorgi, A., et al., 2020. A probiotic preparation hydrolyzes gliadin and protects intestinal cells from the toxicity of pro-inflammatory peptides. Nutrients 12 (2), 495. https://doi.org/10.3390/nu12020495 (MDPI AG).

Gopalakrishnan, S., et al., 2012. Larazotide acetate promotes tight junction assembly in epithelial cells. Peptides 35 (1), 95–101. https://doi.org/10.1016/j.peptides.2012.02.016.

Günzel, D., Yu, A.S.L., 2013. Claudins and the modulation of tight junction permeability. Physiol. Rev. 93 (2), 525–569. https://doi.org/10.1152/physrev.00019.2012.

Hagander, B., et al., 1977. Hepatic injury in adult coeliac disease. Lancet 310 (8032), 270–272. https://doi.org/10.1016/S0140-6736(77)90954-0.

Hall, E.J., Batt, R.M., 1992. Dietary modulation of gluten sensitivity in a naturally occurring enteropathy of Irish setter dogs. Gut 33 (2), 198–205, (BMJ Publishing Group). https://doi.org/10.1136/gut.33.2.198.

Hamazaki, Y., et al., 2002. Multi-PDZ domain protein 1 (MUPP1) is concentrated at tight junctions through its possible interaction with claudin-1 and junctional adhesion molecule. J. Biol. Chem. 277 (1), 455–461. https://doi.org/10.1074/jbc.M109005200.

Hamilton, I., et al., 1982. Intestinal permeability in coeliac disease: the response to gluten withdrawal and single-dose gluten challenge. Gut 23 (3), 202–210. https://doi.org/10.1136/gut.23.3.202.

Holmes, J.L., et al., 2006. Claudin profiling in the mouse during postnatal intestinal development and along the gastrointestinal tract reveals complex expression patterns. Gene Expr. Patterns 6 (6), 581–588. https://doi.org/10.1016/j.modgep.2005.12.001.

Ide, N., 1999. Localization of membrane-associated guanylate kinase (MAGI)-1/BAI-associated protein (BAP) 1 at tight junctions of epithelial cells. Oncogene 18, 7810–7815.

Ikari, A., et al., 2014. Nuclear distribution of claudin-2 increases cell proliferation in human lung adenocarcinoma cells. BBA-Mol. Cell. Res. 1843 (9), 2079–2088, (Elsevier). https://doi.org/10.1016/j.bbamcr.2014.05.017.

Ireton, R.C., et al., 2002. A novel role for p120 catenin in E-cadherin function. J. Cell Biol. 159 (3), 465–476. https://doi.org/10.1083/jcb.200205115.

Ivarsson, A., et al., 2000. Epidemic of coeliac disease in Swedish children. Acta Paediatr. 89 (2), 165–171, (Wiley). https://doi.org/10.1080/080352500750028771.

Jauregi-Miguel, A., et al., 2019. MAGI2 gene region and celiac disease. Front. Nutr. 6, 187, (Frontiers Media S.A.). https://doi.org/10.3389/fnut.2019.00187.

Just, I., et al., 1995. Glucosylation of Rho proteins by Clostridium difficile toxin B. Nature 375 (6531), 500–503. https://doi.org/10.1038/375500a0.

Kagnoff, M.F., et al., 1987. Evidence for the role of a human intestinal adenovirus in the pathogenesis of coeliac disease. Gut 28 (8), 995–1001. https://doi.org/10.1136/gut.28.8.995.

Kaukinen, K., et al., 2002. Celiac disease in patients with severe liver disease: gluten-free diet may reverse hepatic failure. Gastroenterology 122 (4), 881–888. https://doi.org/10.1053/gast.2002.32416.

Kelly, C.P., et al., 2013. Larazotide acetate in patients with coeliac disease undergoing a gluten challenge: a randomised placebo-controlled study. Aliment. Pharmacol. Ther. 37 (2), 252–262, (Blackwell Publishing Ltd). https://doi.org/10.1111/apt.12147.

Kominsky, S.L., et al., 2003. Loss of the tight junction protein claudin-7 correlates with histological grade in both ductal carcinoma in situ and invasive ductal carcinoma of the breast. Oncogene 22 (13), 2021–2033. https://doi.org/10.1038/sj.onc.1206199.

Krause, G., et al., 2008. Structure and function of claudins. Biochim. Biophys. Acta 1778 (3), 631–645, (Elsevier). https://doi.org/10.1016/j.bbamem.2007.10.018.

Kumar, V., et al., 2015. Systematic annotation of celiac disease loci refines pathological pathways and suggests a genetic explanation for increased interferon-gamma levels. Hum. Mol. Genet. 24 (2), 397–409. https://pubmed.ncbi.nlm.nih.gov/25190711/.

Lammers, K.M., et al., 2008. Gliadin induces an increase in intestinal permeability and zonulin release by binding to the chemokine receptor CXCR3. Gastroenterology 135 (1), 194–204.e3. https://doi.org/10.1053/j.gastro.2008.03.023.

Laparra, J.M., et al., 2012. Bifidobacterium longum CECT 7347 modulates immune responses in a gliadin-induced enteropathy animal model. PLoS One 7 (2), e30744. https://doi.org/10.1371/journal.pone.0030744.

Leffler, D.A., et al., 2012. A randomized, double-blind study of larazotide acetate to prevent the activation of celiac disease during gluten challenge. Am. J. Gastroenterol. 107 (10), 1554–1562. https://doi.org/10.1038/ajg.2012.211.

Leffler, D.A., et al., 2015. Larazotide acetate for persistent symptoms of celiac disease despite a gluten-free diet: a randomized controlled trial. Gastroenterology 148 (7), 1311–1319.e6 (W.B. Saunders) https://doi.org/10.1053/j.gastro.2015.02.008.

Lemmers, C., 2002. hINADl/PATJ, a homolog of discs lost, interacts with crumbs and localizes to tight junctions in human epithelial cells. J. Biol. Chem. 277, 25408–25415.

Maiers, J.L., et al., 2013. ZO-1 recruitment to α-catenin -a novel mechanism for coupling the assembly of tight junctions to adherens junctions. J. Cell Sci. 126 (17), 3904–3915. https://doi.org/10.1242/jcs.126565.

Maiuri, L, et al., 2003. Association between innate response to gliadin and activation of pathogenic T cells in coeliac disease. Lancet 362 (9377), 30–37. https://doi.org/10.1016/S0140-6736(03)13803-2.

Marchiando, A.M., et al., 2010. Caveolin-1-dependent occludin endocytosis is required for TNF-induced tight junction regulation in vivo. J. Cell Biol. 189 (1), 111–126. https://doi.org/10.1083/jcb.200902153.

Mårild, K., et al., 2015. Infections and risk of celiac disease in childhood: a prospective nationwide cohort study. Am. J. Gastroenterol. 110 (10), 1475–1484, (Nature Publishing Group). https://doi.org/10.1038/ajg.2015.287.

Márquez, L., et al., 2012. Effects of haptoglobin polymorphisms and deficiency on susceptibility to inflammatory bowel disease and on severity of murine colitis. Gut 61 (4), 528–534. https://doi.org/10.1136/gut.2011.240978.

Matsuda, M., et al., 2004. A peculiar internalization of claudins, tight junction-specific adhesion molecules, during the intercellular movement of epithelial cells. J. Cell Sci. 117 (7), 1247–1257, (The Company of Biologists Ltd). https://doi.org/10.1242/jcs.00972.

Matter, K., Balda, M.S., 2003. Signalling to and from tight junctions. Nat. Rev. Mol. Cell Biol. 4 (3), 225–236. Available at: https://www.google.com/search?client=firefox-b-d&q=Signalling+to+and+from+tight+junctions.+Nat+Rev+Mol+Cell+Biol+2003%3B4%283%29%3A225-36. (Accessed 28 June 2020).

Matysiak-Budnik, T., et al., 2008. Secretory IgA mediates retrotranscytosis of intact gliadin peptides via the transferrin receptor in celiac disease. J. Exp. Med. 205 (1), 143–154, (The Rockefeller University Press). https://doi.org/10.1084/jem.20071204.

McCaffrey, L.M., Macara, I.G., 2012. Signaling pathways in cell polarity. Cold Spring Harb. Perspect. Biol. 4 (6), 1–15, (Cold Spring Harbor Laboratory Press). https://doi.org/10.1101/cshperspect.a009654.

McCormack, J.J., et al., 2017. The scaffold protein Ajuba suppresses CdGAP activity in epithelia to maintain stable cell-cell contacts. Sci. Rep. 7 (1), 1–14, (Nature Publishing Group). https://doi.org/10.1038/s41598-017-09024-4.

Meddings, J., 2008. The significance of the gut barrier in disease. Gut 57 (4), 438–440. https://doi.org/10.1136/gut.2007.143172.

Menzies, I.S., et al., 1979. Abnormal intestinal permeability to sugars in villous atrophy. Lancet 314 (8152), 1107–1109. https://doi.org/10.1016/S0140-6736(79)92507-8.

Miranda-Ribera, A., et al., 2019. Exploiting the Zonulin mouse model to establish the role of primary impaired gut barrier function on microbiota composition and immune profiles. Front. Immunol. 10, 2233, (Frontiers Media S.A.). https://doi.org/10.3389/fimmu.2019.02233.

Mohanan, V., et al., 2018. C1orf106 is a colitis risk gene that regulates stability of epithelial adherens junctions. Science 359 (6380), 1161–1166, (American Association for the Advancement of Science). https://doi.org/10.1126/science.aan0814.

Monsuur, A.J., et al., 2005. Myosin IXB variant increases the risk of celiac disease and points toward a primary intestinal barrier defect. Nat. Genet. 37 (12), 1341–1344. https://doi.org/10.1038/ng1680.

Montalto, M., et al., 2002. Immunohistochemical analysis of ZO-1 in the duodenal mucosa of patients with untreated and treated celiac disease. Digestion 65 (4), 227–233. https://doi.org/10.1159/000063817.

Nagy Szakál, D., et al., 2010. Mucosal expression of claudins 2, 3 and 4 in proximal and distal part of duodenum in children with coeliac disease. Virchows Arch. 456 (3), 245–250, (Springer Verlag). https://doi.org/10.1007/s00428-009-0879-7.

Orlando, A., et al., 2018. Lactobacillus rhamnosus GG protects the epithelial barrier of wistar rats from the pepsin-trypsin-digested gliadin (PTG)-induced enteropathy. Nutrients 10 (11), 1698, (MDPI AG). https://doi.org/10.3390/nu10111698.

Ozaki, H., et al., 1999. Cutting edge: combined treatment of TNF-alpha and IFN-gamma causes redistribution of junctional adhesion molecule in human endothelial cells. J. Immunol. 163 (2), 553–557. https://pubmed.ncbi.nlm.nih.gov/10395639/. (Accessed: June 29, 2020).

Papp, M., et al., 2008. Haptoglobin polymorphism: a novel genetic risk factor for celiac disease development and its clinical manifestations. Clin. Chem. 54 (4), 697–704. https://doi.org/10.1373/clinchem.2007.098780.

Paterson, B.M., et al., 2007. The safety, tolerance, pharmacokinetic and pharmacodynamic effects of single doses of AT-1001 in coeliac disease subjects: a proof of concept study. Aliment. Pharmacol. Ther. 26 (5), 757–766. https://doi.org/10.1111/j.1365-2036.2007.03413.x.

Pearce, S.C., et al., 2018. Marked differences in tight junction composition and macromolecular permeability among different intestinal cell types. BMC Biol. 16 (1), 19, (BioMed Central Ltd.). https://doi.org/10.1186/s12915-018-0481-z.

Pearson, A.D.J., et al., 1982. Intestinal permeability in children with Crohn's disease and coeliac disease. Br. Med. J. 285 (6334), 20–21, (Br Med J (Clin Res Ed)). https://doi.org/10.1136/bmj.285.6334.20.

Perez-Moreno, M., Jamora, C., Fuchs, E., 2003. Sticky business: orchestrating cellular signals at adherens junctions. Cell 112 (4), 535–548, (Cell Press). https://doi.org/10.1016/S0092-8674(03)00108-9.

Pizzuti, D., et al., 2004. Transcriptional downregulation of tight junction protein ZO-1 in active coeliac disease is reversed after a gluten-free diet. Dig. Liver Dis. 36 (5), 337–341, (W.B. Saunders Ltd). https://doi.org/10.1016/j.dld.2004.01.013.

Poliak, S., et al., 2002. Distinct claudins and associated PDZ proteins form different autotypic tight junctions in myelinating Schwann cells. J. Cell Biol. 159 (2), 361–372. https://doi.org/10.1083/jcb.200207050.

Quiros, M., Nusrat, A., 2014. RhoGTPases, actomyosin signaling and regulation of the epithelial apical junctional complex. Semin. Cell Dev. Biol. 36, 194–203, (Academic Press). https://doi.org/10.1016/j.semcdb.2014.09.003.

Raleigh, D.R., et al., 2010. Tight junction-associated MARVEL proteins marvelD3, tricellulin, and occludin have distinct but overlapping functions. Mol. Biol. Cell 21 (7), 1200–1213, (The American Society for Cell Biology). https://doi.org/10.1091/mbc.E09-08-0734.

Rampertab, S.D., et al., 2006. Trends in the presentation of celiac disease. Am. J. Med. 119 (4), 355.e9–355.e14. https://doi.org/10.1016/j.amjmed.2005.08.044.

Riddle, M.S., et al., 2013. Pathogen-specific risk of celiac disease following bacterial causes of foodborne illness: a retrospective cohort study. Dig. Dis. Sci. 58 (11), 3242–3245. https://doi.org/10.1007/s10620-013-2733-7.

Rodgers, L.S., et al., 2013. Epithelial barrier assembly requires coordinated activity of multiple domains of the tight junction protein ZO-1. J. Cell Sci. 126 (7), 1565–1575. https://doi.org/10.1242/jcs.113399.

Rodriguez, P., et al., 1995. Tumour necrosis factor-α induces morphological and functional alterations of intestinal ht29 cl.19a cell monolayers. Cytokine 7 (5), 441–448. https://doi.org/10.1006/cyto.1995.0060.

Roh, M., 2002. The Maguk protein, Pals1, functions as an adapter, linking mammalian homologues of crumbs and discs lost. J. Cell Biol. 157, 161–172.

Rubio-Tapia, A., Murray, J.A., 2007. The liver in celiac disease. Hepatology 46 (5), 1650–1658. https://doi.org/10.1002/hep.21949.

Saitoh, Y., et al., 2015. Structural insight into tight junction disassembly by Clostridium perfringens enterotoxin. Science 347 (6223), 775–778, (American Association for the Advancement of Science). https://doi.org/10.1126/science.1261833.

Sander, G.R., et al., 2005. Rapid disruption of intestinal barrier function by gliadin involves altered expression of apical junctional proteins. FEBS Lett. 579 (21), 4851–4855. https://doi.org/10.1016/j.febslet.2005.07.066. Elsevier.

Sandle, G.I., 2005. Pathogenesis of diarrhea in ulcerative colitis: new views on on old problem. J. Clin. Gastroenterol. 39 (4), S49–S52. https://doi.org/10.1097/01.mcg.0000155520.04253.37.

Scheffler, L, et al., 2018. Widely used commercial ELISA does not detect precursor of haptoglobin2, but recognizes properdin as a potential second member of the zonulin family. Front. Endocrinol. (Lausanne) 9 (FEB), 22. https://doi.org/10.3389/fendo.2018.00022.

Schmitz, H., et al., 1999. Tumor necrosis factor-alpha (TNFalpha) regulates the epithelial barrier in the human intestinal cell line HT-29/B6. J. Cell Sci. 112 (1), 137–146. https://pubmed.ncbi.nlm.nih.gov/9841910/. (Accessed: July 11, 2020).

Schneeberger, E.E., Lynch, R.D., 1992. Structure, function, and regulation of cellular tight junctions. Am. J. Physiol. 262 (6 Pt 1), L647–L661. https://doi.org/10.1152/ajplung.1992.262.6.l647.

Schulzke, J.D., et al., 1995. Epithelial barrier and ion transport in coeliac sprue: electrical measurements on intestinal aspiration biopsy specimens. Gut 37 (6), 777–782, (BMJ Publishing Group). https://doi.org/10.1136/gut.37.6.777.

Schulzke, J.D., et al., 1998. Epithelial tight junction structure in the jejunum of children with acute and treated celiac sprue. Pediatr. Res. 43 (4 I), 435–441, (Nature Publishing Group). https://doi.org/10.1203/00006450-199804000-00001.

Schumann, M., et al., 2008a. Mechanisms of epithelial translocation of the α2-gliadin-33mer in coeliac sprue. Gut 57 (6), 747–754. https://doi.org/10.1136/gut.2007.136366.

Schumann, M., et al., 2008b. Mechanisms of epithelial translocation of the α2-gliadin- 33mer in coeliac sprue. Gut 57 (6), 747–754. https://doi.org/10.1136/gut.2007.136366.

Schumann, M., et al., 2012. Cell polarity-determining proteins Par-3 and PP-1 are involved in epithelial tight junction defects in coeliac disease. Gut 61 (2), 220–228 (BMJ Publishing Group) https://doi.org/10.1136/gutjnl-2011-300123.

Schumann, M., et al., 2017. Celiac disease: role of the epithelial barrier. Cell. Mol. Gastroenterol. Hepatol. 3, 150–162 (Elsevier Inc) https://doi.org/10.1016/j.jcmgh.2016.12.006.

Shen, L., et al., 2006. Myosin light chain phosphorylation regulates barrier function by remodeling tight junction structure. J. Cell Sci. 119 (10), 2095–2106. https://doi.org/10.1242/jcs.02915.

Sollid, L.M., Khosla, C., 2011. Novel therapies for coeliac disease. J. Intern. Med. 269 (6), 604–613 (NIH Public Access) https://doi.org/10.1111/j.1365-2796.2011.02376.x.

Stene, L.C., et al., 2006. Rotavirus infection frequency and risk of celiac disease autoimmunity in early childhood: a longitudinal study. Am. J. Gastroenterol. 101 (10), 2333–2340. https://doi.org/10.1111/j.1572-0241.2006.00741.x.

Stevenson, B.R., et al., 1986. Identification of ZO-1: a high molecular weight polypeptide associated with the tight junction (Zonula Occludens) in a variety of epithelia. J. Cell Biol. 103 (3), 755–766. https://doi.org/10.1083/jcb.103.3.755.

Stuart, R.O., Nigam, S.K., 1995. Regulated assembly of tight junctions by protein kinase C. Proc. Natl. Acad. Sci. U. S. A. 92 (13), 6072–6076, (National Academy of Sciences). https://doi.org/10.1073/pnas.92.13.6072.

Sturgeon, C., Fasano, A., 2016. Zonulin, a regulator of epithelial and endothelial barrier functions, and its involvement in chronic inflammatory diseases. Tissue Barriers 4 (4), e1251384. https://doi.org/10.1080/21688370.2016.1251384 (Taylor and Francis Inc.).

Sturgeon, C., Lan, J., Fasano, A., 2017. Zonulin transgenic mice show altered gut permeability and increased morbidity/mortality in the DSS colitis model. Ann. N. Y. Acad. Sci. 1397 (1), 130–142, (Blackwell Publishing Inc). https://doi.org/10.1111/nyas.13343.

Suzuki, A., Ohno, S., 2006. The PAR-aPKC system: lessons in polarity. J. Cell Sci. 119 (6), 979–987, (The Company of Biologists Ltd). https://doi.org/10.1242/jcs.02898.

Tamura, A., et al., 2011. Loss of claudin-15, but not claudin-2, causes Na+ deficiency and glucose malabsorption in mouse small intestine. Gastroenterology 140 (3), 913–923, (W.B. Saunders). https://doi.org/10.1053/j.gastro.2010.08.006.

Thiery, J.P., Sleeman, J.P., 2006. Complex networks orchestrate epithelial-mesenchymal transitions. Nat. Rev. Mol. Cell Biol. 7 (2), 131–142. https://doi.org/10.1038/nrm1835.

Traweger, A., et al., 2008. Protein phosphatase 1 regulates the phosphorylation state of the polarity scaffold Par-3. Proc. Natl. Acad. Sci. U. S. A. 105 (30), 10402–10407. https://doi.org/10.1073/pnas.0804102105.

Tripathi, A., et al., 2009. Identification of human zonulin, a physiological modulator of tight junctions, as prehaptoglobin-2. Proc. Natl. Acad. Sci. U. S. A. 106 (39), 16799–16804, (National Academy of Sciences). https://doi.org/10.1073/pnas.0906773106.

Trynka, G., et al., 2011. Dense genotyping identifies and localizes multiple common and rare variant association signals in celiac disease. Nat. Genet. 43 (12), 1193–1201. https://doi.org/10.1038/ng.998.

Tsukita, S., Furuse, M., Itoh, M., 2001. Multifunctional strands in tight junctions. Nat. Rev. Mol. Cell Biol. 2 (4), 285–293, (Nature Publishing Group). https://doi.org/10.1038/35067088.

Umeda, K., et al., 2006. ZO-1 and ZO-2 independently determine where claudins are polymerized in tight-junction strand formation. Cell 126 (4), 741–754, (Elsevier). https://doi.org/10.1016/j.cell.2006.06.043.

Utech, M., et al., 2005. Mechanism of IFN-γ-induced endocytosis of tight junction proteins: Myosin II-dependent vacuolarization of the apical plasma membrane. Mol. Biol. Cell 16 (10), 5040–5052. https://doi.org/10.1091/mbc.E05-03-0193.

Van Elburg, R.M., et al., 1993. Intestinal permeability in patients with coeliac disease and relatives of patients with coeliac disease. Gut 34 (3), 354–357. https://doi.org/10.1136/gut.34.3.354.

Van Heel, D.A., et al., 2007. A genome-wide association study for celiac disease identifies risk variants in the region harboring IL2 and IL21. Nat. Genet. 39 (7), 827–829. https://doi.org/10.1038/ng2058.

Van Itallie, C.M., Anderson, J.M., 1997. Occludin confers adhesiveness when expressed in fibroblasts. J. Cell Sci. 110 (Pt 9), 1113–1121. https://pubmed.ncbi.nlm.nih.gov/9175707/. (Accessed: June 19, 2020).

Van Itallie, C.M., Anderson, J.M., 2006. Claudins and epithelial paracellular transport. Annu. Rev. Physiol. 68 (1), 403–429. https://doi.org/10.1146/annurev.physiol.68.040104.131404.

Van Itallie, C.M., et al., 2014. Biotin ligase tagging identifies proteins proximal to E-cadherin, including lipoma preferred partner, a regulator of epithelial cell-cell and cell-substrate adhesion. J. Cell Sci. 127 (4), 885–895. https://doi.org/10.1242/jcs.140475.

Van Itallie, C.M., Anderson, J.M., 2014. Architecture of tight junctions and principles of molecular composition. Semin. Cell Dev. Biol. 36, 157–165. https://doi.org/10.1016/j.semcdb.2014.08.011.

van Meer, G., Simons, K., 1986. The function of tight junctions in maintaining differences in lipid composition between the apical and the basolateral cell surface domains of MDCK cells. EMBO J. 5 (7), 1455–1464, (European Molecular Biology Organization). https://doi.org/10.1002/j.1460-2075.1986.tb04382.x.

Vanuytsel, T., Vermeire, S., Cleynen, I., 2013. The role of Haptoglobin and its related protein, Zonulin, in inflammatory bowel disease. Tissue Barriers 1 (5), e27321. https://doi.org/10.4161/tisb.27321. (Landes Bioscience).

Wang, W., et al., 2000. Human Zonulin, a potential modulator of intestinal tight junctions. J. Cell Sci. 113 (Pt 24), 4435–4440. https://pubmed.ncbi.nlm.nih.gov/11082037/. (Accessed: June 14, 2020).

Wapenaar, M.C., et al., 2008. Associations with tight junction genes PARD3 and MAGI2 in Dutch patients point to a common barrier defect for coeliac disease and ulcerative colitis. Gut 57 (4), 463–467. https://doi.org/10.1136/gut.2007.133132.

Wiest, R., Lawson, M., Geuking, M., 2014. Pathological bacterial translocation in liver cirrhosis. J. Hepatol. 60 (1), 197–209, (Elsevier). https://doi.org/10.1016/j.jhep.2013.07.044.

Williams, J.M., et al., 2015. Epithelial cell shedding and barrier function: a matter of life and death at the small intestinal villus tip. Vet. Pathol. 52 (3), 445–455, (SAGE Publications Inc.). https://doi.org/10.1177/0300985814559404.

Yap, A.S., Niessen, C.M., Gumbiner, B.M., 1998. The juxtamembrane region of the cadherin cytoplasmic tail supports lateral clustering, adhesive strengthening, and interaction with p120(ctn). J. Cell Biol. 141 (3), 779–789, (The Rockefeller University Press). https://doi.org/10.1083/jcb.141.3.779.

Yu, L., et al., 2012. Host-microbial interactions and regulation of intestinal epithelial barrier function: from physiology to pathology. World J. Gastrointest. Pharmacol. 3 (1), 27, (Baishideng Publishing Group Inc). https://doi.org/10.4291/wjgp.v3.i1.27.

Yu, M., et al., 2015. Par-3 modulates intestinal epithelial barrier function through regulating intracellular trafficking of occludin and myosin light chain phosphorylation. J. Gastroenterol. 50 (11), 1103–1113, (Springer-Verlag Tokyo). https://doi.org/10.1007/s00535-015-1066-z.

Yuhan, R., et al., 1997. Enteropathogenic Escherichia coli-induced myosin light chain phosphorylation alters intestinal epithelial permeability. Gastroenterology 113 (6), 1873–1882, (W.B. Saunders). https://doi.org/10.1016/S0016-5085(97)70006-4.

Zeissig, S., et al., 2007. Changes in expression and distribution of claudin 2, 5 and 8 lead to discontinuous tight junctions and barrier dysfunction in active Crohn's disease. Gut 56 (1), 61–72. https://doi.org/10.1136/gut.2006.094375.

Zhu, Y., et al., 2006. Differences in expression patterns of the tight junction proteins, claudin 1, 3, 4 and 5, in human ovarian surface epithelium as compared to epithelia in inclusion cysts and epithelial ovarian tumours. Int. J. Cancer 118 (8), 1884–1891. https://doi.org/10.1002/ijc.21506.

Zihni, C., et al., 2016. Tight junctions: from simple barriers to multifunctional molecular gates. Nat. Rev. Mol. Cell Biol. 17 (9), 564–580, (Nature Publishing Group). https://doi.org/10.1038/nrm.2016.80.

Zihni, C., Balda, M.S., Matter, K., 2014. Signalling at tight junctions during epithelial differentiation and microbial pathogenesis. J. Cell Sci. 127, 3401–3413, (Company of Biologists Ltd). https://doi.org/10.1242/jcs.145029.

Zimmer, K.P., et al., 2010. Endocytotic segregation of gliadin peptide 31-49 in enterocytes. Gut 59 (3), 300–310. https://doi.org/10.1136/gut.2008.169656.

Zolotarevsky, Y., et al., 2002. A membrane-permeant peptide that inhibits MLC kinase restores barrier function in in vitro models of intestinal disease. Gastroenterology 123 (1), 163–172. https://doi.org/10.1053/gast.2002.34235.

CHAPTER FIVE

Epithelial cell dysfunction in coeliac disease

Celia Escudero-Hernández*,†
Department of Biomedical and Clinical Sciences (BKV), Linköping University, Linköping, Sweden
*Corresponding author: e-mail address: celia.escher@gmail.com

Contents

Abstract

The intestinal epithelium limits host-luminal interactions and maintains gut homeostasis. Breakdown of the epithelial barrier and villous atrophy are hallmarks of coeliac disease. Besides the well characterized immune-mediated epithelial damage induced in coeliac mucosa, constitutional changes and early gluten direct effects disturb intestinal epithelial cells. The subsequent modifications in key epithelial signaling pathways leads to outnumbered immature epithelial cells that, in turn, facilitate epithelial dysfunction, promote crypt hyperplasia, and increase intestinal permeability. Consequently, underlying immune cells have a greater access to gluten, which boosts the proinflammatory immune response against gluten and positively feedback the epithelial damage loop. Gluten-free diet is an indispensable treatment for coeliac disease patients, but additional therapies are under development, including those that reinforce intestinal epithelial healing. In this chapter, we provide an overview of intestinal epithelial cell disturbances that develop during gluten intake in coeliac disease mucosa.

† Current address: University Hospital Schleswig Holstein and Institute of Clinical Molecular Biology Christian-Albrechts-University Kiel, Kiel, Germany.

International Review of Cell and Molecular Biology, Volume 358
ISSN 1937-6448
https://doi.org/10.1016/bs.ircmb.2020.09.007

1. Epithelial homeostasis

Intestinal epithelial cells (IECs) are specialized, polarized cells that line the gastrointestinal tract and constitute the largest and most diverse mucosal surface in the body. The IEC dynamic physical barrier separates the gut lumen from the underlying tissue and gut-associated lymphoid tissue (GALT). Despite this crucial barrier function, IECs also uptake nutrients and receive cues from immune cells, mesenchymal cells and microbiota (Peterson and Artis, 2014). To withstand the continuous mechanical, chemical and biological stress, IECs are constantly replenished every four to five days by highly proliferative pluripotent intestinal stem cells (ISC) (Barker, 2014; Koch and Nusrat, 2012).

The intestinal barrier is organized into protrusions (villi) and tubular invaginations (crypts) of the intestinal epithelium (Fig. 1A). Villi increase the surface area and connect to the circulatory system, hence, improving the absorptive capacity of the intestine. Villus length decreases along the intestinal tract, being more than 1 mm length in the duodenum and lacking in the colon (Gehart and Clevers, 2019). On the other side, crypts protect ISCs from the hazards of the digestive process with physical restrains (crypts are connected to the intestinal lumen only by a ~6 μm opening that limits diffusion), and chemical protection (mucus and antimicrobial product secretion) (Gehart and Clevers, 2019). In addition, the stem cell niche at the bottom of the crypt, composed by ISCs, secretory Paneth cells, and the surrounding mesenchyme, ensure the metabolic environment for ISC proliferation. There, ISCs continuously divide to give rise to progenitor or transit amplifying cells. These cells proliferate and become mature intestinal epithelial cells (IEC), which will be pushed out from the crypt to the tip of the villus, where they eventually undergo apoptosis and are released into the lumen (Barker et al., 2007; Gehart and Clevers, 2019) (Fig. 1A).

Crypt base columnar (CBC) cells are the main ISCs, which are long-lived, continuously cycling cells that generate multiple lineages and self-renew. In addition, +4 ($Bmi1^+$, polycomb ring finger) cells are candidate ISCs that locate above the uppermost Paneth cell of the stem cell niche (Fig. 1A). Although their contribution to tissue homeostasis is still under debate, +4 cells might comprise a reserve stem cell able to replenish CBC cells under injury conditions (Li et al., 2014, 2016). Still, the intestinal epithelium is far more plastic and dynamic since lineage-committed cells can also mobilize, de-differentiate and fall back into damaged stem cell niches

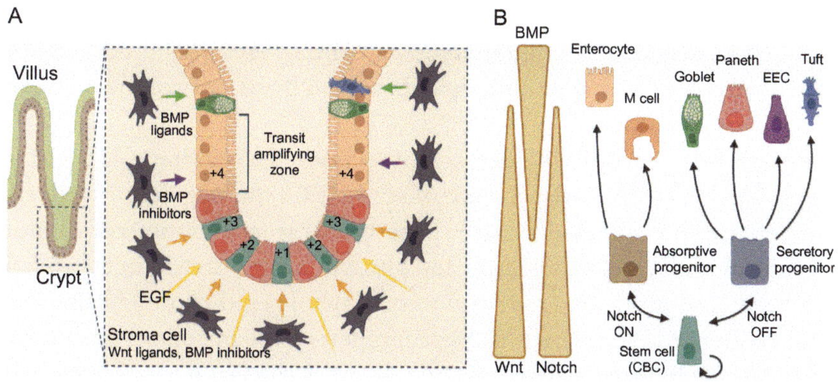

Fig. 1 Intestinal homeostasis and cell differentiation. (A) The intestinal epithelium is organized into villi and crypts. In the crypts, crypt base columnar (CBC) cells or intestinal stem cells (ISC) proliferate continuously with the protection and nourishment of Paneth cells (secretion of antimicrobial peptides, provision of lactate, Wnt and Notch ligands). Above the stem cell zone, lineage-committed progenitor cells divide in the transit amplifying zone and migrate upstream. Fully mature epithelial cells keep migrating toward the villus tip. Stroma or mesenchymal cells contribute to the stem cell niche with the production of Wnt ligands, epithelial growth factor (EGF, yellow arrows), and bone morphogenic protein (BMP) inhibitors (purple arrows). BMP ligands are provided by stroma cells upper in the crypt and villus (not shown) to promote cell differentiation (green arrows). +4 cells constitute a reserve stem cell able to replenish CBC cells under injury conditions. (B) Wnt and Notch signaling through cell-to-cell contact are abundant at the bottom of the crypt, which creates a gradient of Wnt and Notch availability. Intercrypt and intervillus mesenchymal cells secrete BMP ligands that promote an increasing gradient of BMP activity toward the villi. Crypt base columnar (CBC) cells differentiate once they are pushed out the stem cell niche at the bottom of the crypt. Notch signaling favors CBC cell differentiation toward absorptive cells (Notch ON), which will further mature along the increasing BMP gradient, whereas the secretory fate requires a lack of Notch signals (Notch OFF). In the proximities of Peyer's patches, the receptor activator of NF-κB ligand (RANKL) is presented by stroma cells and promotes absorptive progenitors to differentiate into microfold (M) cells. On the other side, secretory progenitors mostly differentiate into goblet cells in the absence of Notch and Wnt signaling. When Wnt signals and fibroblast growth factor (FGF) are high, CBC cells turn into Paneth cells instead. Low Wnt levels, lack of Notch ligands, and limited EGF lead secretory progenitors to differentiate into diverse enteroendocrine cells (EEC). Although not well understood, Tuft and goblet cell profiles are promoted in an IL-4/IL-13 rich environment. *Adapted from Gehart, H., Clevers H. (2019). Tales from the crypt: new insights into intestinal stem cells. Nat. Rev. Gastroenterol. Hepatol. 16(1): 19–34 and created with BioRender.com.*

to function as *bona fide* CBC cells (Gehart and Clevers, 2019). These intestinal cell reprogramming and plasticity might be possible due to permissive open chromatin states that maintain throughout the differentiation process (i.e., similar DNA methylation and histone modifications have been described between ISCs and differentiated IECs). As a result, the differentiation status of cells can be entirely determined by its environment (Gehart and Clevers, 2019).

Paneth cells—or REG4$^+$ deep crypt secretory cells in the colon—, nurture and protect CBC cells, and are the only IECs moving downwards after differentiation (Sasaki et al., 2016). Paneth cells secrete antimicrobial products such as lysozyme, α-defensins and phospholipase A2 that mix with goblet cell-derived protective mucus along the whole epithelium (Allaire et al., 2018; Gassler, 2017). Every CBC cell competes to be in contact with at least one Paneth cell because these provide essential paracrine maintenance signals, including WNT and Notch ligands, epidermal growth factor (EGF), and transforming growth factor (TGF)-α (Sato et al., 2011). Since CBC cells depend heavily on mitochondrial oxidative phosphorylation, Paneth cells undergo glycolysis that provides CBC cells with lactate (Rodriguez-Colman et al., 2017). CBC cell oxidative metabolism and production of reactive oxygen species (ROS) can lead to DNA damage. However, accumulation of mutations in the gut is slow, possibly due to the combination of lactate antioxidant effects, the high spatial competition of CBC cells for Paneth cell contact, and the chance of a CBC cell clone to dominate if it proliferates quickly (Gehart and Clevers, 2019). Consequently, only mutations that provide niche independence and increases proliferation succeed, as it occurs in 93% of all colon cancers (e.g., activating mutations in WNT signaling) (The Cancer Genome Atlas Network, 2012).

The underlying mesenchyme of myofibroblasts, fibroblasts, pericytes, endothelial cells, neural cells and smooth muscle cells provide structural support, WNT signals, bone morphogenetic protein (BMP) ligands, BMP antagonists (only at the bottom of the crypts), TGF-α and cytokines that regulate IEC behavior (Fig. 1A). In return, the intestinal epithelium favors mesenchymal cell growth and maintenance via hedgehog signaling pathway (Gehart and Clevers, 2019). Among immune cells, group 3 innate lymphoid cells (ILC3) reside close to intestinal crypts, lack antigen receptors and, upon injury, produce interleukin (IL)-22 that promote IEC regeneration (Lindemans et al., 2015).

Cell differentiation is promoted as WNT and Notch signals decrease and BMP signals increase during progenitor cell progression along the crypt-villus axis (He et al., 2004) (Fig. 1B). Among the 15 identified IEC subtypes, absorptive cells are particularly favored by Notch signaling (Gehart and Clevers, 2019; Haber et al., 2017). Enterocytes represent up to 80% of

IECs and are responsible for ion, water, sugar, peptide and lipid transport; whereas microfold or M-cells sample the intestinal lumen and transport antigens to lymphoid follicular cells underneath, in the Peyer's patches (Gehart and Clevers, 2019) (Fig. 1B). On the other side, lack of Notch signals determines the secretory fate toward Paneth cells when the environment is rich in WNT signals and fibroblast growth factor (FGF), or to goblet cells when WNT signals are low and IL-4 and IL-13 are produced upon ILC2 mediated immunity (Gehart and Clevers, 2019; Gerbe et al., 2016; von Moltke et al., 2016) (Fig. 1B). Similarly to goblet cells, Tuft cells, a rare cell type specialized to fight against helminths and to promote neuronal development, are favored after exposure to IL-4 and IL-13 (Gerbe et al., 2016; von Moltke et al., 2016) (Fig. 1B). Lastly, differentiation to sensory and hormone-secretory enteroendocrine cells is promoted when signaling pathways toward other secretory cell types are inhibited and the transcription factor NEUROG3 is activated (Fig. 1B).

The intestinal barrier depends on IEC polarization and strict separation of the luminal contents from the underlying mucosa. To effectively comply with that function, maintain IEC attached and seal paracellular transport, the molecular fence built by tight-junctions (TJs) is crucial. Since TJ were discussed in details in the previous chapter, below we will focus on reviewing the effects of gluten peptides in epithelial biology and major signaling pathways in coeliac disease (CD).

2. Genetic and epigenetic modifications in the coeliac epithelium

Genome-wide association studies (GWAS) and Immunochip analyses have identified genetic risk variants outside the human leukocyte antigen (*HLA*) region in CD, including genes related to oxidative stress, protein trafficking and cell cycle. Besides the apoptotic genes from the family of tumor necrosis factor (TNF) ligand superfamily (TNFSF) such as Fas ligand (*FASL*), *TNFSF4* and *TNFSF18*, the only genes associated with epithelial cells are linked to cytoskeletal reorganization, barrier function and tight junction organization (discussed in detail in the previous chapter) (Dubois et al., 2010; Garner et al., 2014; Hunt et al., 2008; Kumar et al., 2015; Trynka, et al., 2011). A distinct, less studied genetic polymorphism is sequence copy number variation (CNV), which is the result of complex sequence rearrangements. In CD, an increased CNV of the epithelial-derived antimicrobial peptide β-defensin gene cluster is underrepresented in the diseased mucosa, which could result in increased bacterial infiltration (Fernandez-Jimenez et al., 2010). Still, the main candidate for immune-mediated disorders in this cluster, gene *DEFB4*, encodes a highly inducible protein, which prevents the correlation between gene expression and copy number (Fernandez-Jimenez et al., 2010).

In recent years, modulation of gene expresion by DNA sequence-independent events has gained attention and, together with *in vitro* culture of indefinitely maintained ISC as mini-guts or organoids, (Sato et al., 2009, 2011) specific epithelial cell modifications in CD start to be unraveled. The epigenetic changes that have been associated with epithelial cells in CD so far include methylation of DNA or histones, and gene regulation by non-coding RNAs.

DNA methylation often represses gene expression, and in active CD duodenal mucosa, DNA is generally hypomethylated in regions containing CD-associated single nucleotide polymorphisms (SNPs) (Fernandez-Jimenez et al., 2019). This hypomethylation pattern is genotype-independent and cell type-specific because epithelial and immune cell associated profiles differ. In epithelial cells, most of the regions are related to immune response genes that are co-expressed, suggesting common regulatory mechanisms (Fernandez-Jimenez et al., 2019). In contrast, histone methylation is increased in CD due to overexpression of polycomb repressive complex-2 (PRC2), which catalyzes histone 3 methylation in CpG-rich DNA regions (Oittinen et al., 2017). PRC2 normally represses ISC differentiation toward absorptive IEC in the crypts, and inhibits cell proliferation in mature enterocytes by regulating the expression of Wnt-related genes. Thus, PRC2 overexpression might result in outnumbered immature transit amplifying cells, which is characteristic of gluten-induced crypt hyperplasia (Oittinen et al., 2017).

Gene expression can also be post-transcriptionally prevented by non-coding micro RNA (miRNA) interactions with messenger RNA (mRNA), which promotes mRNA degradation and protein translation inhibition. Consequently, miRNA dysregulation has severe consequences for homeostasis. Indeed, lack of the double-stranded RNA-specific endoribonuclease Dicer1 that processes pre-miRNA provokes intestinal barrier dysfunction and crypt extension in mice, possibly due to a decrease in BMP2 (McKenna et al., 2010). The first CD-associated miRNA, miR-449a, is highly expresssed in pediatric active and gluten-free diet (GFD)-treated CD mucosa (Capuano et al., 2011). miR-449a upregulation in CD might constitutively diminish Notch signals and increase nuclear β-catenin (and thus Wnt signaling induction) in coeliac IECs which, in turn, result in a decreased number of goblet cells (Capuano et al., 2011). Subsequent studies linked a generalized downregulation of the miR-192/194 miRNA cluster on chromosome 11, and miR-192 precursor homologue miR-215, with active CD (Vaira et al., 2014). Downregulation of these miRNA clusters, as well miR-31-5p and miR-551a, is involved in mitogen-associated protein kinase (MAPK) signaling pathway, cytoskeletal and extracellular matrix remodeling, and Wnt signaling (Buoli Comani et al., 2015; Vaira et al., 2014). Although these signaling pathways are key in IEC biology, only miR-192-5p has been particularly associated with epithelial cells. miR-192-5p downregulation in coeliac IECs seems to upregulate the macrophage activator C-X-C motif ligand (CXCL) 2 and the nucleotide-binding oligomerization domain-containing protein (NOD) 2 that recognizes bacterial peptidoglycan, activates NF-κB and induces secretion of TNF, IL-6, and IL-18 (Magni et al., 2014). Interestingly, miR-192-5p and miR-31-5p are also downregulated in active CD plasma, and only miR-192 levels remain downregulated despite GFD-treatment and response, including GFD-refractory CD patients (Buoli Comani et al., 2015; Vaira et al., 2014, 2020).

Autophagy mechanisms are essential to digest organelles, large aggregates of proteins, and microorganisms; hence, fundamental in host defense against pathogens. The autophagy-related proteins Atg7 and Beclin 1 (Atg6 homolog) are overexpressed after early exposure to gluten on the IEC line Caco-2 (Comincini et al., 2017). Their negative regulators miR-17 and miR-30 decrease, possibly due to their release in exosomes as these miRNAs are increased in blood of CD patients. Subsequently, miRNA uptake by IECs repress autophagy, which might disrupt gluten transport and increase its toxicity (Comincini et al., 2017). However, alternative autophagy activators seem to counteract the effects of these miRNAs, such as starvation, or the mammalian target of rapamycin (mTOR)-independent autophagy activator disaccharide trehalose (Manai et al., 2018, 2019).

Long non-coding RNAs (lncRNA) are cell-type specific RNA molecules longer than 200 base pairs able to silence gene expression through chromatic structure regulation, or control RNA maturation, transport and protein synthesis. The first CD associated lcnRNA was *lcn13*, which overlaps with *IL18RAP* (IL-18 receptor accessory protein) gene, and has been linked to the nuclear factor kappa-light-chain-enhancer of activated B-cells B (NF-kB) pathway (Castellanos-Rubio et al., 2016). In CD epithelium, the lncRNA HCG14 is slightly downregulated and polymorphisms in this region are associated with decreased NOD1 expression. Thus, coeliac IEC might develop an altered innate immune response that could assist in CD onset (Santin et al., 2018).

3. Epithelial cell profile and structural changes in coeliac disease

In line with genetic and epigenetic findings, Wnt, Notch, and EGF receptor (EGFR) signaling pathways are disrupted in CD mucosa (Capuano et al., 2011; Juuti-Uusitalo et al., 2007; Oittinen et al., 2017). As a result, a decreased number of secretory goblet and Tuft cells has been associated with active CD (Capuano et al., 2011; Di Sabatino et al., 2008; Huh et al., 2020) (Fig. 2). Consistently, the secretion of some Paneth and

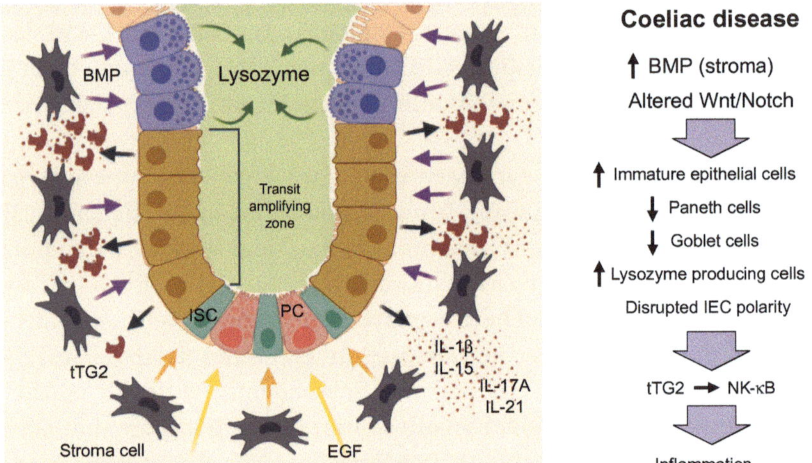

Fig. 2 Changes in intestinal epithelial cells during coeliac disease. Intestinal stem cells (ISC) re-programme toward immature intestinal epithelial cells that expand the transit amplifying zone (brown cells), and lysozyme-producing mucus cells (blue cells). Bone morphogenic protein (BMP) antagonist secretion by stroma cells is increased (purple arrows), and Wnt and Notch signals are altered. Thus, genuine Paneth (PC) and goblet cells are scarce and crypt hyperplasia is promoted. The epithelial-mesenchymal transition pathway is disrupted in intestinal epithelial cells, which causes loss of cell polarity, induces morphological changes, and promotes tissue transglutaminase (tTG) 2 expression and secretion (black arrows). Active tTG2 induces NF-κB activation that results in dysregulation of extracellular matrix components and secretion of proinflammatory cytokines IL-1β, IL-15, IL-17A, and IL-21 (black arows). Secreted tTG2 crosslinks with extracellular matrix proteins and causes protein deposition, fibrosis and tissue damage. *Created with BioRender.com.*

goblet cell-derived compounds that protect the mucosa and promote repair, such as β-defensin (hBD) 1 and intestinal trefoil factor, is decreased in active CD (Ciacci et al., 2002; Di Sabatino et al., 2008; Frye et al., 2000; Intrieri et al., 2010; Taha et al., 2005). In contrast, the antimicrobial lysozyme is remarkably produced due to highly abundant lysozyme-producing mucus cells in CD (Rubio, 2011) (Fig. 2). Thus, ISC re-programme toward less differentiated secretory IECs (Rubio, 2011). The increased expression of the ISC specific marker and Wnt co-receptor leucine-rich repeat-containing G protein-coupled receptor (Lgr) 5 in active CD indicates crypt hyperplasia (Senger et al., 2015). This, together with a downregulation of hedgehog signaling molecules and upregulation of BMP antagonists by mesenchymal cells, altered Notch signals, and decreased numbers of Paneth cells, further promotes an increased proliferation of immature transit amplifying epithelial cells (Capuano et al., 2011; Di Sabatino et al., 2008; Senger et al., 2015) (Fig. 2).

Transcriptomic analyses of intestinal CD-derived organoids further support the above mentioned processes, and reveal an intense downregulation of epithelial-mesenchymal transition (EMT) gene pathway (Dieterich et al., 2020; Freire et al., 2019). This pathway is involved in the trans-differentiation of ISCs into motile mesenchymal cells and *vice versa* (i.e., mesenchymal-epithelial transition, MET), and is crucial in wound healing processes (Lamouille et al., 2014). Since EMT induces loss of cell polarity and morphological changes, genes related to the TGF-β signaling family, cytoskeletal and cellular adhesion reassembly, and extracellular matrix are also dysregulated in CD (Dieterich et al., 2020; Lamouille et al., 2014; Senger et al., 2015). In particular, a decreased production of extracellular matrix components disrupt the accumulation of growth factors and lymphocytes, which impairs tissue scaffolding (Dieterich et al., 2020). Initially, these extracellular matrix alterations were attributed to CD-associated antibodies against gluten peptides and tissue transglutaminase (tTG) 2 or primary defects in fibroblasts, but the latest reports on organoids demonstrate an additional role for epithelial cells (Dieterich et al., 2020; Roncoroni et al., 2013). Interestingly, tTG2, the CD autoantigen involved in tissue repair, stabilizes extracellular matrix proteins by binding to integrins (Dieterich et al., 1997; Mehta et al., 2010; Wang and Griffin, 2012).

Tissue transglutaminases are multifunctional proteins that are expressed by IECs. In particular, tTG2 transduce signals from hormones and growth factors by acting as a GPTase/ATPase, or as a threonine protein kinase, respectively, in a Ca^{2+}-independent manner (Wang and Griffin, 2012).

When translocated to the cell nuclei, tTG2 increases the transcription of survival genes (Wang and Griffin, 2012). Gluten peptides p31-43 and p57-68 increase intracellular Ca^{2+} concentration in IEC lines, thus, promoting Ca^{2+}-dependent activation of cytosolic and nuclear tTG2, which results in deamidation of gluten peptides, MAPK-related kinases, and transcription factors (Caputo et al., 2012; Esposito et al., 2003; Stamnaes and Sollid, 2015) (Fig. 3). This enzymatic activity is particularly important in CD since it greatly enhances gluten peptide affinity for HLA molecules in antigen presenting cells (Stamnaes and Sollid, 2015). In addition, active tTG2 maintains NF-κB activation in IECs which, in turn, sustains gene expression of extracellular matrix components and tTG2 itself, and induce proinflammatory cytokine expression (IL-1β, IL-15, IL-17A, IL-21) (Caputo et al., 2012, Stamnaes and Sollid, 2015) (Figs. 2 and 3). During injury, tTG2 is continuously released from damaged cells. Secreted tTG2 then crosslinks with extracellular matrix proteins, which causes protein deposition and subsequent fibrosis and tissue damage (Wang and Griffin, 2012) (Fig. 2).

4. Direct effects of gluten in the intestinal epithelium

Gluten peptides from the α-gliadin fraction can be detected by C-X-C motif chemokine receptor (CXCR) 3 in IECs and activate MyD88 (myeloid differentiation primary response 88) pathway and NF-κB signaling (Freire et al., 2019, Lammers et al., 2008, Lerner and Matthias, 2015, Maiuri et al., 2003a,b). As a result, the tight junction regulator zonulin is released and epithelial impermeability is compromised, allowing passage of luminal contents, including toxic gluten peptides (Lammers et al., 2008; Zimmermann et al., 2014) (Fig. 3).

Recently, the chloride ion channel CFTR (cystic fibrosis transmembrane conductance regulator) has been associated with p31-43 detection, which locks CFTR in its closed state after binding and reducing the ATPase activity of the nuclear binding domain 1 (NBD1) (Villella et al., 2019). Due to CFTR implication in epithelial adaptation to oxidative stress, tTG2 and the inflammasome are activated, NF-κB is translocated to cell nuclei, where cyclooxygenase (COX)-2, inducible nitric oxide synthase (iNOS), and IL-15 expression is stimulated (Maiuri et al., 2003a, b; Villella et al., 2019) (Fig. 3). IEC activation induces the production of prostaglandins and nitric oxide metabolites, which can be detected in blood

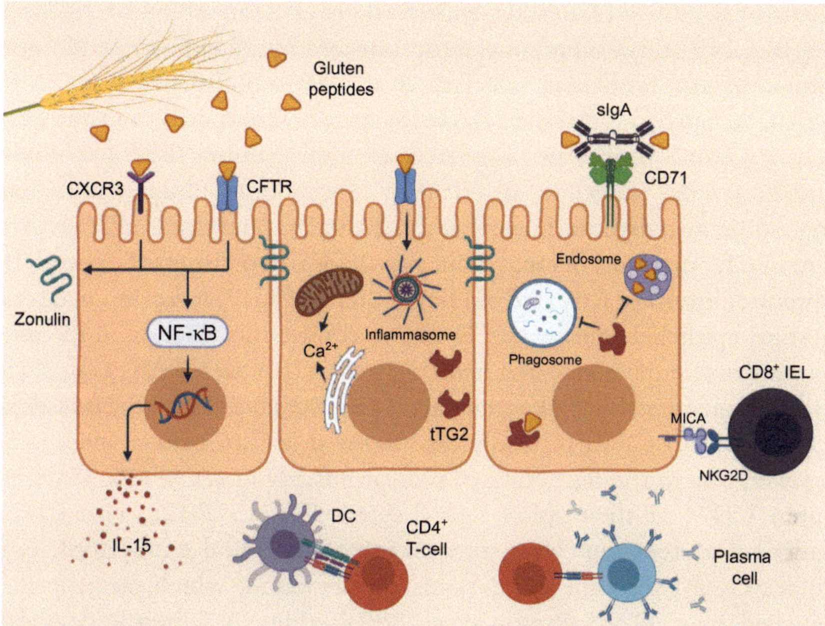

Fig. 3 Epithelial damage in coeliac disease. Intestinal epithelial cells (IEC) recognize gluten peptides through CXCR3 and CFTR which, in turn, increases barrier permeability through zonulin release (a tight junction protein), and promotes IL-15 secretion through NF-κB activation. CFTR blockage in its closed form also promotes inflammasome and tissue transglutaminase (tTG2) activation. Transcytosis of gluten peptides occurs after peptide recognition by secretory immunoglobulin (sIg) A, and then transferrin receptor (CD71)-dependent endocytosis. The concentration of reactive oxygen species increases after exposure to gluten and alters Wnt and Notch signaling pathways, and impedes tTG2 degradation (not shown). Reactive oxygen species and active tTG2 inhibit autophagosome, peroxisome and endosome maturation, which causes epithelial growth factor receptor to exert a prolonged action (not shown). Inside IECs, gluten peptides induce calcium release from the mitochondria and the endoplasmic reticulum, which, if sustained, induces cell death due to protein misfolding inside the endoplasmic reticulum. In addition, active tTG2 deamidates gluten peptides, which become easily recognizable by underlying dendritic cells (DC), thus, contributing to develop a T-cell mediated proinflammatory immune response. These $CD4^+$ T helper cells can synergistically activate gluten-primed B-cells, hence, inducing plasma cell differentiation and antibody production (anti-tTG2 and anti-gluten peptides). Ultimately, the internal stress and the proinflammatory environment induces MICA expression in IECs, which is recognized by NKG2D receptor expressed in IL-15/IL-21 re-programmed cytotoxic intraepithelial T-cells. This grants cytotoxic $CD8^+$ T-cells license to kill IECs and contribute to tissue atrophy. *Created with BioRender.com.*

(Maiuri et al., 2003a,b). In addition, activated tTG2 can inhibit Beclin 1 protein (BECN1), thus, suppresing autophagy and interfeting with endosomal maturation and trafficking (Luciani et al., 2010a,b, Villella et al., 2019) (Fig. 3). Reactive oxygen species (ROS) and tTG2 also drives inflammation via peroxisome proliferator-activated receptor gamma (PPARγ) downregulation after induction of PPARγ crosslinking, ubiquitination and proteasome degradation (Luciani et al., 2010a,b). Due to PPARγ involvement in CD, transport of long-chain fatty acids is also disrupted. Indeed, the transporter intestinal fatty acid binding protein (IFABP) serves as a marker of intestinal epithelial damage in CD due to consistent reduction in the duodenal mucosa and increase in circulating plasma (Bottasso Arias et al., 2015).

A complementary uptake route involves gluten binding to the transferrin receptor CD71. CD71 is usually expresseed at IEC basolateral membrane; however, due to the disruption of cellular polarity in active CD, IECs also express CD71 at their apical side (Lebreton et al., 2012). This CD71-dependent gluten transcytosis process implicates apical endocytosis, early and recycling endosomes, and basolateral secretion, which, in general, is facilitated by tTG2 (Lebreton et al., 2012) (Fig. 3). Upon endocytosis, p31-43 remains in early endocytic vesicles and prevents their maturation through disruption in the localization of the hepatocyte growth factor-regulated tyrosine kinase substrate (Barone et al., 2010) (Fig. 3). After 24h, IECs can still retain p31-43 in perinuclear vesicles, which leads to increased levels of ROS, including $O_2^{-\cdot}$, H_2O_2, peroxynitrite, and lipid peroxides (Luciani et al., 2010a,b; Perez et al., 2017). ROS indirectly increase tTG2 levels through inhibition of tTG2 ubiquitination (Luciani et al., 2010a,b). Epithelial oxidative stress also interferes in cellular growth and morphology through a decrease of F-actin expression and cytoskeletal restructuration (Dolfini et al., 2005; Hudson et al., 1976). Moreover, Wnt and Notch pathways are redox-sensitive since they may be modulated by members of the nicotinamide adenine dinucleotide phosphate (NADPH) oxidase (NOX) and Tp53-inducible glycolysis and apoptosis regulator (TIGAR) families (Perez et al., 2017). Finally, glutathione (GSH), a dietary-derived antioxidant that balance redox status and is connected to DNA methylation events, and GSH peroxidase and reductase, are depleted in CD (Perez et al., 2017; Trivedi et al., 2014). This, together with the opioid activity of certain proline-rich gluten-derived peptides, which can inhibit cysteine uptake, can further perturb the redox balance and DNA methylation profile in IECs (Trivedi et al., 2014).

By intercepting endocytic vesicle maturation, p31-43 also promotes cell proliferation due to a prolonged EGFR activation (Barone et al., 2011; Maiuri et al., 2003a,b). Insterestingly, this activity is dependent on IL-15, which is also expressed by IECs after exposure to gluten, and is a key cytokine in innate-adaptive immune response crosstalk (Barone et al., 2011, Maiuri et al., 2003a,b). On the other side, p57-68 reaches late endocytic vesicles (Caputo et al., 2012). Both p31-43 and p57-68 peptides mobilize Ca^{2+} ions from the mitochondria, and p31-43 also does from the endoplasmic reticulum (ER), which results in intracellular tTG2 activation and subsequent gluten peptide deamidation *in situ* (Caputo et al., 2012; Esposito et al., 2003) (Fig. 3). Ca^{2+} depletion from the ER can also provoke protein misfolding and ER-stress, thus, activating the unfolded protein response. This mechanism induces chaperone production, e.g., GRP78, and can be used as a ER-stress marker (McGuckin et al., 2010). GRP78 expression in increased active CD mucosa, and p31-43-stimulated IECs also moderately increase CHOP (C/EBP homologous protein) transcription factor. Consequently, stress-linked apoptosis can be triggered due to irrecoverable ER-stress (Zinszner et al., 1998).

Gluten directly induces IEC apoptosis by increasing Fas and Fas ligand (FasL) expression, FasL secretion, and decreasing anti-apoptotic Bcl-2 (B-cell lymphoma 2 protein) expression (Giovannini et al., 2003; Shalimar et al., 2013). In combination with IEC-derived, gluten-induced IL-15 secretion, Fas is further upregulated in IECs, and FasL in upregulated in intraepithelial lymphocytes (IELs) (Di Sabatino et al., 2006; Maiuri et al., 2000). IL-15 can also cooperate with α-gliadin p31-49 to induce expression of major histocompatibility complex class I chain-related gene-A (MICA) in IECs, which is an unconventional HLA class I molecule that acts as an epithelial stress signal (Groh et al., 1996; Hue et al., 2004) (Fig. 3).

Intestinal epithelial cells can function as non-professional antigen presenting cells due to the expression of major histocompatibility complex (MHC) type I molecules CD1d, MICA/B, and HLA-E, especially outside intestinal crypts (Perera et al., 2007). Therefore, gluten peptides that escape from endocytic vesicles and reach the cytosol for a subsequent delivery to the ER, can be loaded into MHC-I molecules in combination with tTG2 or alone. Afterward, MHC-peptide(-tTG2) complexes are transported to the basolateral membrane, where they cross-present gluten-derived peptides to $CD8^+$ intraepithelial T-cells (Gromme and Neefjes, 2002; Kovacsovics-Bankowski and Rock, 1995; Mazzarella et al., 2008).

In CD, IECs fail to express CD1d, which could affect the priming of a particular T regulatory cell subset identified as $CD8^+$ $CD28^-$ (Perera et al., 2007).

5. Epithelial damage driven by immune cells

Intraepithelial lymphocytes (IELs), gut-associated lymphoid tissue, and additional underlying immune cells maintain and defend the intestinal epithelium, and promote sustained tolerance against dietary and microbial antigens (Tordesillas and Berin, 2018). In CD, non-immunodominant gluten peptides trigger innate immune responses in IEC, macrophages and dendritic cells by inducing cell activation, IL-15 and IL-21 secretion, and cyclooxygenase (COX)-2 expression (Maiuri et al., 2003a,b) (Fig. 3). Early epithelial barrier defects increase intestinal permeability, thus, enabling immunodominant peptides to develop a T helper (Th)1/Th17 adaptive immune response in the lamina propria, which is enhanced by tTG2 (Escudero-Hernandez et al., 2020; Maiuri et al., 2003a,b; Serena et al., 2019; Stamnaes and Sollid, 2015) (Fig. 3).

Activated Th cells vastly produce proinflammatory cytokines interferon (IFN)-γ, TNF-α, IL-18 and IL-21, that attract more immune cells to the intestine and establishes a positive proinflammatory feedback that perpetuates lymphocyte infiltration (Serena et al., 2019; Stamnaes and Sollid, 2015). These cytokines increase the nascent epithelial permeability, and reprogramme and activate cytotoxicity in natural killer (NK) cells and IELs (Al-Sadi et al., 2016; Ciszewski et al., 2020; Ma et al., 2004). IFN-γ response is particularly important in IECs because IEL-secreted IFN-γ, acting through Janus kinase/signal transducers and activators of transcription (JAK/STAT) signaling and IFN regulatory factor (IRF) 1 transcription factor activation, promotes inflammasome activation by pre-caspase 1 cleavage and activation of inflammasome sensor nucleotide-binding oligomerization domain, leucine rich repeat and pyrin domain (NLRP) 6 in enterocytes. In addition, NLRP8 is activated in goblet cells by IL-17A—also upregulated in active CD—and, together, promote IL-18 secretion by IECs (Pietz et al., 2017). Such environment ultimately increases the epithelial stress, and triggers MICA overexpression in IECs during active CD (Groh et al., 1996, Hue et al., 2004). MICA is recognized by NK receptor D (NKG2D) molecules expressed by IL-15/IL-21-dependent reprogrammed cytotoxic $CD8^+$ IELs (Ciszewski et al., 2020) (Fig. 3). This bound, which is reinforced by HLA-E (IEC)—NKG2C (IEL) interaction, prompts IEC elimination by

IELs (Hue et al., 2004; Meresse et al., 2006). Other immune cells with similar profile and action over IECs are also increased in CD active mucosa, including natural killer (NK), NK T-cells, and T T-cell receptor (TCR)$\gamma\delta^+$ cells (Calleja et al., 2011).

B-cells are antigen presenting cells that can internalize tTG2–gluten complexes via B-cell receptor (BCR) (du Pre and Sollid, 2015). Gluten-reactive $CD4^+$ T-cells cooperate to activate tTG2- and gluten-reactive B-cells by TCR—MHC-II interaction, which results in activation of both T-cells and B-cells. In addition, B-cells differentiate into antibody-producing plasma cells (du Pre and Sollid, 2015) (Fig. 3). As mentioned before, antibody accumulation in the basement membrane region of the mucosa can damage the extracellular matrix by reducing IEC attachment to the tTG2-fibronectin matrix, thus, causing mechanical stress (Rauhavirta et al., 2019; Roncoroni et al., 2013). Anti-tTG2 also inhibit IEC differentiation and induce proliferation by intercepting TGF-β-mediated fibroblast–IEC crosstalk (Halttunen and Maki, 1999; Rauhavirta et al., 2019). Similarly, antibodies that play a role in intestinal homeostasis are involved in CD pathogenesis, such as secretory immunoglobulin (sIg) A that neutralizes antigens from the lumen and its secretion from Peyer's patches cells is induced by M cells (Rios et al., 2016). Since gluten peptides are recognized by sIgA and these can subsequently bind the apically expressed transferrin receptor CD71, peptides further surpass the epithelial barrier by transcytosis (Lebreton et al., 2012) (Fig. 3).

6. Consequences of microbiota–epithelium interactions for coeliac disease

The human enteric compartment is an ideal niche for the coevolution of numerous microbes cohabiting the gut in a bi-directional symbiotic environment. However, numerous chronic human diseases, as CD, are associated with dysbiosis (Valitutti et al., 2019). Toll-like receptors (TLRs), common receptors that recognize structurally conserved microbe-derived molecules that are expressed by immune cells and IECs, are dysregu lated in CD. In particular, TLR2 and TLR4 that detect parasitic (TLR2) and bacterial (TLR2 and TLR4) components are increased in active CD (Kalliomaki et al., 2012). On the contrary, expression of the TLR inhibitor and Toll-interacting protein (TOLLIP) is decreased, and has been correlated with luminal bacterial load (Kalliomaki et al., 2012).

Antimicrobial peptides shape the composition of gut microbiota and protect from bacterial translocation, but both Paneth cells and antimicrobial peptide secretions are altered in CD (Di Sabatino et al., 2008; Intrieri et al., 2010; Taha et al., 2005). Indeed, Paneth and goblet cells are replaced by lysozyme-producing mucus cells (Rubio, 2011) (Fig. 2). This, and the generalized immature nature of enterocytes in CD mucosa, could facilitate gram-negative bacterial colonization (e.g., proinflammatory *Bacteroides-Prevotella* and *Escherichia coli*), which could contribute to loss of gluten tolerance in genetically predisposed individuals (Schippa et al., 2010). Still, ISC re-programming toward a lysozyme-producing mucus cell profile could result from an adaptation to signals from pathogenic duodenal bacteria and microbial transglutaminases (mTG) (Rubio, 2011). Normal microbiota, especially bacteria from the *Streptomyces* genus, produce mTG (Lerner et al., 2017). This Ca^{2+}-independent transglutaminase is crucial for bacterial survival and can deamidate gluten peptides; hence, enhancing the immunogenic effects of gluten in CD mucosa (Lerner et al., 2017; Stricker et al., 2019). Due to mTG emulsifing properties, it is increasingly being used in food industry, so mTG association with the induction of autoimmune diseases, especially CD, is under debate (Lerner et al., 2017).

Altogether, whether bacterial misbalance is a cause or consequence of CD remains unclear (Valitutti et al., 2019). The only study that explored the microbiome and included asymptomatic CD patients identified after screening of family members found that these patients had a microbial diversity similar to controls, while classical active CD microbiota was less diverse (Wacklin et al., 2013). Interestingly, asymptomatic CD histopathological profile was comparable to active CD mucosa (i.e., lymphocyte infiltration, increase of $TCR\gamma\delta^{+}$ T-cells, and decreased villous height to crypt depth ratio) (Wacklin et al., 2013).

Infections have been associated with CD due to the potential to depolarize IECs and interfere with transcytosis processes, as it happens after infection with *Pseudomonas aeruginosa* (Kierbel et al., 2007). Recently, the suspected role of reoviruses in CD was confirmed (Bouziat et al., 2017). Reoviruses are avirulent, non-enveloped double-stranded RNA viruses that cause frequent early childhood infections and elicit protective immunity. The first contact of a reovirus with an IEC is through infectious subviral particles (ISVPs) produced after virions are digested by serine proteases present in the intestinal lumen (Stanifer et al., 2016). ISVPs are endocytosed but quickly escape from the vesicles, whereas an intact virion encountering a cell would follow the endocytic route until proteases cathepsins, among others, in late endosomes maturate them into ISVPs and their genetic material is delivered to the cytoplasm (Stanifer et al., 2016). ISVPs are more infectious than virions, but induce a more moderate innate immune response in IECs (i.e., IFN-β1 is much more expressed after infection with virions). Interestingly, the virion-induced innate immune response is linked to cell death, and does not require viral replication. On the other hand, ISVPs induce IECs to secrete TGF-β, which promotes IEC survival. Therefore, reoviruses evolved to use ISVPs as a mechanism to enhance the initial infection of cells, and to limit stimulation of the cellular immune system (Stanifer et al., 2016). Once the infection is established, reoviruses have access to the underlying dendritic cells, and disrupt oral tolerance by suppressing regulatory T-cell conversion and promoting a Th1 response (Bouziat et al., 2017). Furthermore, undigested gluten p31-43 enhances the IFN-mediated viral response by interfering with endocytic trafficking and cooperating with the virus to activate TLR7 signaling pathway, including NF-κB activation (Nanayakkara et al., 2018). Therefore, alimentary proteins able to mimic and potentiate viral immune responses could trigger CD. Indeed, anti-tTG IgA antibodies can also recognize proteins from rotavirus, a virus from Reoviridae family (Dolcino et al., 2013). In IEC lines, anti-rotavirus antibodies induce apoptosis, inflammation and epithelial barrier dysfunction. Interestingly, the production of these antibodies precedes CD onset and anti-tTG2 production (Dolcino et al., 2013).

7. Diet and intestinal epithelial health in coeliac disease

The nutritional environment can influence ISC functioning. For example, the western-style diet (high fat, and low vitamin D_3, calcium, methyl donors and fiber), perturbs Wnt signaling and IEC maturation through imbalance in ROS metabolism, DNA methylation and gene expression in mice which, in turn, promotes tumorigenesis (Li et al., 2020). The most common nutritional deficiences in CD involve iron and vitamins B_9 (folic acid), B_{12} (cobalamin), and D, which are associated with the development of complications such as anemia, osteoporosis, type I diabetes, and neurological and skin disorders (Malterre, 2009; Ponziani et al., 2012; Vici et al., 2016). These defetcs are often related to the reduced absorptive area in CD mucosa, which improves after following a GFD, so the specific causes have been underinvestigated (Freeman, 2015; Ponziani et al., 2012).

Iron and iron heme complex biosynthesis are important to IEC cellular respiration, oxidative stress, regulation of cell growth and differentiation, and apoptosis (Oates and West, 2006). Expression of iron transporters divalent metal transporter (DMT) 1 and ferroportin or metal transporter (MTP) 1, the ferroxidase hephaestin, the storage protein ferritin and the transferrin receptor CD71 are increased in active CD, while body iron stores are shortened (Freeman, 2015). The increased secretion of TNF-α in CD, in particular, inhibits IEC iron absorption due to its capability to modify the expression and/or cellular localization of ferritin (Sharma et al., 2009).

Vitamins B_9 and B_{12} are closely related to DNA, RNA and protein metabolism since these are required for DNA methylation: folates are the base substrate to replenish methyl donor products (S-adenosylmethionine, SAMe), and vitamin B_{12} is a cofactor of methionine synthase that converts methyl donors into methionine for subsequent use in DNA methylation. These processes are also involved in DNA metabolism since the resulting products are used for purine and pyrimidine base formation. Consequently, DNA synthesis is reduced by B_9 or B_{12} deficiency, which has severe consequences on cell replication, especially in cells with a high proliferation rate (e.g., IECs) (Ponziani et al., 2012). As discussed before, hypomethylation is generalized in IECs from active CD mucosa.[26] And, interestingly, intestinal mucosa can recover DNA methylation after stimulation with folate (Pufulete et al., 2005). Imbalance in the absorption mechanisms of vitamins B_9 and B_{12} have not been explored in CD. On the one side, vitamin B_9 derived folates are absorbed after conversion into monoglutamic folate by enzymes of the intestinal juice and of the enterocyte brush border and, after absorption, enterocytes further reduce them. Since CD mucosa lacks brush border formations in enterocytes, folates are malabsorbed (Ponziani et al., 2012). On the other side, vitamin B_{12} binds to intrinsic factor (secreted by the stomach) in the duodenum and it is mostly absorbed in this form in the distal ileum through transcobalamin I, II and III dependent endocytosis. A minor free vitamin B_{12} portion is also absorbed along the entire intestine by passive diffusion (Fernandez-Banares et al., 2009).

Vitamin D regulates the expression of over 900 genes through the nuclear vitamin D receptor (VDR), modulating blood pressure, calcium absorption, and immune responses. Indeed, vitamin D deficiency contributes to increase the rates of autoimmune diseases, including CD (Vici et al., 2020). Both vitamin D_2 and D_3 forms are found in food and supplements, but most of vitamin D derives from vitamin D_3 synthesis in the skin through the action of ultraviolet light B (UVB) (Vici et al., 2020). In CD, no differences in VDR gene expression have been found, but some genetic polymorphisms could be involved in the susceptibility to the disease (Colston et al., 1994; San-Pedro et al., 2005). The active vitamin D form (1,25-dihydroxyvitamin D, 1,25$(OH)_2D_3$) can downregulate BMP gene expression by increasing DNA methylation and histone modifications in the BMP promoter gene. However, this mechanism has been studied in bone marrow stromal cells, and it remains unknown whether it also acts on IECs (Fu et al., 2013). Still, a recent study confirms ISC dysfunction due to downregulation of oxidative phosphorylation and to increase in DNA mismatch repair activity induced by diets deficient in vitamin D and/or calcium (Li et al., 2020). Intestinal epithelial VDR also regulates autophagy and antimicrobial secretions from Paneth cells, and ameliorates dysbiosis; thus, the altered cellular differentiation and microbiota in CD could cause defects on vitamin D positive impacts (Wu et al., 2015). It has been shown that 1,25$(OH)_2D_3$ also blocks TNF-α-induced intestinal permeability. Specifically, 1,25$(OH)_2D_3$ disrupts NF-κB binding to the promoter of myosin light chain kinase (*MLCK*) gene in IECs, which blocks zonulin release, preserves mucosal barrier integrity and alleviates intestinal permeability in gluten-sensitized mice (Dong et al., 2018; Du et al., 2015). In addition, vitamin D exerts immune regulatory functions in dendritic and T-cells expressing high levels of VDR (Baeke et al., 2010; Ferreira et al., 2011).

Strict gluten exlusion from the diet is the only reliable treatment for CD patients. However, gluten-free products are scarce in fiber, vitamins, and some minerals such as iron, zinc, magnesium and calcium. Oppositely, the GFD is usually rich in energy, fatty acids, sodium, sugars and simple carbohydrates, which increases the risk of cardiovascular disease and obesity in GFD consumers (Naik et al., 2018; Vici et al., 2016). Particularly, a high fat diet increases bile acid secretion that favors toward increased deoxycholic acid (DCA) levels. In turn, DCA causes defective intestinal barrier function by activating WNT signaling pathway, induces low-grade inflammation, and promotes a decrease of *Lactobacillus* and an increase of *Clostridium* bacteria (Di Ciaula et al., 2017). Therefore, alimentary education, evaluation of nutritional quality of the diet, and regular nutritional screening are crucial to avoid sustained intestinal barrier imbalance and other complications in CD patients (Naik et al., 2018, Vici et al., 2016). Alternatively, micronutrient supplementation or gluten-free product enrichment could complement the GFD, especially in cases where overdosing or mutations do not negatively affect the patient (i.e., *TMPRSS6 mutations alter iron absorption,* and vitamin D overexposure upregulate CD immune-mediators) (Martin-Masot et al., 2019; Vici et al., 2016, 2020).

8. Treatments targeting intestinal epithelial cells

Adherence to a strict GFD limits nutritional variety, which can impede the full recovery of the intestinal mucosa (Naik et al., 2018, Vici et al., 2016). Given the social burden of GFD, additional treatment options are under investigation. Among novel treatments with a focus on IECs, oral enzymes or neutralizing antibodies against gluten would reduce the contact of gluten-derived peptides with mucosal cells (Gujral et al., 2015; Malamut et al., 2019). As discussed before, gluten p31-43 blocks CFTR in IECs and triggers the epithelial stress response. Thus, CFTR potentiators as the approved drug Ivacaftor could represent a novel therapeutic option for CD patients (Esposito et al., 2019; Luciani et al., 2010a,b). Anti-IL-15, JAK (part of cytokine downstream signaling pathways) inhibitors and tTG2 blocking therapies also seem encouraging, but their efficacy and safety are still under assessment (Malamut et al., 2019). Alternatively, combined IL-15 and tTG2 gene silencing in IECs could be considered (Attarwala et al., 2017).

A promising drug is the synthetic peptide larazotide acetate (AT-1001), which prevents zonulin release from tight junctions, thus, structurally preserving the intestinal barrier and limiting gluten paracellular transport (Khaleghi et al., 2016). If this could be combined with the epithelial mitogen R-spondin-1, which stimulates crypt cell growth, accelerates mucosal regeneration and restores intestinal architecture in experimental colitis in mice, coeliac mucosa could be restored (Zhao et al., 2007). Still, none of these treatments will cure CD or replace the GFD, but it is likely that they will be used as maintenance therapies or be combined with additional complementary options. For instance, treatment with *Bifidobacterium bifidum* IATA-ES2 recovers the decreased number of goblet cells caused by gluten or IFN-γ stimuli, as observed using rat small intestines (Cinova et al., 2011). Microbiota-derived bioproducts butyrate, lactate, and polysaccharide A also improve intestinal barrier function and reduce gluten-induced cytokine secretion in human organoids (Freire et al., 2019). Although effects on IECs have not been explored, parasitic helminth infection with *Necator americanus* induces gluten tolerance in CD (Croese et al., 2015). In a similar infection model with the nematode *Nippostrongylus brasiliensis,* it has been shown that *hookworm-secreted extracellular vesicles alone can be uptaken by intestinal organoids and prevent colitis in mice* (Eichenberger et al., 2018).

The constant conflict between apoptotic and survival signals leads to excessive cell death in active CD and induces the mobilization of bone marrow multipotent stem cells. These cells seem to function as a supplementary source of ISCs that promotes repair; hence, regenerative medicine could mirror ISC replacement to treat CD (Mastrandrea et al., 2008). The current experience is limited to isolated case reports of different severe CD complications treated with autologous or allogenic hematopoietic stem cell transplantation (Al-toma et al., 2007; Bishton and Haynes, 2007; Ciccocioppo et al., 2013; Hoekstra et al., 2010; Kline et al., 2007; Tack et al., 2011). Still, these results seem promising since some of the patients re-introduced gluten intake without developing clinical symptoms after follow-up for several years (Bishton and Haynes, 2007; Ciccocioppo et al., 2013; Kline et al., 2007). On the other side, intestinal organoids or miniguts are the result of *in vitro* ISC growing that spontaneously rebuild the crypt-villus structure (Sato et al., 2009). In 2017, human colon organoids were successfully transplanted into recipient mice (Sugimoto et al., 2018). In combination with genome editing techniques, disorders with a genetic ground (i.e., HLA-DQ2/DQ8 in CD) could be corrected with patient-derived tissue, as it has been done in organoids derived from cystic fibrosis patients (Schwank et al., 2013). Therefore, intestinal organoids represent a new instrument for individual-based therapy for the treatment of CD and other chronic gut disorders.

9. Concluding remarks

Coeliac disease is an autoimmune disorder triggered by dietary gluten. Despite the central role of immune cells in the development of proinflammatory responses and intestinal epithelial damage, it is increasingly obvious that epithelial cell dysfunction drives CD pathogenesis before gluten is overtly available to underlying immune cells (Dieterich et al., 2020; du Pre and Sollid, 2015; Freire et al., 2019). Consistently, gluten causes epigenetic, endosomal, oxidative, and metabolic stress, which reprogramme ISCs toward immature epithelial cell profiles (Barone et al., 2010; Dieterich et al., 2020; Fernandez-Jimenez et al., 2019; Luciani et al., 2010a,b; Nanayakkara et al., 2018). The activation and mislocalization of tTG2 in IECs promotes extracellular instability and physical tensions, besides an enhance of gluten peptide recognition by antigen presenting HLA molecules (Caputo et al., 2012; du Pre and Sollid, 2015). IECs also develop an immune response against gluten that induces IL-15 secretion and cross-presentation of gluten-derived peptides to IELs (Hue et al., 2004; Maiuri et al., 2000). Given the burden of gluten-induced modifications, IECs express stress signals that ultimately attract cytotoxic cells to engage into IEC clearance (Hue et al., 2004).

Changes in microbiome and diet affect IEC profile and destabilize the epithelial barrier, which promotes intestinal permeabilization (Li et al., 2020; Rubio, 2011; Stanifer et al., 2016). Thus, immunodominant gluten peptides cross the IEC fence and develop an adaptive immune response that boost inflammation and villus atrophy (du Pre and Sollid, 2015). Although the GFD is an indispensable treatment for CD patients, its nutritional deficiencies require regular evaluation and nutritional supplementation to achieve full epithelial recovery (Naik et al., 2018, Vici et al., 2016).

Organoids represent a unique and invaluable technologic breakthrough into the foundations of IEC biology (Gehart and Clevers, 2019). Indefinite ISC culture has been proved to facilitate drug testing, regenerative and personalized medicine. Therefore, once organoid transplantation can be routinely performed on a larger scale, they will aid CD treatment, especially in cases with severe complications and poor prognosis. Integrating our knowledge on epithelial damage and repair, immune responses and treatment deficiencies will be key to better understand CD and develop more advanced, individualized and efficient treatments.

References

Al-Sadi, R., Guo, S., Ye, D., Rawat, M., Ma, T.Y., 2016. TNF-alpha modulation of intestinal tight junction permeability is mediated by NIK/IKK-alpha axis activation of the canonical NF-kappaB pathway. Am. J. Pathol. 186 (5), 1151–1165.

Al-toma, A., Visser, O.J., van Roessel, H.M., von Blomberg, B.M., Verbeek, W.H., Scholten, P.E., Ossenkoppele, G.J., Huijgens, P.C., Mulder, C.J., 2007. Autologous hematopoietic stem cell transplantation in refractory celiac disease with aberrant T cells. Blood 109 (5), 2243–2249.

Allaire, J.M., Crowley, S.M., Law, H.T., Chang, S.Y., Ko, H.J., Vallance, B.A., 2018. The intestinal epithelium: central coordinator of mucosal immunity. Trends Immunol. 39 (9), 677–696.

Attarwala, H., Clausen, V., Chaturvedi, P., Amiji, M.M., 2017. Cosilencing intestinal transglutaminase-2 and interleukin-15 using gelatin-based nanoparticles in an in vitro model of celiac disease. Mol. Pharm. 14 (9), 3036–3044.

Baeke, F., Korf, H., Overbergh, L., van Etten, E., Verstuyf, A., Gysemans, C., Mathieu, C., 2010. Human T lymphocytes are direct targets of 1,25-dihydroxyvitamin D3 in the immune system. J. Steroid Biochem. Mol. Biol. 121 (1-2), 221–227.

Barker, N., 2014. Adult intestinal stem cells: critical drivers of epithelial homeostasis and regeneration. Nat. Rev. Mol. Cell Biol. 15 (1), 19–33.

Barker, N., van Es, J.H., Kuipers, J., Kujala, P., van den Born, M., Cozijnsen, M., Haegebarth, A., Korving, J., Begthel, H., Peters, P.J., Clevers, H., 2007. Identification of stem cells in small intestine and colon by marker gene Lgr5. Nature 449 (7165), 1003–1007.

Barone, M.V., Nanayakkara, M., Paolella, G., Maglio, M., Vitale, V., Troiano, R., Ribecco, M.T., Lania, G., Zanzi, D., Santagata, S., Auricchio, R., Troncone, R., Auricchio, S., 2010. Gliadin peptide P31-43 localises to endocytic vesicles and interferes with their maturation. PLoS One 5 (8), e12246.

Barone, M.V., Zanzi, D., Maglio, M., Nanayakkara, M., Santagata, S., Lania, G., Miele, E., Ribecco, M.T., Maurano, F., Auricchio, R., Gianfrani, C., Ferrini, S., Troncone, R., Auricchio, S., 2011. Gliadin-mediated proliferation and innate immune activation in celiac disease are due to alterations in vesicular trafficking. PLoS One 6 (2), e17039.

Bishton, M.J., Haynes, A.P., 2007. Combination chemotherapy followed by autologous stem cell transplant for enteropathy-associated T cell lymphoma. Br. J. Haematol. 136 (1), 111–113.

Bottasso Arias, N.M., Garcia, M., Bondar, C., Guzman, L., Redondo, A., Chopita, N., Corsico, B., Chirdo, F.G., 2015. Expression Pattern of Fatty Acid Binding Proteins in Celiac Disease Enteropathy. Mediators Inflamm. 2015, 738563.

Bouziat, R., Hinterleitner, R., Brown, J.J., Stencel-Baerenwald, J.E., Ikizler, M., Mayassi, T., Meisel, M., Kim, S.M., Discepolo, V., Pruijssers, A.J., Ernest, J.D., Iskarpatyoti, J.A., Costes, L.M., Lawrence, I., Palanski, B.A., Varma, M., Zurenski, M.A., Khomandiak, S., McAllister, N., Aravamudhan, P., Boehme, K.W., Hu, F., Samsom, J.N., Reinecker, H.C., Kupfer, S.S., Guandalini, S., Semrad, C.E., Abadie, V., Khosla, C., Barreiro, L.B., Xavier, R.J., Ng, A., Dermody, T.S., Jabri, B., 2017. Reovirus infection triggers inflammatory responses to dietary antigens and development of celiac disease. Science 356 (6333), 44–50.

Buoli Comani, G., Panceri, R., Dinelli, M., Biondi, A., Mancuso, C., Meneveri, R., Barisani, D., 2015. miRNA-regulated gene expression differs in celiac disease patients according to the age of presentation. Genes Nutr. 10 (5), 482.

Calleja, S., Vivas, S., Santiuste, M., Arias, L., Hernando, M., Nistal, E., Casqueiro, J., Ruiz de Morales, J.G., 2011. Dynamics of non-conventional intraepithelial lymphocytes-NK, NKT, and gammadelta T-in celiac disease: relationship with age, diet, and histopathology. Dig. Dis. Sci. 56 (7), 2042–2049.

The Cancer Genome Atlas Network, 2012. Comprehensive molecular characterization of human colon and rectal cancer. Nature 487 (7407), 330–337.

Capuano, M., Iaffaldano, L., Tinto, N., Montanaro, D., Capobianco, V., Izzo, V., Tucci, F., Troncone, G., Greco, L., Sacchetti, L., 2011. MicroRNA-449a overexpression, reduced NOTCH1 signals and scarce goblet cells characterize the small intestine of celiac patients. PLoS One 6 (12), e29094.

Caputo, I., Secondo, A., Lepretti, M., Paolella, G., Auricchio, S., Barone, M.V., Esposito, C., 2012. Gliadin peptides induce tissue transglutaminase activation and ER-stress through Ca2+ mobilization in Caco-2 cells. PLoS One 7 (9), e45209.

Castellanos-Rubio, A., Fernandez-Jimenez, N., Kratchmarov, R., Luo, X., Bhagat, G., Green, P.H., Schneider, R., Kiledjian, M., Bilbao, J.R., Ghosh, S., 2016. A long non-coding RNA associated with susceptibility to celiac disease. Science 352 (6281), 91–95.

Ciacci, C., Di Vizio, D., Seth, R., Insabato, G., Mazzacca, G., Podolsky, D.K., Mahida, Y.R., 2002. Selective reduction of intestinal trefoil factor in untreated coeliac disease. Clin. Exp. Immunol. 130 (3), 526–531.

Ciccocioppo, R., Bernardo, M.E., Russo, M.L., Vanoli, A., Franco, C., Martinetti, M., Catenacci, L., Giorgiani, G., Zecca, M., Piralla, A., Baldanti, F., Locatelli, F., Corazza, G.R., 2013. Allogeneic hematopoietic stem cell transplantation may restore gluten tolerance in patients with celiac disease. J. Pediatr. Gastroenterol. Nutr. 56 (4), 422–427.

Cinova, J., De Palma, G., Stepankova, R., Kofronova, O., Kverka, M., Sanz, Y., Tuckova, L., 2011. Role of intestinal bacteria in gliadin-induced changes in intestinal mucosa: study in germ-free rats. PLoS One 6 (1), e16169.

Ciszewski, C., Discepolo, V., Pacis, A., Doerr, N., Tastet, O., Mayassi, T., Maglio, M., Basheer, A., Al-Mawsawi, L.Q., Green, P.H., Auricchio, R., Troncone, R., Waldmann, T.A., Azimi, N., Tagaya, Y., Barreiro, L.B., Jabri, B., 2020. Identification of a γc receptor antagonist that prevents reprogramming of human tissue-resident cytotoxic T cells by IL15 and IL21. Gastroenterology 158 (3), 625–637. e613.

Colston, K.W., Mackay, A.G., Finlayson, C., Wu, J.C., Maxwell, J.D., 1994. Localisation of vitamin D receptor in normal human duodenum and in patients with coeliac disease. Gut 35 (9), 1219–1225.

Comincini, S., Manai, F., Meazza, C., Pagani, S., Martinelli, C., Pasqua, N., Pelizzo, G., Biggiogera, M., Bozzola, M., 2017. Identification of autophagy-related genes and their regulatory miRNAs associated with celiac disease in children. Int. J. Mol. Sci. 18 (2).

Croese, J., Giacomin, P., Navarro, S., Clouston, A., McCann, L., Dougall, A., Ferreira, I., Susianto, A., O'Rourke, P., Howlett, M., McCarthy, J., Engwerda, C., Jones, D., Loukas, A., 2015. Experimental hookworm infection and gluten microchallenge promote tolerance in celiac disease. J. Allergy Clin. Immunol. 135 (2), 508–516.

Di Ciaula, A., Garruti, G., Lunardi Baccetto, R., Molina-Molina, E., Bonfrate, L., Wang, D.Q., Portincasa, P., 2017. Bile Acid Physiology. Ann. Hepatol. 16 (Suppl. 1: s3–105). s4–s14.

Di Sabatino, A., Ciccocioppo, R., Cupelli, F., Cinque, B., Millimaggi, D., Clarkson, M.M., Paulli, M., Cifone, M.G., Corazza, G.R., 2006. Epithelium derived interleukin 15 regulates intraepithelial lymphocyte Th1 cytokine production, cytotoxicity, and survival in coeliac disease. Gut 55 (4), 469–477.

Di Sabatino, A., Miceli, E., Dhaliwal, W., Biancheri, P., Salerno, R., Cantoro, L., Vanoli, A., De Vincenzi, M., Blanco Cdel, V., MacDonald, T.T., Corazza, G.R., 2008. Distribution, proliferation, and function of Paneth cells in uncomplicated and complicated adult celiac disease. Am. J. Clin. Pathol. 130 (1), 34–42.

Dieterich, W., Ehnis, T., Bauer, M., Donner, P., Volta, U., Riecken, E.O., Schuppan, D., 1997. Identification of tissue transglutaminase as the autoantigen of celiac disease. Nat. Med. 3 (7), 797–801.

Dieterich, W., Neurath, M.F., Zopf, Y., 2020. Intestinal ex vivo organoid culture reveals altered programmed crypt stem cells in patients with celiac disease. Sci. Rep. 10 (1), 3535.

Dolcino, M., Zanoni, G., Bason, C., Tinazzi, E., Boccola, E., Valletta, E., Contreas, G., Lunardi, C., Puccetti, A., 2013. A subset of anti-rotavirus antibodies directed against the viral protein VP7 predicts the onset of celiac disease and induces typical features of the disease in the intestinal epithelial cell line T84. Immunol. Res. 56 (2-3), 465–476.

Dolfini, E., Elli, L., Roncoroni, L., Costa, B., Colleoni, M.P., Lorusso, V., Ramponi, S., Braidotti, P., Ferrero, S., Falini, M.L., Bardella, M.T., 2005. Damaging effects of gliadin on three-dimensional cell culture model. World J. Gastroenterol. 11 (38), 5973–5977.

Dong, S., Singh, T.P., Wei, X., Yao, H., Wang, H., 2018. Protective effect of 1,25-dihydroxy vitamin D3 on pepsin-trypsin-resistant gliadin-induced tight junction injuries. Dig. Dis. Sci. 63 (1), 92–104.

Du, J., Chen, Y., Shi, Y., Liu, T., Cao, Y., Tang, Y., Ge, X., Nie, H., Zheng, C., Li, Y.C., 2015. 1,25-Dihydroxyvitamin D protects intestinal epithelial barrier by regulating the myosin light chain kinase signaling pathway. Inflamm. Bowel Dis. 21 (11), 2495–2506.

du Pre, M.F., Sollid, L.M., 2015. T-cell and B-cell immunity in celiac disease. Best Pract. Res. Clin. Gastroenterol. 29 (3), 413–423.

Dubois, P.C., Trynka, G., Franke, L., Hunt, K.A., Romanos, J., Curtotti, A., Zhernakova, A., Heap, G.A., Adany, R., Aromaa, A., Bardella, M.T., van den Berg, L.H., Bockett, N.A., de la Concha, E.G., Dema, B., Fehrmann, R.S., Fernandez-Arquero, M., Fiatal, S., Grandone, E., Green, P.M., Groen, H.J., Gwilliam, R., Houwen, R.H., Hunt, S.E., Kaukinen, K., Kelleher, D., Korponay-Szabo, I., Kurppa, K., MacMathuna, P., Maki, M., Mazzilli, M.C., McCann, O.T., Mearin, M.L., Mein, C.A., Mirza, M.M., Mistry, V., Mora, B., Morley, K.I., Mulder, C.J., Murray, J.A., Nunez, C., Oosterom, E., Ophoff, R.A., Polanco, I., Peltonen, L., Platteel, M., Rybak, A., Salomaa, V., Schweizer, J.J., Sperandeo, M.P., Tack, G.J., Turner, G., Veldink, J.H., Verbeek, W.H., Weersma, R.K., Wolters, V.M., Urcelay, E., Cukrowska, B., Greco, L., Neuhausen, S.L., McManus, R., Barisani, D., Deloukas, P., Barrett, J.C., Saavalainen, P., Wijmenga, C., van Heel, D.A., 2010. Multiple common variants for celiac disease influencing immune gene expression. Nat. Genet. 42 (4), 295–302.

Eichenberger, R.M., Ryan, S., Jones, L., Buitrago, G., Polster, R., Montes de Oca, M., Zuvelek, J., Giacomin, P.R., Dent, L.A., Engwerda, C.R., Field, M.A., Sotillo, J., Loukas, A., 2018. Hookworm secreted extracellular vesicles interact with host cells and prevent inducible colitis in mice. Front. Immunol. 9, 850.

Escudero-Hernandez, C., Martin, A., de Pedro Andres, R., Fernandez-Salazar, L., Garrote, J.A., Bernardo, D., Arranz, E., 2020. Circulating dendritic cells from celiac disease patients display a gut-homing profile and are differentially modulated by different gliadin-derived peptides. Mol. Nutr. Food Res. 64 (6), e1900989.

Esposito, C., Paparo, F., Caputo, I., Porta, R., Salvati, V.M., Mazzarella, G., Auricchio, S., Troncone, R., 2003. Expression and enzymatic activity of small intestinal tissue transglutaminase in celiac disease. Am. J. Gastroenterol. 98 (8), 1813–1820.

Esposito, S., Villella, V.R., Ferrari, E., Monzani, R., Tosco, A., Rossin, F., D'Eletto, M., Castaldo, A., Luciani, A., Silano, M., Bona, G., Marseglia, G.L., Romani, L., Piacentini, M., Raia, V., Kroemer, G., Maiuri, L., 2019. Genistein antagonizes gliadin-induced CFTR malfunction in models of celiac disease. Aging 11 (7), 2003–2019.

Fernandez-Banares, F., Monzon, H., Forne, M., 2009. A short review of malabsorption and anemia. World J. Gastroenterol. 15 (37), 4644–4652.

Fernandez-Jimenez, N., Castellanos-Rubio, A., Plaza-Izurieta, L., Gutierrez, G., Castano, L., Vitoria, J.C., Bilbao, J.R., 2010. Analysis of beta-defensin and Toll-like receptor gene copy number variation in celiac disease. Hum. Immunol. 71 (8), 833–836.

Fernandez-Jimenez, N., Garcia-Etxebarria, K., Plaza-Izurieta, L., Romero-Garmendia, I., Jauregi-Miguel, A., Legarda, M., Ecsedi, S., Castellanos-Rubio, A., Cahais, V., Cuenin, C., Degli Esposti, D., Irastorza, I., Hernandez-Vargas, H., Herceg, Z., Bilbao, J.R., 2019. The methylome of the celiac intestinal epithelium harbours genotype-independent alterations in the HLA region. Sci. Rep. 9 (1), 1298.

Ferreira, G.B., van Etten, E., Verstuyf, A., Waer, M., Overbergh, L., Gysemans, C., Mathieu, C., 2011. 1,25-Dihydroxyvitamin D3 alters murine dendritic cell behaviour in vitro and in vivo. Diabetes Metab. Res. Rev. 27 (8), 933–941.

Freeman, H.J., 2015. Iron deficiency anemia in celiac disease. World J. Gastroenterol. 21 (31), 9233–9238.

Freire, R., Ingano, L., Serena, G., Cetinbas, M., Anselmo, A., Sapone, A., Sadreyev, R.I., Fasano, A., Senger, S., 2019. Human gut derived-organoids provide model to study gluten response and effects of microbiota-derived molecules in celiac disease. Sci. Rep. 9 (1), 7029.

Frye, M., Bargon, J., Lembcke, B., Wagner, T.O., Gropp, R., 2000. Differential expression of human alpha- and beta-defensins mRNA in gastrointestinal epithelia. Eur. J. Clin. Invest. 30 (8), 695–701.

Fu, B., Wang, H., Wang, J., Barouhas, I., Liu, W., Shuboy, A., Bushinsky, D.A., Zhou, D., Favus, M.J., 2013. Epigenetic regulation of BMP2 by 1,25-dihydroxyvitamin D3 through DNA methylation and histone modification. PLoS One 8 (4), e61423.

Garner, C., Ahn, R., Ding, Y.C., Steele, L., Stoven, S., Green, P.H., Fasano, A., Murray, J.A., Neuhausen, S.L., 2014. Genome-wide association study of celiac disease in North America confirms FRMD4B as new celiac locus. PLoS One 9 (7), e101428.

Gassler, N., 2017. Paneth cells in intestinal physiology and pathophysiology. World J Gastrointest Pathophysiol 8 (4), 150–160.

Gehart, H., Clevers, H., 2019. Tales from the crypt: new insights into intestinal stem cells. Nat. Rev. Gastroenterol. Hepatol. 16 (1), 19–34.

Gerbe, F., Sidot, E., Smyth, D.J., Ohmoto, M., Matsumoto, I., Dardalhon, V., Cesses, P., Garnier, L., Pouzolles, M., Brulin, B., Bruschi, M., Harcus, Y., Zimmermann, V.S., Taylor, N., Maizels, R.M., Jay, P., 2016. Intestinal epithelial tuft cells initiate type 2 mucosal immunity to helminth parasites. Nature 529 (7585), 226–230.

Giovannini, C., Matarrese, P., Scazzocchio, B., Vari, R., D'Archivio, M., Straface, E., Masella, R., Malorni, W., De Vincenzi, M., 2003. Wheat gliadin induces apoptosis of intestinal cells via an autocrine mechanism involving Fas-Fas ligand pathway. FEBS Lett. 540 (1-3), 117–124.

Groh, V., Bahram, S., Bauer, S., Herman, A., Beauchamp, M., Spies, T., 1996. Cell stress-regulated human major histocompatibility complex class I gene expressed in gastrointestinal epithelium. Proc. Natl. Acad. Sci. U. S. A. 93 (22), 12445–12450.

Gromme, M., Neefjes, J., 2002. Antigen degradation or presentation by MHC class I molecules via classical and non-classical pathways. Mol. Immunol. 39 (3-4), 181–202.

Gujral, N., Suh, J.W., Sunwoo, H.H., 2015. Effect of anti-gliadin IgY antibody on epithelial intestinal integrity and inflammatory response induced by gliadin. BMC Immunol. 16, 41.

Haber, A.L., Biton, M., Rogel, N., Herbst, R.H., Shekhar, K., Smillie, C., Burgin, G., Delorey, T.M., Howitt, M.R., Katz, Y., Tirosh, I., Beyaz, S., Dionne, D., Zhang, M., Raychowdhury, R., Garrett, W.S., Rozenblatt-Rosen, O., Shi, H.N., Yilmaz, O., Xavier, R.J., Regev, A., 2017. A single-cell survey of the small intestinal epithelium. Nature 551 (7680), 333–339.

Halttunen, T., Maki, M., 1999. Serum immunoglobulin A from patients with celiac disease inhibits human T84 intestinal crypt epithelial cell differentiation. Gastroenterology 116 (3), 566–572.

He, X.C., Zhang, J., Tong, W.G., Tawfik, O., Ross, J., Scoville, D.H., Tian, Q., Zeng, X., He, X., Wiedemann, L.M., Mishina, Y., Li, L., 2004. BMP signaling inhibits intestinal stem cell self-renewal through suppression of Wnt-beta-catenin signaling. Nat. Genet. 36 (10), 1117–1121.

Hoekstra, J.H., Groot-Loonen, J.J., van der Weij, A., Hoogerbrugge, P.M., Kooy, Y., Koning, F., 2010. Successful treatment of coeliac disease by allogeneic haematopoietic stem cell transplantation. J. Pediatr. Gastroenterol. Nutr. 51 (6), 793–794.

Hudson, D.A., Cornell, H.J., Purdham, D.R., Rolles, C.J., 1976. Non-specific cytotoxicity of wheat gliadin components towards cultured human cells. Lancet 1 (7955), 339–341.

Hue, S., Mention, J.J., Monteiro, R.C., Zhang, S., Cellier, C., Schmitz, J., Verkarre, V., Fodil, N., Bahram, S., Cerf-Bensussan, N., Caillat-Zucman, S., 2004. A direct role for NKG2D/MICA interaction in villous atrophy during celiac disease. Immunity 21 (3), 367–377.

Huh, W.J., Roland, J.T., Asai, M., Kaji, I., 2020. Distribution of duodenal tuft cells is altered in pediatric patients with acute and chronic enteropathy. Biomed. Res. 41 (2), 113–118.

Hunt, K.A., Zhernakova, A., Turner, G., Heap, G.A., Franke, L., Bruinenberg, M., Romanos, J., Dinesen, L.C., Ryan, A.W., Panesar, D., Gwilliam, R., Takeuchi, F.,

McLaren, W.M., Holmes, G.K., Howdle, P.D., Walters, J.R., Sanders, D.S., Playford, R.J., Trynka, G., Mulder, C.J., Mearin, M.L., Verbeek, W.H., Trimble, V., Stevens, F.M., O'Morain, C., Kennedy, N.P., Kelleher, D., Pennington, D.J., Strachan, D.P., McArdle, W.L., Mein, C.A., Wapenaar, M.C., Deloukas, P., McGinnis, R., McManus, R., Wijmenga, C., van Heel, D.A., 2008. Newly identified genetic risk variants for celiac disease related to the immune response. Nat. Genet. 40 (4), 395–402.

Intrieri, M., Rinaldi, A., Scudiero, O., Autiero, G., Castaldo, G., Nardone, G., 2010. Low expression of human beta-defensin 1 in duodenum of celiac patients is partially restored by a gluten-free diet. Clin. Chem. Lab. Med. 48 (4), 489–492.

Juuti-Uusitalo, K., Maki, M., Kainulainen, H., Isola, J., Kaukinen, K., 2007. Gluten affects epithelial differentiation-associated genes in small intestinal mucosa of coeliac patients. Clin. Exp. Immunol. 150 (2), 294–305.

Kalliomaki, M., Satokari, R., Lahteenoja, H., Vahamiko, S., Gronlund, J., Routi, T., Salminen, S., 2012. Expression of microbiota, Toll-like receptors, and their regulators in the small intestinal mucosa in celiac disease. J. Pediatr. Gastroenterol. Nutr. 54 (6), 727–732.

Khaleghi, S., Ju, J.M., Lamba, A., Murray, J.A., 2016. The potential utility of tight junction regulation in celiac disease: focus on larazotide acetate. Therap. Adv. Gastroenterol. 9 (1), 37–49.

Kierbel, A., Gassama-Diagne, A., Rocha, C., Radoshevich, L., Olson, J., Mostov, K., Engel, J., 2007. Pseudomonas aeruginosa exploits a PIP3-dependent pathway to transform apical into basolateral membrane. J. Cell Biol. 177 (1), 21–27.

Kline, R.M., Neudorf, S.M., Baron, H.I., 2007. Correction of celiac disease after allogeneic hematopoietic stem cell transplantation for acute myelogenous leukemia. Pediatrics 120 (4), e1120–e1122.

Koch, S., Nusrat, A., 2012. The life and death of epithelia during inflammation: lessons learned from the gut. Annu. Rev. Pathol. 7, 35–60.

Kovacsovics-Bankowski, M., Rock, K.L., 1995. A phagosome-to-cytosol pathway for exogenous antigens presented on MHC class I molecules. Science 267 (5195), 243–246.

Kumar, V., Gutierrez-Achury, J., Kanduri, K., Almeida, R., Hrdlickova, B., Zhernakova, D.V., Westra, H.J., Karjalainen, J., Ricano-Ponce, I., Li, Y., Stachurska, A., Tigchelaar, E.F., Abdulahad, W.H., Lahdesmaki, H., Hofker, M.H., Zhernakova, A., Franke, L., Lahesmaa, R., Wijmenga, C., Withoff, S., 2015. Systematic annotation of celiac disease loci refines pathological pathways and suggests a genetic explanation for increased interferon-gamma levels. Hum. Mol. Genet. 24 (2), 397–409.

Lammers, K.M., Lu, R., Brownley, J., Lu, B., Gerard, C., Thomas, K., Rallabhandi, P., Shea-Donohue, T., Tamiz, A., Alkan, S., Netzel-Arnett, S., Antalis, T., Vogel, S.N., Fasano, A., 2008. Gliadin induces an increase in intestinal permeability and zonulin release by binding to the chemokine receptor CXCR3. Gastroenterology 135 (1), 194–204. e193.

Lamouille, S., Xu, J., Derynck, R., 2014. Molecular mechanisms of epithelial-mesenchymal transition. Nat. Rev. Mol. Cell Biol. 15 (3), 178–196.

Lebreton, C., Menard, S., Abed, J., Moura, I.C., Coppo, R., Dugave, C., Monteiro, R.C., Fricot, A., Traore, M.G., Griffin, M., Cellier, C., Malamut, G., Cerf-Bensussan, N., Heyman, M., 2012. Interactions among secretory immunoglobulin A, CD71, and transglutaminase-2 affect permeability of intestinal epithelial cells to gliadin peptides. Gastroenterology 143 (3), 698–707. e694.

Lerner, A., Aminov, R., Matthias, T., 2017. Transglutaminases in Dysbiosis As Potential Environmental Drivers of Autoimmunity. Front. Microbiol. 8, 66.

Lerner, A., Matthias, T., 2015. Changes in intestinal tight junction permeability associated with industrial food additives explain the rising incidence of autoimmune disease. Autoimmun. Rev. 14 (6), 479–489.

Li, N., Nakauka-Ddamba, A., Tobias, J., Jensen, S.T., Lengner, C.J., 2016. Mouse label-retaining cells are molecularly and functionally distinct from reserve intestinal stem cells. Gastroenterology 151 (2), 298–310. e297.

Li, N., Yousefi, M., Nakauka-Ddamba, A., Jain, R., Tobias, J., Epstein, J.A., Jensen, S.T., Lengner, C.J., 2014. Single-cell analysis of proxy reporter allele-marked epithelial cells establishes intestinal stem cell hierarchy. Stem Cell Rep. 3 (5), 876–891.

Li, W., Peregrina, K., Houston, M., Augenlicht, L.H., 2020. Vitamin D and the nutritional environment in functions of intestinal stem cells: Implications for tumorigenesis and prevention. J. Steroid Biochem. Mol. Biol. 198, 105556.

Lindemans, C.A., Calafiore, M., Mertelsmann, A.M., O'Connor, M.H., Dudakov, J.A., Jenq, R.R., Velardi, E., Young, L.F., Smith, O.M., Lawrence, G., Ivanov, J.A., Fu, Y.Y., Takashima, S., Hua, G., Martin, M.L., O'Rourke, K.P., Lo, Y.H., Mokry, M., Romera-Hernandez, M., Cupedo, T., Dow, L., Nieuwenhuis, E.E., Shroyer, N.F., Liu, C., Kolesnick, R., van den Brink, M.R.M., Hanash, A.M., 2015. Interleukin-22 promotes intestinal-stem-cell-mediated epithelial regeneration. Nature 528 (7583), 560–564.

Luciani, A., Villella, V.R., Esposito, S., Brunetti-Pierri, N., Medina, D., Settembre, C., Gavina, M., Pulze, L., Giardino, I., Pettoello-Mantovani, M., D'Apolito, M., Guido, S., Masliah, E., Spencer, B., Quaratino, S., Raia, V., Ballabio, A., Maiuri, L., 2010a. Defective CFTR induces aggresome formation and lung inflammation in cystic fibrosis through ROS-mediated autophagy inhibition. Nat. Cell Biol. 12 (9), 863–875.

Luciani, A., Villella, V.R., Vasaturo, A., Giardino, I., Pettoello-Mantovani, M., Guido, S., Cexus, O.N., Peake, N., Londei, M., Quaratino, S., Maiuri, L., 2010b. Lysosomal accumulation of gliadin p31-43 peptide induces oxidative stress and tissue transglutaminase-mediated PPARgamma downregulation in intestinal epithelial cells and coeliac mucosa. Gut 59 (3), 311–319.

Ma, T.Y., Iwamoto, G.K., Hoa, N.T., Akotia, V., Pedram, A., Boivin, M.A., Said, H.M., 2004. TNF-alpha-induced increase in intestinal epithelial tight junction permeability requires NF-kappa B activation. Am. J. Physiol. Gastrointest. Liver Physiol. 286 (3), G367–G376.

Magni, S., Buoli Comani, G., Elli, L., Vanessi, S., Ballarini, E., Nicolini, G., Rusconi, M., Castoldi, M., Meneveri, R., Muckenthaler, M.U., Bardella, M.T., Barisani, D., 2014. miRNAs affect the expression of innate and adaptive immunity proteins in celiac disease. Am. J. Gastroenterol. 109 (10), 1662–1674.

Maiuri, L., Ciacci, C., Auricchio, S., Brown, V., Quaratino, S., Londei, M., 2000. Interleukin 15 mediates epithelial changes in celiac disease. Gastroenterology 119 (4), 996–1006.

Maiuri, L., Ciacci, C., Ricciardelli, I., Vacca, L., Raia, V., Auricchio, S., Picard, J., Osman, M., Quaratino, S., Londei, M., 2003a. Association between innate response to gliadin and activation of pathogenic T cells in coeliac disease. Lancet 362 (9377), 30–37.

Maiuri, M.C., De Stefano, D., Mele, G., Iovine, B., Bevilacqua, M.A., Greco, L., Auricchio, S., Carnuccio, R., 2003b. Gliadin increases iNOS gene expression in interferon-gamma-stimulated RAW 264.7 cells through a mechanism involving NF-kappa B. Naunyn Schmiedebergs Arch. Pharmacol. 368 (1), 63–71.

Malamut, G., Cording, S., Cerf-Bensussan, N., 2019. Recent advances in celiac disease and refractory celiac disease. F1000Res 8, F1000 Faculty Rev-969. https://doi.org/10.12688/f1000research.18701.1. eCollection 2019.

Malterre, T., 2009. Digestive and nutritional considerations in celiac disease: could supplementation help? Altern. Med. Rev. 14 (3), 247–257.

Manai, F., Azzalin, A., Gabriele, F., Martinelli, C., Morandi, M., Biggiogera, M., Bozzola, M., Comincini, S., 2018. The in vitro effects of enzymatic digested Gliadin on the functionality of the autophagy process. Int. J. Mol. Sci. 19 (2).

Manai, F., Azzalin, A., Morandi, M., Riccardi, V., Zanoletti, L., Dei Giudici, M., Gabriele, F., Martinelli, C., Bozzola, M., Comincini, S., 2019. Trehalose modulates autophagy process to counteract gliadin cytotoxicity in an in vitro celiac disease model. Cell 8 (4).

Martin-Masot, R., Nestares, M.T., Diaz-Castro, J., Lopez-Aliaga, I., Alferez, M.J.M., Moreno-Fernandez, J., Maldonado, J., 2019. Multifactorial etiology of anemia in celiac disease and effect of gluten-free diet: a comprehensive review. Nutrients 11 (11), 348.

Mastrandrea, F., Semeraro, F.P., Coradduzza, G., Manelli, M., Scarcia, G., Pezzuto, F., Serio, G., 2008. CD34+ hemopoietic precursor and stem cells traffic in peripheral blood of celiac patients is significantly increased but not directly related to epithelial damage severity. Eur. Ann. Allergy Clin. Immunol. 40 (3), 90–103.

Mazzarella, G., Stefanile, R., Camarca, A., Giliberti, P., Cosentini, E., Marano, C., Iaquinto, G., Giardullo, N., Auricchio, S., Sette, A., Troncone, R., Gianfrani, C., 2008. Gliadin activates HLA class I-restricted CD8+ T cells in celiac disease intestinal mucosa and induces the enterocyte apoptosis. Gastroenterology 134 (4), 1017–1027.

McGuckin, M.A., Eri, R.D., Das, I., Lourie, R., Florin, T.H., 2010. ER stress and the unfolded protein response in intestinal inflammation. Am. J. Physiol. Gastrointest. Liver Physiol. 298 (6), G820–G832.

McKenna, L.B., Schug, J., Vourekas, A., McKenna, J.B., Bramswig, N.C., Friedman, J.R., Kaestner, K.H., 2010. MicroRNAs control intestinal epithelial differentiation, architecture, and barrier function. Gastroenterology 139 (5), 1654–1664. e1651.

Mehta, K., Kumar, A., Kim, H.I., 2010. Transglutaminase 2: a multi-tasking protein in the complex circuitry of inflammation and cancer. Biochem. Pharmacol. 80 (12), 1921–1929.

Meresse, B., Curran, S.A., Ciszewski, C., Orbelyan, G., Setty, M., Bhagat, G., Lee, L., Tretiakova, M., Semrad, C., Kistner, E., Winchester, R.J., Braud, V., Lanier, L.L., Geraghty, D.E., Green, P.H., Guandalini, S., Jabri, B., 2006. Reprogramming of CTLs into natural killer-like cells in celiac disease. J. Exp. Med. 203 (5), 1343–1355.

Naik, R.D., Seidner, D.L., Adams, D.W., 2018. Nutritional Consideration in Celiac Disease and Nonceliac Gluten Sensitivity. Gastroenterol. Clin. North Am. 47 (1), 139–154.

Nanayakkara, M., Lania, G., Maglio, M., Auricchio, R., De Musis, C., Discepolo, V., Miele, E., Jabri, B., Troncone, R., Auricchio, S., Barone, M.V., 2018. P31-43, an undigested gliadin peptide, mimics and enhances the innate immune response to viruses and interferes with endocytic trafficking: a role in celiac disease. Sci. Rep. 8 (1), 10821.

Oates, P.S., West, A.R., 2006. Heme in intestinal epithelial cell turnover, differentiation, detoxification, inflammation, carcinogenesis, absorption and motility. World J. Gastroenterol. 12 (27), 4281–4295.

Oittinen, M., Popp, A., Kurppa, K., Lindfors, K., Maki, M., Kaikkonen, M.U., Viiri, K., 2017. Polycomb Repressive Complex 2 Enacts Wnt Signaling in Intestinal Homeostasis and Contributes to the Instigation of Stemness in Diseases Entailing Epithelial Hyperplasia or Neoplasia. Stem Cells 35 (2), 445–457.

Perera, L., Shao, L., Patel, A., Evans, K., Meresse, B., Blumberg, R., Geraghty, D., Groh, V., Spies, T., Jabri, B., Mayer, L., 2007. Expression of nonclassical class I molecules by intestinal epithelial cells. Inflamm. Bowel Dis. 13 (3), 298–307.

Perez, S., Talens-Visconti, R., Rius-Perez, S., Finamor, I., Sastre, J., 2017. Redox signaling in the gastrointestinal tract. Free Radic. Biol. Med. 104, 75–103.

Peterson, L.W., Artis, D., 2014. Intestinal epithelial cells: regulators of barrier function and immune homeostasis. Nat. Rev. Immunol. 14 (3), 141–153.
Pietz, G., De, R., Hedberg, M., Sjoberg, V., Sandstrom, O., Hernell, O., Hammarstrom, S., Hammarstrom, M.L., 2017. Immunopathology of childhood celiac disease-Key role of intestinal epithelial cells. PLoS One 12 (9), e0185025.
Ponziani, F.R., Cazzato, I.A., Danese, S., Fagiuoli, S., Gionchetti, P., Annicchiarico, B.E., D'Aversa, F., Gasbarrini, A., 2012. Folate in gastrointestinal health and disease. Eur. Rev. Med. Pharmacol. Sci. 16 (3), 376–385.
Pufulete, M., Al-Ghnaniem, R., Khushal, A., Appleby, P., Harris, N., Gout, S., Emery, P.W., Sanders, T.A., 2005. Effect of folic acid supplementation on genomic DNA methylation in patients with colorectal adenoma. Gut 54 (5), 648–653.
Rauhavirta, T., Hietikko, M., Salmi, T., Lindfors, K., 2019. Transglutaminase 2 and Transglutaminase 2 Autoantibodies in Celiac Disease: a Review. Clin Rev Allergy Immunol 57 (1), 23–38.
Rios, D., Wood, M.B., Li, J., Chassaing, B., Gewirtz, A.T., Williams, I.R., 2016. Antigen sampling by intestinal M cells is the principal pathway initiating mucosal IgA production to commensal enteric bacteria. Mucosal Immunol. 9 (4), 907–916.
Rodriguez-Colman, M.J., Schewe, M., Meerlo, M., Stigter, E., Gerrits, J., Pras-Raves, M., Sacchetti, A., Hornsveld, M., Oost, K.C., Snippert, H.J., Verhoeven-Duif, N., Fodde, R., Burgering, B.M., 2017. Interplay between metabolic identities in the intestinal crypt supports stem cell function. Nature 543 (7645), 424–427.
Roncoroni, L., Elli, L., Bardella, M.T., Perrucci, G., Ciulla, M., Lombardo, V., Tomba, C., Conte, D., Doneda, L., 2013. Extracellular matrix proteins and displacement of cultured fibroblasts from duodenal biopsies in celiac patients and controls. J. Transl. Med. 11, 91.
Rubio, C.A., 2011. Lysozyme-rich mucus metaplasia in duodenal crypts supersedes Paneth cells in celiac disease. Virchows Arch. 459 (3), 339–346.
San-Pedro, J.I., Bilbao, J.R., Perez de Nanclares, G., Vitoria, J.C., Martul, P., Castano, L., 2005. Heterogeneity of vitamin D receptor gene association with celiac disease and type 1 diabetes mellitus. Autoimmunity 38 (6), 439–444.
Santin, I., Jauregi-Miguel, A., Velayos, T., Castellanos-Rubio, A., Garcia-Etxebarria, K., Romero-Garmendia, I., Fernandez-Jimenez, N., Irastorza, I., Castano, L., Bilbao, J.R., 2018. Celiac diasease-associated lncRNA named HCG14 regulates NOD1 expression in intestinal cells. J. Pediatr. Gastroenterol. Nutr. 67 (2), 225–231.
Sasaki, N., Sachs, N., Wiebrands, K., Ellenbroek, S.I., Fumagalli, A., Lyubimova, A., Begthel, H., van den Born, M., van Es, J.H., Karthaus, W.R., Li, V.S., Lopez-Iglesias, C., Peters, P.J., van Rheenen, J., van Oudenaarden, A., Clevers, H., 2016. Reg4+ deep crypt secretory cells function as epithelial niche for Lgr5+ stem cells in colon. Proc. Natl. Acad. Sci. U. S. A. 113 (37), E5399–E5407.
Sato, T., van Es, J.H., Snippert, H.J., Stange, D.E., Vries, R.G., van den Born, M., Barker, N., Shroyer, N.F., van de Wetering, M., Clevers, H., 2011. Paneth cells constitute the niche for Lgr5 stem cells in intestinal crypts. Nature 469 (7330), 415–418.
Sato, T., Vries, R.G., Snippert, H.J., van de Wetering, M., Barker, N., Stange, D.E., van Es, J.H., Abo, A., Kujala, P., Peters, P.J., Clevers, H., 2009. Single Lgr5 stem cells build crypt-villus structures in vitro without a mesenchymal niche. Nature 459 (7244), 262–265.
Schippa, S., Iebba, V., Barbato, M., Di Nardo, G., Totino, V., Checchi, M.P., Longhi, C., Maiella, G., Cucchiara, S., Conte, M.P., 2010. A distinctive 'microbial signature' in celiac pediatric patients. BMC Microbiol. 10, 175.
Schwank, G., Koo, B.K., Sasselli, V., Dekkers, J.F., Heo, I., Demircan, T., Sasaki, N., Boymans, S., Cuppen, E., van der Ent, C.K., Nieuwenhuis, E.E., Beekman, J.M., Clevers, H., 2013. Functional repair of CFTR by CRISPR/Cas9 in intestinal stem cell organoids of cystic fibrosis patients. Cell Stem Cell 13 (6), 653–658.

Senger, S., Sapone, A., Fiorentino, M.R., Mazzarella, G., Lauwers, G.Y., Fasano, A., 2015. Celiac disease histopathology recapitulates hedgehog downregulation, consistent with wound healing processes activation. PLoS One 10 (12), e0144634.

Serena, G., Huynh, D., Lima, R.S., Vise, L.M., Freire, R., Ingano, L., Leonard, M.M., Senger, S., Fasano, A., 2019. Intestinal epithelium modulates macrophage response to gliadin in celiac disease. Front. Nutr. 6, 167.

Shalimar, D.M., Das, P., Sreenivas, V., Gupta, S.D., Panda, S.K., Makharia, G.K., 2013. Mechanism of villous atrophy in celiac disease: role of apoptosis and epithelial regeneration. Arch. Pathol. Lab. Med. 137 (9), 1262–1269.

Sharma, N., Begum, J., Eksteen, B., Elagib, A., Brookes, M., Cooper, B.T., Tselepis, C., Iqbal, T.H., 2009. Differential ferritin expression is associated with iron deficiency in coeliac disease. Eur. J. Gastroenterol. Hepatol. 21 (7), 794–804.

Stamnaes, J., Sollid, L.M., 2015. Celiac disease: Autoimmunity in response to food antigen. Semin. Immunol. 27 (5), 343–352.

Stanifer, M.L., Rippert, A., Kazakov, A., Willemsen, J., Bucher, D., Bender, S., Bartenschlager, R., Binder, M., Boulant, S., 2016. Reovirus intermediate subviral particles constitute a strategy to infect intestinal epithelial cells by exploiting TGF-beta dependent pro-survival signaling. Cell. Microbiol. 18 (12), 1831–1845.

Stricker, S., de Laffolie, J., Rudloff, S., Komorowski, L., Zimmer, K.P., 2019. Intracellular localization of microbial transglutaminase and its influence on the transport of gliadin in enterocytes. J. Pediatr. Gastroenterol. Nutr. 68 (3), e43–e50.

Sugimoto, S., Ohta, Y., Fujii, M., Matano, M., Shimokawa, M., Nanki, K., Date, S., Nishikori, S., Nakazato, Y., Nakamura, T., Kanai, T., Sato, T., 2018. Reconstruction of the human colon epithelium in vivo. Cell Stem Cell 22 (2), 171–176. e175.

Tack, G.J., Wondergem, M.J., Al-Toma, A., Verbeek, W.H., Schmittel, A., Machado, M.V., Perri, F., Ossenkoppele, G.J., Huijgens, P.C., Schreurs, M.W., Mulder, C.J., Visser, O.J., 2011. Auto-SCT in refractory celiac disease type II patients unresponsive to cladribine therapy. Bone Marrow Transplant. 46 (6), 840–846.

Taha, A.S., Faccenda, E., Angerson, W.J., Balsitis, M., Kelly, R.W., 2005. Natural antibiotic expression in celiac disease- -correlation with villous atrophy and response to a gluten-free diet. Dig. Dis. Sci. 50 (4), 791–795.

Tordesillas, L., Berin, M.C., 2018. Mechanisms of oral tolerance. Clin Rev Allergy Immunol 55 (2), 107–117.

Trivedi, M.S., Shah, J.S., Al-Mughairy, S., Hodgson, N.W., Simms, B., Trooskens, G.A., Van Criekinge, W., Deth, R.C., 2014. Food-derived opioid peptides inhibit cysteine uptake with redox and epigenetic consequences. J. Nutr. Biochem. 25 (10), 1011–1018.

Trynka, G., Hunt, K.A., Bockett, N.A., Romanos, J., Mistry, V., Szperl, A., Bakker, S.F., Bardella, M.T., Bhaw-Rosun, L., Castillejo, G., de la Concha, E.G., de Almeida, R.C., Dias, K.R., van Diemen, C.C., Dubois, P.C., Duerr, R.H., Edkins, S., Franke, L., Fransen, K., Gutierrez, J., Heap, G.A., Hrdlickova, B., Hunt, S., Plaza Izurieta, L., Izzo, V., Joosten, L.A., Langford, C., Mazzilli, M.C., Mein, C.A., Midah, V., Mitrovic, M., Mora, B., Morelli, M., Nutland, S., Nunez, C., Onengut-Gumuscu, S., Pearce, K., Platteel, M., Polanco, I., Potter, S., Ribes-Koninckx, C., Ricano-Ponce, I., Rich, S.S., Rybak, A., Santiago, J.L., Senapati, S., Sood, A., Szajewska, H., Troncone, R., Varade, J., Wallace, C., Wolters, V.M., Zhernakova, A., Spanish Consortium on the Genetics of Coeliac Disease (CEGEC), PreventCD Study Group, Wellcome Trust Case Control Consortium (WTCCC), Thelma, B.K., Cukrowska, B., Urcelay, E., Bilbao, J.R., Mearin, M.L., Barisani, D., Barrett, J.C., Plagnol, V., Deloukas, P., Wijmenga, C., van Heel, D.A., 2011. Dense genotyping identifies and localizes multiple common and rare variant association signals in celiac disease. Nat. Genet. 43 (12), 1193–1201.

Vaira, V., Gaudioso, G., Laginestra, M.A., Terrasi, A., Agostinelli, C., Bosari, S., Di Sabatino, A., Vanoli, A., Paulli, M., Ferrero, S., Roncoroni, L., Lombardo, V., Perera, L.P., Fabris, S., Vecchi, M., Pileri, S., Elli, L., 2020. Deregulation of miRNAs-cMYC circuits is a key event in refractory celiac disease type-2 lymphomagenesis. Clin. Sci. (Lond.) 134 (10), 1151–1166.

Vaira, V., Roncoroni, L., Barisani, D., Gaudioso, G., Bosari, S., Bulfamante, G., Doneda, L., Conte, D., Tomba, C., Bardella, M.T., Ferrero, S., Locatelli, M., Elli, L., 2014. microRNA profiles in coeliac patients distinguish different clinical phenotypes and are modulated by gliadin peptides in primary duodenal fibroblasts. Clin. Sci. (Lond.) 126 (6), 417–423.

Valitutti, F., Cucchiara, S., Fasano, A., 2019. Celiac disease and the microbiome. Nutrients 11 (10).

Vici, G., Belli, L., Biondi, M., Polzonetti, V., 2016. Gluten free diet and nutrient deficiencies: A review. Clin. Nutr. 35 (6), 1236–1241.

Vici, G., Camilletti, D., Polzonetti, V., 2020. Possible role of vitamin D in celiac disease onset. Nutrients 12 (4).

Villella, V.R., Venerando, A., Cozza, G., Esposito, S., Ferrari, E., Monzani, R., Spinella, M.C., Oikonomou, V., Renga, G., Tosco, A., Rossin, F., Guido, S., Silano, M., Garaci, E., Chao, Y.K., Grimm, C., Luciani, A., Romani, L., Piacentini, M., Raia, V., Kroemer, G., Maiuri, L., 2019. A pathogenic role for cystic fibrosis transmembrane conductance regulator in celiac disease. EMBO J. 38 (2).

von Moltke, J., Ji, M., Liang, H.E., Locksley, R.M., 2016. Tuft-cell-derived IL-25 regulates an intestinal ILC2-epithelial response circuit. Nature 529 (7585), 221–225.

Wacklin, P., Kaukinen, K., Tuovinen, E., Collin, P., Lindfors, K., Partanen, J., Maki, M., Matto, J., 2013. The duodenal microbiota composition of adult celiac disease patients is associated with the clinical manifestation of the disease. Inflamm. Bowel Dis. 19 (5), 934–941.

Wang, Z., Griffin, M., 2012. TG2, a novel extracellular protein with multiple functions. Amino Acids 42 (2-3), 939–949.

Wu, S., Zhang, Y.G., Lu, R., Xia, Y., Zhou, D., Petrof, E.O., Claud, E.C., Chen, D., Chang, E.B., Carmeliet, G., Sun, J., 2015. Intestinal epithelial vitamin D receptor deletion leads to defective autophagy in colitis. Gut 64 (7), 1082–1094.

Zhao, J., de Vera, J., Narushima, S., Beck, E.X., Palencia, S., Shinkawa, P., Kim, K.A., Liu, Y., Levy, M.D., Berg, D.J., Abo, A., Funk, W.D., 2007. R-spondin1, a novel intestinotrophic mitogen, ameliorates experimental colitis in mice. Gastroenterology 132 (4), 1331–1343.

Zimmermann, C., Rudloff, S., Lochnit, G., Arampatzi, S., Maison, W., Zimmer, K.P., 2014. Epithelial transport of immunogenic and toxic gliadin peptides in vitro. PLoS One 9 (11), e113932.

Zinszner, H., Kuroda, M., Wang, X., Batchvarova, N., Lightfoot, R.T., Remotti, H., Stevens, J.L., Ron, D., 1998. CHOP is implicated in programmed cell death in response to impaired function of the endoplasmic reticulum. Genes Dev. 12 (7), 982–995.

CHAPTER SIX

The gliadin p31–43 peptide: Inducer of multiple proinflammatory effects

Fernando Gabriel Chirdo[a,*], Salvatore Auricchio[b], Riccardo Troncone[b,c], and Maria Vittoria Barone[b,c]

[a]Departamento de Ciencias Biológicas, Facultad de Ciencias Exactas, Instituto de Estudios Inmunológicos y Fisiopatológicos—IIFP (UNLP-CONICET), La Plata, Argentina
[b]European Laboratory for the Investigation of Food Induced Diseases (ELFID), University Federico II, Naples, Italy
[c]Department of Translational Medical Science, University Federico II, Naples, Italy
*Corresponding author: e-mail address: fchirdo@gmail.com

Contents

International Review of Cell and Molecular Biology, Volume 358
ISSN 1937-6448
https://doi.org/10.1016/bs.ircmb.2020.10.003

Abstract

Coeliac disease (CD) is the prototype of an inflammatory chronic disease induced by food. In this context, gliadin p31–43 peptide comes into the spotlight as an important player of the inflammatory/innate immune response to gliadin in CD. The p31–43 peptide is part of the p31–55 peptide from α-gliadins that remains undigested for a long time, and can be present in the small intestine after ingestion of a gluten-containing diet. Different biophysical methods and molecular dynamic simulations have shown that p31–43 spontaneously forms oligomeric nanostructures, whereas experimental approaches using *in vitro* assays, mouse models, and human duodenal tissues have shown that p31–43 is able to induce different forms of cellular stress by driving multiple inflammatory pathways. Increased proliferative activity of the epithelial cells in the crypts, enterocyte stress, activation of TG2, induction of Ca^{2+}, IL-15, and NFκB signaling, inhibition of CFTR, alteration of vesicular trafficking, and activation of the inflammasome platform are some of the biological effects of p31–43, which, in the presence of appropriate genetic susceptibility and environmental factors, may act together to drive CD.

1. Introduction

Coeliac disease (CD) is an immune-mediated enteropathy triggered in genetically susceptible individuals by a group of wheat proteins (commonly called gluten) and related prolamins from toxic cereals (Sollid and Jabri, 2013). It is characterized by a variable combination of gluten-induced symptoms, generation of CD-associated autoantibodies, and enteropathy (Husby et al., 2020; Lindfors et al., 2019). A central role in the pathogenesis of CD is played by the HLA-restricted gliadin-specific intestinal T cell response (Iversen and Sollid, 2020; Shan et al., 2002). The molecular basis of the interaction between gliadin peptides and HLA-DQ2.5 or HLA-DQ8 molecules, (*HLA-DQA1*05/HLA-DQB1*02:01* and *HLA-DQA1*03:01/HLA-DQB1*03:02*, respectively), as well as the role of transglutaminase 2 (TG2) in converting neutral glutamine into negatively charged residues, and consequently, strongly enhancing the peptide affinity to the HLA susceptibility alleles, have been well established (Molberg et al., 1998). Poor digestion together with deamidation can explain why strongly immunogenic peptides are generated, and several gliadin peptides that are preferentially recognized in the context of HLA molecules have been identified (Sollid et al., 2012, 2020). Furthermore, the T cell receptor expressed by gliadin-specific T cells has been characterized, and the rules governing the molecular interactions with gliadin T cell epitopes have been established (Qiao et al., 2014). The cytokine pattern involved has also been defined, and a Th1 phenotype dominated by IFNγ has been attributed to intestinal gliadin-specific T cell clones expanded from the coeliac mucosa (Jabri and Sollid, 2017).

Despite such progress, a central question remains still unanswered, namely, why a proinflammatory T cell response is generated instead of a regulatory response, which normally promotes oral tolerance to dietary protein antigens. Dendritic cells play a central role in determining the outcome of the immune response (Mowat and Agace, 2014). In an inflamed environment enriched in cytokines such as IL-15 or type I interferons, T cells tend to acquire a proinflammatory phenotype. Recently, elegant studies have been conducted in mice demonstrating that mucosal inflammation due to reovirus infection may disrupt oral tolerance by suppressing the regulatory T cell (Treg) conversion and promoting Th1 immunity against gliadin (Bouziat et al., 2017).

The factors that create a proinflammatory environment in the CD intestine, leading to an expansion of gliadin-specific T cells in genetically susceptible individuals and further shifting them toward a proinflammatory phenotype, remain to be identified. Many mice-based studies as mentioned above as well as epidemiological data have suggested viral infections as one of such factors (Stene et al., 2006). However, a number of other factors may contribute to the generation of a "sterile" inflammation, such as metabolic, chemical, and physical factors, drugs, alterations of the mucosal barrier, and changes in the microbiota. Dietary proteins are also a candidate for this role: wheat proteins, such as ATIs (amylase trypsin inhibitors), have been shown to activate innate immunity through TLRs (Junker et al., 2012), but most of the evidence points to gliadin itself and in particular, a peptide from the N-terminal portion of α-gliadin, named peptide p31–43. In addition, the capacity to "stress" epithelial cells could play a decisive role for the initiation of the effector destructive phase of the disease. However, it remains to be explained why CD patients are particularly "sensitive" to gliadin. Both genetic and environmental factors have been advocated.

This review aims at discussing: (1) the available evidence in relation to the biological activity of p31–43, and particularly its proinflammatory properties and its ability to trigger innate immunity, (2) the mechanisms underlying such activity at the cellular level, and (3) the specific susceptibility of coeliac patients to such biological activities.

2. Wheat prolamins: A very particular group of proteins

Wheat is one of the most popular cereal crops worldwide, both in terms of production and utilization. Wheat flour contains the storage proteins present in the endosperm of the grain and is a major source of protein and dietary fibers important for human nutrition (Shewry, 2019).

As originally proposed by Osborne (1908), wheat proteins can be classified in groups according to their solubility in different solvents: albumins in water, globulins in saline solutions, and gliadins in 60–70% aqueous ethanol. The molecular mass of gliadins varies from about 30 to 75 kDa. The proteins that require more aggressive denaturing solvents in order to be solubilized are collectively termed glutenins. They have bigger structures with molecular masses ranging from 30 to 90 kDa, and can be divided into high (HMW) and low (LMW) molecular weight glutenins (Shewry and Halford, 2002) (Fig. 1A). Both gliadins and glutenins have an uncommonly high content of proline and glutamine residues, and for this reason, they are called prolamins. One of the most used classification of gliadins is based on electrophoresis in acidic conditions. α-, β-, γ-, and ω-gliadins are defined in decreasing order of mobility at pH = 3 (Bernardin et al., 1967).

Pioneering studies from Kasarda et al. (1967) characterized the specific ability of α-gliadins to self-assemble and form fibrils. After laborious biochemical techniques, Kasarda et al. (1967) and Bernardin et al. (1967) optimized a procedure to purify α-gliadins, and finally described the first complete sequence of a member of the α-gliadin type, namely, A-gliadin (Kasarda et al., 1984). This work was instrumental for our current knowledge on the molecular basis of the functional properties of wheat proteins. The gliadin sequence showed well-defined regions containing unique sequences, poly-glutamine domains, and repetitive sequences (Fig. 1B). The repetitive sequences in the N-terminal end are formed by motifs containing proline and glutamine residues, such as PQQPFPQQ (P: proline, Q: glutamine, F: phenylalanine) found in γ-gliadins. These sequences are responsible for many of the biochemical and immunochemical properties of these proteins (Shewry and Halford, 2002). As we discuss later, this region also contains the most studied peptides related with CD pathogenesis.

In addition to their solubility in different solvents, prolamins can be divided into gliadins and glutenins also based on their polymerization properties. Glutenins can polymerize by the formation of intermolecular disulfide bonds, whereas gliadins remain as monomers. In the presence of water,

A

Prolamin type	Molecular mass (kDa)	Aggregation
S-rich gliadins (α/β, γ)	30-45	monomers
S-poor gliadins (ω)	30-75	monomers
HMW-glutenins	65-90	polymers
LMW-glutenins	30-75	polymers

B

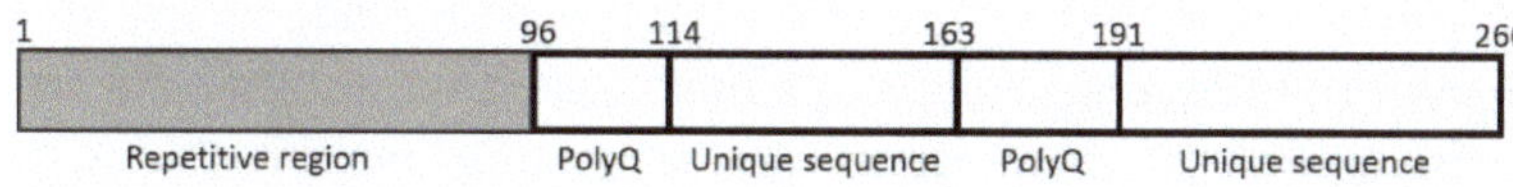

C

		Sequence
p31-43 and related peptides	p31-43	LGQQQPFPPQQPY
	p31-49	LGQQQPFPPQQPYPQPQPF
	p31-55	LGQQQPFPPQQPYPQPQPFPSQQPY
33mer	p57-89	LQLQPFPQPQLPYPQPQLPYPQPQLPYPQPQPF

D

Fig. 1 Molecular characteristics of prolamins. (A) Classification of prolamins. (B) Structure of an α-type gliadin. The gray box account for the repetitive region where, the most studied peptides related to CD pathogenesis are allocated. (C) Sequence of common model peptides used in the study of pathogenic mechanisms in CD. (D) p31–55 has an almost duplicate sequence of p31–43. Panel (A): Adapted from Kasarda, D.D., et al., 1984. Nucleic acid (cDNA) and amino acid sequences of alpha-type gliadins from wheat (Triticum aestivum*). Proc. Natl. Acad. Sci. U. S. A. 81 (15), 4712–4716. doi: 10.1073/pnas.81.15.4712.*

gliadins and glutenins form the viscoelastic structure called gluten, which is stabilized by non-covalent interactions and disulfide bond exchange, and determines the bread-making quality of wheat flour (Shewry, 2019; Shewry and Halford, 2002).

Furthermore, the purification of other components from wheat, barley, and rye (Field et al., 1982) allowed us to draw parallels between the proteins derived from these cereals: gliadins, hordeins and secalins, respectively. Studies from the group of Shewry PR showed the relationship between the seed storage proteins from wheat, barley, and rye at the molecular level (Kreis et al., 1985) and in their immunochemical reactivity (Festenstein et al., 1984).

The product of gliadin digestion by pepsin and trypsin (commonly named as PT-gliadin) has been largely used in the investigation of the pathogenic mechanisms of CD. This enzymatic treatment, representing the gastric-pancreatic digestion, generates relatively large fragments, in accordance with the low digestibility of gluten compared to other dietary proteins. In order to characterize the remaining peptides after gluten digestion, Shan et al., (2002) analyzed the products generated after the treatment of α-gliadin with pepsin and trypsin by liquid chromatography coupled with tandem mass spectroscopy. The most representative fragment found was the one comprising residues 57–89, commonly referred to as the 33-mer. This peptide has been extensively used in research because it is highly stable to proteolytic digestion and contains a total of six T cell epitopes belonging to three distinct types: PFPQPQLPY, PQPQLPYPQ (three copies), and PYPQPQLPY (two copies). Additionally, as discussed above, the short peptide comprising residues 31–43 from α-gliadin (p31–43), which is not recognized by T cells, has also been studied. In conclusion, PT-gliadin and the synthetic peptides 33-mer and p31–43, as well as other related peptides (Fig. 1C), have been used for decades in the research on CD pathogenesis.

3. Early evidence of the toxic effects of gliadin peptides

The studies aimed at investigating the toxic effects of gliadin fragments were instrumental in generating the first insights on the mechanism of mucosal damage by gluten peptides (Browning and Trier, 1969). Intestinal biopsies were placed on a stainless-steel mesh on a Petri dish with the villus surface of the tissue specimens facing upwards. Using this approach, human mucosal biopsies could be cultivated for up to 24 h or even longer, in a medium under controlled atmosphere (95% O_2/5% CO_2) at 37 ° C. These studies were conducted to evaluate the role of gliadin fractions, PT-gliadin, and later, synthetic peptides. Treated tissue pieces could be then used for histological, electron microscopy, immunochemical or physiological analysis. Although organ culture has some limitations, primarily due to the loss of connection with the vascular and lymphatic systems and because gliadin fragments may access the interior of the tissue by a non-physiological manner, this technique allowed major advances in our understanding of CD pathology.

Falchuk et al. (1974) incubated intestinal biopsies from CD patients and healthy individuals with PT-gliadin or A-gliadin for 24–72 h and analyzed them by electron microscopy, finding that gluten fragments were toxic on microvilli at the ultrastructural level and produced changes in brush border enzymes.

By peptic/tryptic digestion of whole gliadin and following several purification steps, Wieser et al. (1984, 1986) isolated a 53 amino acid long peptide (named B3142), which corresponds to the N-terminal sequence of A-gliadin comprising residues 3–55. This peptide showed toxic activity in organ-culture tests when samples from healthy individuals and CD patients were compared. Interestingly, a proline instead of a leucine residue was found at position 31. This is a known variation in the α-gliadin type sequence, and the one with leucine residue is most commonly used in the studies (Fig. 1C). Altogether, these studies showed that sequences with high content of proline and glutamine residues seem to be directly involved in gluten toxicity, and some motifs, such as QQQP, were initially described as necessary for the toxic effects of larger fragments.

Using organ cultures with peptides obtained after the digestion of A-gliadin with chymotrypsin, de Ritis et al. (1988) identified A-gliadin fragments comprising residues 1–30, 1–55, and 31–55 that had toxic effects, as assessed by the observed reduction in enterocyte height. On the contrary, the fragment comprising residues 56–68 did not show any toxic effects in that experimental system. However, while most of the peptides were isolated to a 98–99% purity, the chromatographic separation caused the partial contamination of p1–30 with p31–55. Therefore, the possibility that the toxic effects of p1–30 were due to the presence of p31–55 could not be excluded. These reports suggest that in organ culture assays, some of the peptides from the N-terminal region of α-type gliadins, particularly p31–55, exert toxic effects on the small intestine.

Moreover, the toxic activity of gliadin fractions has also been demonstrated *in vivo*, after instilling pure α-, β-, γ-, and ω-gliadins into the proximal small intestine in CD patients, which caused alterations in morphometric parameters of the mucosa after 6 h (Ciclitira et al., 1984). Further studies by the group of Ciclitira, using the same experimental approach, were the first to investigate the effects of synthetic peptides with sequences corresponding to amino acids 3–21, 31–49, and 202–220 of A-gliadin.

The assessment of morphometric parameters of enterocytes and the number of intraepithelial lymphocytes (IELs) 6h after the instillation of 200mg of each peptide showed significant histological changes only for p31–49 (Sturgess et al., 1994). In a second study, the authors compared the enterocyte height in duodenal samples from CD patients to that of healthy individuals and reported a significant reduction upon *in vivo* treatment with p31–49, but not with the other synthetic peptides (p3–21 or p202–220) (Shidrawi et al., 1995). These studies showed that, after instillation into the small intestine of CD patients, the p31–49 peptide is toxic *in vivo*.

The toxicity of the p31–55 fragment was also observed by Maiuri et al. (1996a) using organ cultures of intestinal biopsies from CD patients. In order to investigate the toxic activity of smaller synthetic fragments, two synthetic peptides covering the p31–55, p31–43, and p44–55 sequences were tested. While p31–43 produced the same effects (reduction of enterocyte height and induction of expression of HLA-DR in cells from the crypts) as p31–55 after 24h incubation, the concentration of p44–55 needed to obtain equivalent results higher. It is interesting to notice that the sequence of fragment p31–55 is almost identical to that of p31–43 (9 out of 12 residues are the same in the p44–55 sequence) (Fig. 1D). This work provided the first piece of evidence that p31–43, a shorter sequence (13-mer) from the originally recognized p31–55 (25-mer), was able to produce toxic effects on the small intestinal mucosa.

4. Biochemistry of p31–43

4.1 Structural analysis of p31–43

Flour from wheat has become one of the main dietary sources in human nutrition because wheat storage proteins have the unique property of forming a viscoelastic dough when they are combined with water and kneaded. During dough making, gluten proteins are the main determinants of dough properties, which eventually determine the structure and texture of the wheat flour-based baked foods. Therefore, the need to produce better products led to the crossing and selection of the better wheat varieties for each purpose for thousands of years, but more recently, molecular and structural studies have provided basic knowledge that can help us to understand the biochemical properties of these proteins. The conformation of prolamins has been studied by different techniques, but the first description of their secondary structure was provided by circular dichroism. Accordingly, the large repetitive regions of prolamins show an equilibrium between β-turn and poly-proline II conformations, whereas the non-repetitive regions present α-helix or random conformations (Kasarda et al., 1967, 1968; Tatham et al., 1985). Using circular dichroism, Tatham et al. (1990) showed that peptides from the N-terminal region of α-gliadin (p3–55, p3–19, and p39–45) present β-reverse turns, which in water and isoosmotic saline buffers are in equilibrium with a poly-proline II structure.

Our group has recently analyzed the conformation of p31–43 using different biophysical approaches. Transmission electron microscopy showed that p31–43 spontaneously forms oligomeric nanostructures, whose size depends on the peptide concentration, whereas circular dichroism analysis showed that p31–43 presents a poly-proline II secondary structure (Gómez Castro et al., 2019). This peptide has hypochromic properties suggesting that it may undergo self-assembly in solution. The relationship between oligomer size and p31–43 concentration was also observed by atomic force microscopy. At lower concentrations (5 μM) (Fig. 2A and B), spherical nanostructures with different heights were observed, while some bigger oligomers, clusters, and precipitation were detected at 50–100 μM. The size of these structures correlated well with that from structures observed by electron microscopy. These findings suggest that p31–43 is able to self-assemble into oligomers, which interact with each other to generate bigger structures (Herrera et al., 2020).

In silico analysis, using coarse-grained molecular dynamics simulations, showed that p31–43 adopts a poly-proline II structure, before assembling spontaneously into small clusters and then, into larger structures. *In silico* modeling showed that isolated p31–43 peptides exhibit broad conformational dynamics with several poly-proline II components. Simulation of multiple replicas showed a spontaneous tendency for aggregation with a concomitant increase in the poly-proline II content (Fig. 2C) (Gómez Castro et al., 2019). The poly-proline II secondary structure of p31–43 was also confirmed by nuclear magnetic resonance experiments (Calvanese et al., 2019). The agreement of the superposition between structures of individual peptides within the oligomer identified by molecular dynamics simulations and experimentally determined structures was remarkable (Fig. 2D). Thus, different approaches support a model of p31–43 monomers in equilibrium with self-assembled structures, whereby isolated p31–43 peptides suffer minor conformational changes when passing from monomeric to oligomeric states. These findings further strengthen the idea that oligomers might work as reservoirs protecting p31–43 from degradation (Barrera et al., 2019).

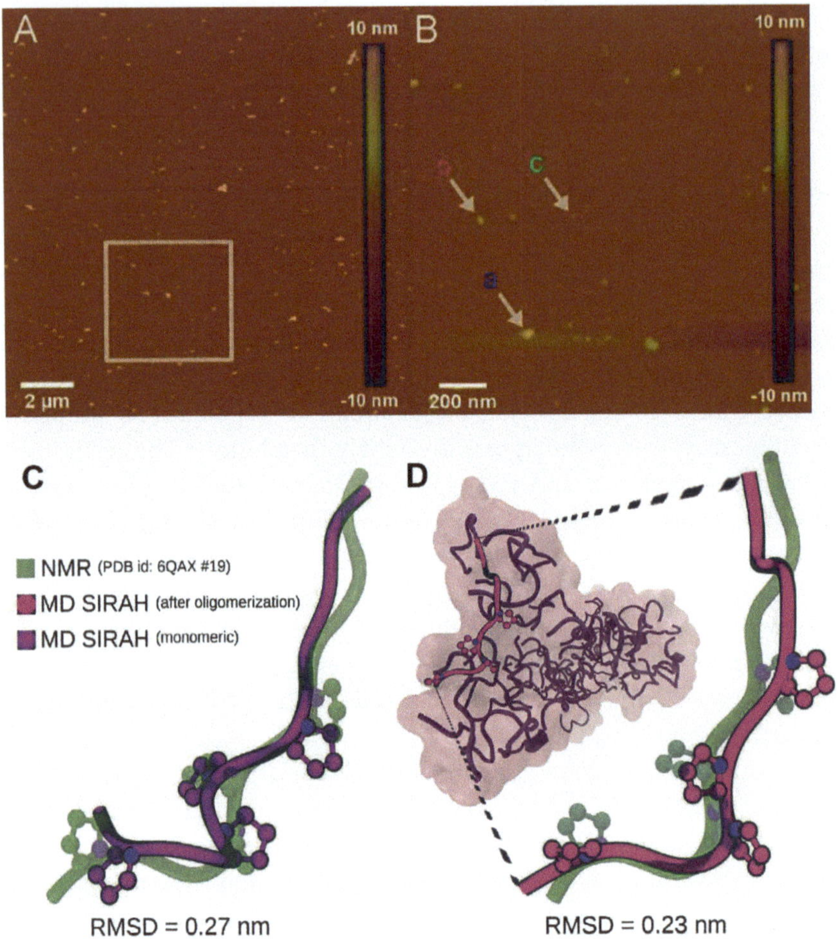

Fig. 2 Characterization of p31–43 structure and oligomer formation. (A) Analysis by Atomic Force Microscopy (AFM) of 5 μM gliadin p31–43 solutions in Milli Q water and deposited on mica surface. AFM images were obtained in tapping mode ($10 \times 10\,\mu m^2$). (B) Cross-section of the oligomers indicated in A showing representative different size oligomers: (a) blue: higher oligomers, (b) red: intermediate oligomers and (c) green: smaller oligomers. *Structural analysis of p31–43 conformation.* Structural alignments for the best matching solution between NMR and coarse-grained molecular dynamics p31–43 conformers. Low root mean square deviation (RMSD) values indicate the high degree of structural similarity observed between experiments and simulations. Similar conformations were obtained for two different simulation scenarios: (C) p31–43 in its monomeric form and (D) p31–43 after its oligomerization. *Panel (B): Adapted from Gomez Castro, M.F., Doctoral Thesis. Facultad de Ciencias Exactas. UNLP. May 2019; Gómez Castro, M.F., et al., 2019. P31-43 gliadin peptide forms oligomers and induces NLRP3 inflammasome/caspase 1-dependent mucosal damage in small intestine. Front. Immunol. 10 (Jan), 1–11. doi: 10.3389/fimmu.2019.00031. Panels (C and D): Prepared by Barrera E., PhD. Institut Pasteur Montevideo, Uruguay.*

Since this model proposes an equilibrium between isolated peptides and oligomers, free monomeric forms may also be able to cause toxic effects. Villella et al. (2019) published the first report describing the molecular effect of p31–43 monomers, which bind to the NBD1 subunit of the anion channel (cystic fibrosis transmembrane conductance regulator, CFTR), inhibit its function, and induce different inflammatory pathways (see Section 5.8).

Circular dichroism studies and molecular dynamics simulations of p31–43 over a wide pH range (3–8) did not show structural transitions, indicating that its conformation is not affected by changes in pH. Since p31–43 is an uncharged peptide, different types of intermolecular forces may be the key drivers for self-assembly. This model suggests that the aggregation process might contribute to the resistance properties of p31–43, and would not be affected by changes in pH along the gastrointestinal tract and in the small intestinal mucosa. Altogether, studies by circular dichroism, transmission electron microscopy, atomic force microscopy, and molecular dynamics simulations demonstrate that p31–43 is able to self-organize and form nanostructures (Herrera et al., 2020). Therefore, the toxic effects induced by p31–43 can be linked to either the monomeric form or to nanostructures, and could be related to the poly-proline II structure and the oligomerization properties of the peptide.

4.2 Gliadins are proteins difficult to digest

Several studies have approached the issue of gliadin digestibility in different systems. Using a sophisticated *in vitro* multi-compartment model that included oral, gastric, and duodenal phases of digestion, 25 peptides were obtained after the treatment of A-gliadin recombinant protein (Mamone et al., 2007). After the supplementation with intestinal BBM (Brush Border Membranes) containing additional enzymes, only two main peptides, a 25-mer (p31–55) and a 33-mer (p57–89), remained undigested. Overnight incubation of the synthetic p31–55 with BBM confirmed its high resistance to proteolytic enzymes. Surprisingly, in the same system, the immuno-active epitope for the adaptive T cell-mediated immune response (the 33-mer), was completely digested after BBM supplementation. This indicates that while A-gliadin can be extensively hydrolyzed *in vitro*, the peptide p31–55 remains unaffected. Interestingly, the digestion of cooked pasta, (*Triticum durum* semolina) in a similar system showed the persistence of several wheat-derived peptides, which were identified by liquid chromatography–mass spectrometry. Some of the peptides were substantially modified, but the peptide p31–55 and its shortened form p31–43 were mostly intact (Gianfrani et al., 2015). These data indicate that p31–55, as well as p31–43, are particularly resistant to digestion by gastric, pancreatic, and BBM enzymes, and therefore, they can come into contact with the intestinal epithelium in our everyday life.

4.3 Passage across the epithelium and mechanisms of cell entry of gliadin peptides

The transport of gliadin peptides from the apical side through the intestinal epithelium occurs by a non-degradative pathway, which results in the release of intact peptides on the basal side. This route seems to be preferentially active in untreated CD patients, since a leakage through the paracellular route due to an increase in intestinal permeability does not explain this passage. In untreated CD patients, p31–49, as well as the 33-mer, appear intact on the basal side, whereas p57–68 is completely degraded during intestinal transport (Matysiak-Budnik et al., 2003).

In untreated CD patients, CD71 is overexpressed in the intestinal epithelium and binds luminal IgA or IgA anti-gliadin/gliadin complexes, which are endocytosed and transported to the basal side without lysosomal degradation (Matysiak-Budnik et al., 2008). This protected passage which is observed for p31–49 and may be active for p31–43 as well, can explain how these gliadin peptides can use a transcellular pathway to get into the tissue with no or minimal degradation.

Additional *in vitro* experiments showed that the undigested p31–43 and p57–68 peptides enter the cells by endocytosis. Their entrance into the cells requires a temperature of 37 °C and the presence of Ca^{2+} in the media (Barone et al., 2014). Experiments with inhibitors of receptor-mediated endocytosis, such as methyl-β-cyclodextrin, showed reduced entrance of p31–43 labeled with a fluorescent tag. On the other side, filipin, an inhibitor of lipid raft/caveolae-mediated endocytosis, was not able to block the entrance of p31–43. These inhibitors had the opposite effect on p57–68, indicating that both peptides enter intestinal epithelial cells by endocytosis, but only p57–68 enters the cells by lipid raft/caveolae-mediated endocytosis (Barone et al., 2014; Zimmermann et al., 2014). An additional difference between these peptides has been observed: whereas p31–43 is segregated in early endosomes after endocytosis, p57–68 is transported to late endosomes (Barone and Zimmer, 2016). Further studies have demonstrated that in a Caco-2 (human intestinal cell line) monolayer culture, p31–43 is translocated to the basal side (Barone and Zimmer, 2016), and the p31–43 transcytosis can be enhanced in the presence of anti-gliadin antibodies bound to the transferrin receptor (Mnard et al., 2012).

Some works have highlighted the role of TG2 (transglutaminase 2) on p31–43 uptake. *In vitro* studies using Caco-2 cells showed that the interaction of anti-TG2 antibodies with TG2 at the cell surface reduced p31–43 uptake, but had no effect on p57–68 uptake (Martucciello et al., 2020). The uptake of EGF by Caco-2 cells was also affected negatively by antibodies against TG2, suggesting a general role of surface TG2 in endocytosis. However, p31–43 enters HEK 293 cells, which do not express TG2, indicating that TG2 may be a regulator of peptide uptake, but its presence is not necessary for internalization. The TG2 enzymatic activity does not seem to be involved in p31–43 uptake, since enzymatic inhibitors, such as the competitive substrate mono-dansyl-cadaverine (MDC) and the inhibitor cystamine, do not interfere with p31–43 internalization (Martucciello et al., 2020). Therefore, different *in vitro* assays showed that the cell entry of p31–43 is fast and protected.

4.4 A receptor for p31–43 is missing

Several experimental approaches have sought a surface receptor for p31–43, but so far, no membrane protein has been identified as a potential p31–43 receptor/carrier (Paolella et al., 2018). This has led to the hypothesis that the entrance of p31–43 into the cell might involve a direct interaction with the cell membrane. This hypothesis is supported by several observations:

- p31–43, but not the immunogenic p57–68, can interact with a membrane-mimetic environment (Vilasi et al., 2010).
- The entrance of fluorescently labeled p31–43 into Caco-2 cells is not reduced by increasing the concentration of the unlabeled peptide, indicating that the peptide does not bind to a receptor (Paolella et al., 2018).
- p31–43 competes with HRS (Hepatocyte growth factor-Regulated tyrosine kinase Substrate), a regulator of endocytosis (see Section 5.7) present on the cytosolic side of endosomes (Vilasi et al., 2010), suggesting that p31–43 is able to penetrate the cell membrane or escape from endocytic vesicles.

This result is not entirely surprising, since many naturally occurring bioactive peptides can interact with the membranes directly and do not need a receptor to enter the cells. These peptides, defined as cell-penetrating peptides (CPPs), use endocytic pathways and then, are able to escape from vesicles (Bechara and Sagan, 2013). As described in Section 4.1, p31–43 presents a poly-proline II helical secondary structure and forms oligomers with a well-defined compact conformation, and these are characteristics shared by other CPPs (Franz et al., 2016; Herrera et al., 2020; Oba et al., 2019). Altogether, these observations strengthen the hypothesis that p31–43, owing to its peculiar sequence and structure, can enter the cells without a receptor.

5. Signaling pathways and pathological effects induced by p31–43

5.1 Activation of innate immunity

Innate immunity has been recognized as a necessary component in the development of CD (Kim et al., 2015). Under homeostatic conditions, regulatory mechanisms in the intestinal mucosa keep an adequate balance between immunity against pathogens and tolerance toward innocuous stimuli (Mowat, 2018). However, several innate pathways could become interconnected and synergize to potentiate local inflammation, crossing the threshold for the full development of an adaptative gluten-specific response through the activation of *lamina propria* $CD4^+$ Th1 cells.

Different experimental approaches, using organ culture assays, *in vitro* assays, and animal models, have demonstrated the biological effects elicited by p31–43 that lead to the induction of innate immunity mechanisms, such as IL-15 and IFNα upregulation (Maiuri et al., 2003; Nanayakkara et al., 2018).

As mentioned before, the study by Maiuri et al. (1996b) was the first demonstration of the toxic effects of p31–43 in an organ culture assay. Incubation of duodenal biopsies for 24 h with 1 mg/mL p31–43 caused a reduction of enterocyte height and an upregulation of HLA class II molecules.

Further investigations from Maiuri et al. (2003), using the same approach, were aimed at assessing the toxicity of p31–43 in comparison with two other immunogenic peptides derived from the sequence of the 33-mer, namely, p57–68 (QLQPFPQPQLPY), and p62–75 (PQPQLPYPQPQLPY), their deamidated forms, and a non-related peptide. Toxic effects were observed when the intestinal tissue from CD patients was treated with 20 μg/mL p31–43, but not with the other gliadin or control peptides tested. The TUNEL reaction showed increased enterocyte death after incubation with p31–43 incubation for 24 h. Evidence for the activation of innate mechanisms was provided by the upregulation of IL-15, ICAM1, and

HLA molecules, and the activation of dendritic cells as determined by the expression of CD83. Some of these effects were further increased when the biopsies were incubated for 3h with p31–43 and then, with one of the other α-gliadin immunogenic peptide (p57–68) for 21h, indicating that the first innate response step primes the system for a subsequent more robust adaptative response (Maiuri et al., 2003).

5.2 p31–43 is linked to IL-15Rα/EGFR expression

IL-15 is at the crossroad between innate immunity and p31–43 effects (Jabri and Abadie, 2015). IL-15 belongs to a family of cytokines (also including IL-2, IL-4, IL-7, IL-9, and IL-21), which share part of a heterotrimeric receptor, providing specificity for IL-15 through the α chain. IL-15 binding to its receptor leads to the phosphorylation of Jak molecules and STAT5, driving T cell and NK cell expansion. In T lymphocytes, IL-15 triggers other signaling pathways as well, including activation of distinct protein kinases, such as Src, Lck, Syk, PI3K, and MAPK (Budagian et al., 2006).

In active CD, IL-15 and its receptor are overexpressed in duodenal tissue (Bernardo et al., 2008; Maiuri et al., 2000). This cytokine plays multiple roles in inducing changes on epithelial cells, IELs, $CD4^+$ and $CD8^+$ T lymphocytes, and dendritic cells from the *lamina propria* of the small intestine (Maiuri et al., 2000; Mention et al., 2003; Meresse et al., 2004). Immunofluorescence analysis of sections of duodenal tissues from CD patients treated with 20 μg/mL p31–43 for 3h showed increased expression of IL-15 (Maiuri et al., 2003). *In vitro* stimulation of Caco-2 cells with gliadin-derived p31–43, but not p57–68, upregulated the mRNA levels of IL-15 after overnight incubation but not after shorter incubation times (3 or 6h). Upregulation of IL-15 surface expression was also observed in p31–43-treated Caco-2 cells by flow cytometry (Barone et al., 2011). Furthermore, western blot analysis of membrane fractions of Caco-2 cells treated with p31–43 showed an increment in the expression of IL-15 and its receptor IL-15Rα (see Section 5.2). This is linked to the biological activity of IL-15 bound to its membrane receptor (Duitman et al., 2008), which, in addition to the already mentioned effects on T and NK cells, affects the expression of other molecules, such as stress markers (see Section 5.5). The *in vitro* induction of IL-15 was also reported by Vincentini et al. (2015), who analyzed the protein extracts of T84 cells (human intestinal epithelial cells) stimulated with 40μg/mL p31–43 for 24h. Furthermore p31–43 activated the IL-15 downstream effector STAT5 in epithelial cells and fibroblasts from

CD patients (Lania et al., 2019; Nanayakkara et al., 2013). Interestingly, IL-15 induced cell proliferation in the intestine, indicating that innate immunity and enterocyte proliferation are closely related (Reinecker et al., 1996). Evidence obtained from Caco-2 cells and intestinal biopsies from CD patients have demonstrated that IL-15 and EGF cooperate in inducing enterocyte proliferation (see Section 5.2).

In untreated CD patients, the intestinal mucosa suffers from chronic and aggressive damage, and an increase in proliferative activity is part of the strategy to remodel the mucosa and control tissue destruction (Przemioslo et al., 1995). Different studies carried on cell lines and biopsies from untreated and potential CD patients showed that among the multiple pathways that might contribute to induce proliferation, p31–43 acts in an EGFR-mediated manner. The effect mediated by p31–43 was specific for epithelial cells found in the crypts of CD patients, and it was not observed in tissues from healthy subjects (Barone and Zimmer, 2016; Barone et al., 2007; Nanayakkara et al., 2013).

5.3 NFκB and MAPK signaling driven by p31–43

The NFκB pathway is considered the masterpiece of the inflammatory response. *In vitro* studies using Caco-2 showed that incubation with 50 μg/mL p31–43 for 45 min increased the expression of p65 NFκB. In parallel, the increased levels of ERK phosphorylation showed that the MAPK pathway was also activated (Capozzi et al., 2013). Thus, these pathways were rapidly induced by p31–43 *in vitro*. The NFκB pathway is activated in the intestinal mucosa of untreated CD patients, and remains active even during the remission phase. This observation suggests that following the induction of the inflammatory response by gluten peptides, other mechanisms maintain an ongoing inflammatory process in the intestinal mucosa of CD patients regardless of the further gluten intake (Castellanos-Rubio and Bilbao, 2018). NFκB and MAPK pathways were also found active in fibroblasts derived from the skin and small intestine of CD patients. This is remarkable since the skin fibroblasts are at a significant distance from the tissue with the active inflammatory response (Lania et al., 2019). Interestingly none of the effects on NFκB or MAPK signaling were induced by the immunogenic peptide p57–68. Remarkably, p31–43 could activate the MAPK and NFκB pathways in cells from control subjects, suggesting that p31–43 "*per se*" has the ability to induce inflammation. Studies using synthetic peptides containing alanine replacements along the p31–43 sequence demonstrated that the reported effects are specific for this peptide (Lania et al., 2019).

5.4 Cellular stress caused by p31–43

The production of reactive oxygen species (ROS) and ER-stress are routes for cellular distress leading to inflammation. The production of ROS by gliadins was demonstrated by different *in vitro* assays (Monguzzi et al., 2019; Orlando et al., 2019; Rivabene et al., 1999), *ex vivo* assays using duodenal biopsies (Monguzzi et al., 2019) and *in vivo* studies using the *C. elegans* model (Lim et al., 2018). Interestingly, Maiuri et al. (2010) found that when T84 cells were treated with 20 μg/mL p31–43 for 24 h, the peptide was accumulated in vesicles, a result that is consistent with the induction of oxidative stress as early as 3 h post-treatment. These effects were not observed when a scrambled-sequence p31–43 peptide, the p57–68 and p62–75 gliadin peptides, or a peptide with unrelated sequence were used instead. These authors also showed that incubation of duodenal biopsies from CD patients but not from healthy controls with p31–43 for 3 h induced ROS through the activation of the peroxisome proliferator-activated receptor (PPARγ).

CD is an example of a pathology driven by specific $CD4^+$ Th1 cells. These T cells preferentially recognize gluten peptides that have been deamidated by the enzyme TG2. TG2 catalyzes peptide deamidation by converting specific glutamine residues to glutamate at certain positions. Negatively charged anchor residues are responsible for the preferential binding of deamidated gluten peptides to the disease-associated HLA alleles, DQ2 and DQ8. This enzymatic activity generates one of the best characterized post-translational modification on neoepitopes as a way to break T cell tolerance to self-antigens and promote autoimmunity (Sollid and Jabri, 2013).

TG2 is a Ca^{2+}-dependent enzyme with multiple physiological roles (mitochondria function, cell adhesion, migration, and phagocytosis, among others), which explains why it appears as a mediator in so many human diseases, such as inflammatory disorders (rheumatoid arthritis, lupus, uveitis), neurological diseases (Alzheimer's disease, Parkinson's disease, Huntington's disease), and cancer (Szondy et al., 2017). The expression of TG2 gene is regulated by the canonical NFκB pathway; consequently, it can be induced by proinflammatory cytokines such as TNFα. In addition, TG2 can activate

NFκB by blocking the inhibitory function of IκBα *via* polymer formation (Lee et al., 2004), leading to a complex cross-regulation between TG2 activity and the NFκB pathway, a circuit that plays a role in inflammation and cancer (Mehta et al., 2010). Importantly, the combination of TNFα and IFNγ produce a strong synergistic effect on TG2 expression in human cell lines and duodenal biopsies of untreated CD patients (Bayardo et al., 2012).

In the enterocytes of untreated CD patients, the levels of TG2 are increased in the brush border, cytoplasm, and also in the extracellular matrix (Caputo et al., 2013; Esposito, 2003; Hansson et al., 2002; Martucciello et al., 2020), TG2 is predominantly localized inside the cells, but following an unconventional secretion route, which depends on TG2 binding to phosphoinositides on endosomal membranes, it is delivered in the recycling endosomes and becomes secreted (Zemskov et al., 2011). Although TG2 has been observed in different cellular locations, the precise site where gliadin deamidation and cross-linking takes place has not been properly defined. Since appropriate Ca^{2+} levels are critical for TG2 activity, the low Ca^{2+} concentration inside the cell, as well as in the extracellular space, keeps intracellular and secreted TG2 in an inactive conformation (Siegel et al., 2008). Recent work from Sollid's group, showed that TG2 can be released into the lumen from shed enterocytes, implying that deamidation can occur in the intestinal lumen prior to peptide internalization to the tissue (Iversen et al., 2020).

There is a complex connection between Ca^{2+} levels and TG2 function, which links mitochondria and the ER. D'Eletto et al. (2018) showed that TG2 interacts with GRP75 (glucose-regulated protein 75) in mitochondria-associated membranes, creating a close contact between the ER and mitochondria. These physical bridges promote the transport of Ca^{2+} from the endoplasmic reticulum to mitochondria. Therefore, TG2 also participates in the regulation of intracellular Ca^{2+}.

Taking into account that Ca^{2+} levels are critical for many aspects of cellular physiology but particularly for TG2 function, the effect of p31–43 on Ca^{2+} mobilization was assessed on Caco-2 cells. *In vitro* assays showed that in Caco-2 cells, low amounts of p31–43, such as 0.2 μg/mL, were able to mobilize Ca^{2+} from the ER and the mitochondria. The consequent increase in cytosolic Ca^{2+} levels led to the activation of the TG2 enzymatic function in the cytosol (Caputo et al., 2012). When Caco-2 cells were treated with 20 μg/mL p31–43 for 24 h, an increase was observed in the expression of TG2, as well as in that of the glucose-regulated protein-(GRP)78 and of the CCAAT/enhancer binding protein-homologous protein, which are both biochemical markers of ER-stress. None of these effects was observed for p57–68 (Caputo et al., 2012). Further work in Caco-2 cells, showed that p31–43, but not p57–68, could bind to Ca^{2+}-activated TG2 (Villella et al., 2019). p31–43 was found covalently bound to TG2, promoting the formation of a supramolecular complex involving TG2 and the NBD1 subunit of CFTR channel (see Section 5.8). Thus, activation of TG2 may sustain the p31–43-mediated inhibition of CFTR function through the formation of covalent links between p31–43, CFTR, and TG2, generating a trimolecular complex. Therefore, p31–43 can cause a cascade of different events, including ROS induction, an ER-response, and the mobilization of intracellular Ca^{2+} from ER store, activating TG2, which would then participate in the inflammatory response, as well in the deamidation of immunogenic gliadin peptides and the formation of TG2/gliadin complexes, inducing the formation of anti-TG2 autoantibodies (Iversen and Sollid, 2020).

5.5 Expression of stress molecules: The case of MICA

MHC class I chain-related A (MICA) proteins are non-classical MHC class I polymorphic proteins, which, in the gastrointestinal epithelium, are conformationally stable without being bound to conventional MHC class I peptides. Therefore, MICA molecules have no role in conventional antigen presentation, but their functions are associated with responses under cellular stress (Groh et al., 1996). Unlike the case of MHC class I genes, the MICA promoter region contains elements found in the genes of heat shock proteins. Under conditions of cellular stress, such as thermal, oxidative stress, viral infections, and tumor transformation, MICA expression is upregulated (Groh et al., 1996; Perera et al., 2007; Raulet, 2003). The surface expression of MICA has been shown to be a signal for enterocyte killing by interacting with the NKG2D receptor constitutively expressed on NK cells, $CD8^+$ α/β T cells, and intestinal intraepithelial γ/δ T cells (Hüe et al., 2004; Meresse et al., 2004; Raulet, 2003). NKG2D expression and signaling, as well as MICA expression (Hüe et al., 2004), are upregulated by IL-15.

In the intestinal mucosa during active CD, MICA associated to the intracellular compartment was redistributed to the cell surface (Hüe et al., 2004; Meresse et al., 2004). Remarkably, the increased MICA levels returned to normal under gluten-free diet, highlighting the importance of the signals derived from gliadin peptides in the direct or indirect, through cellular stress, triggering of MICA expression (Allegretti et al., 2013). It has been reported that the gliadin peptides p31–49 (Hüe et al., 2004) and p31–43 (Allegretti et al., 2013) increase MICA expression. Together, these observations suggest that stress conditions induce the expression of MICA in enterocytes, while inflammation and IL-15 potentiate the cytotoxicity activity of some IEL subsets. These findings support a key role for MICA/NKG2D in the activation of intraepithelial immunity in response to enterocyte damage (Hüe et al., 2004; Meresse et al., 2004). Since p31–43 can promote ER-stress, which may induce MICA overexpression (Allegretti et al., 2013) and IL-15 (Barone et al., 2011; Maiuri et al., 2003), this may be one way by which p31–43 mediate enterocyte cytotoxicity. More recently, Jabri's group reported that in CD, additional enterocyte markers can modulate the specificity of the IEL response, in particular, that of some of the TCRγ/δ$^+$ subsets (Mayassi et al., 2019).

5.6 Induction of mediators of inflammation: *In vitro* findings

It has been found that several proinflammatory cytokines and chemokines that play a key role in the orchestration of the inflammatory response in the intestinal mucosa, promoting the recruitment and activation of immune cells, are produced as consequence of p31–43 stimulation.

The production of key inflammatory mediators was evaluated *in vitro* using T84 cells (human intestinal epithelial cells), stimulated with 40 μg/mL p31–43 for 3 h in a transwell co-culture system in the presence of peripheral blood mononuclear cells (PBMC) (Vincentini et al., 2015). Quantitative ELISA in samples from the culture supernatant showed increased production of IL-1β, and also of the inflammatory chemokines CCL2, CCL3 (MIP1α), and CCL4 (MIP1β), which are major chemoattractant for monocytes, macrophages, and dendritic cells.

The inflammatory effects of p31–43 were also observed in mast cells. Using LAD2 cells, a human mast cell line, Frossi et al. (2017) demonstrated that 100 μg/mL p31–43 was able to induce degranulation, assessed by the release of β-hexosaminidase. Interestingly, flow cytometry analysis showed that p31–43 binds to LAD2 cells and activates the NFκB pathway. After incubation for 24 h, p31–43 was found to induce the expression of cytokines and chemokines in the culture supernatant, particularly that of IL-6, TNFα, IL-17, CXCL8, CCL2, but not of IL-1β, IL12p70, or IL-10. These effects were not observed for other dietary proteins (rice and soya proteins) nor for the immunogenic gliadin peptides p57–68 (QLQPFPQPQLPY) and p62–75 (PQPQLPYPQPQLPY). It is noteworthy that after incubation with p31–43, mast cells isolated from duodenal biopsies of CD patients with severe histological changes (as partial or total villous atrophy) exhibited significantly higher histamine release compared with cells isolated from control patients (Frossi et al., 2017).

Undoubtedly, type I IFNs play a critical role in the amplification of the inflammatory response in the coeliac lesion. In duodenal biopsies of CD patients, who were either untreated or in remission, and under a gluten-free diet, but not in those from healthy controls, incubation with p31–43 upregulated the IFNα pathway (induction of IFNα and MxA proteins, as well as of IFNα mRNA 7 and 17), suggesting that p31–43 may contribute to the expansion of the effect of this pathway in the small intestine of patients with CD (Nanayakkara et al., 2018). Studies with mouse models confirmed the rapid induction of type I IFNs after p31–43 treatment, as well as the fact that type I IFN signaling is relevant for the intestinal pathology, since p31–43-dependent enteropathy was not observed in IFNαR deficient mice (Araya et al., 2016; Gómez Castro et al., 2019). Thus, type I IFNs also play a role in the loss of oral gluten tolerance in CD patients (Kim et al., 2015; Monteleone et al., 2001).

Since a growing body of evidence suggests that gliadin peptides induce an innate response in the intestine, Thomas et al. (2006) performed studies on small intestinal segments in mice. PT-gliadins and the 33-mer and p31–43 peptides induced zonulin release and increased the intestinal permeability, as assessed by a reduction in the transepithelial resistance. These effects were dependent on MyD88, but TLR2 or TLR4 were not required. The authors also observed that incubation of primary macrophages from C57BL/6 mice with 100 μg/mL of p31–43 for 3 h led to the production of proinflammatory cytokines (TNFα, IL-12 p35/p40, IL-6, and IFNβ) and chemokines (CXCL10 and MCP-5). The requirement of MyD88 for some of the biological effects mediated by p31–43 was also documented by Frossi et al. (2017). These authors found that murine mast cells from wild-type, but not from MyD88-deficient mice, challenged *in vitro* with 100 μg/mL p31–43 for 24 h produced proinflammatory cytokines, including IL-6, IL-17, TNFα, and MCP-1.

The induction of a plethora of proinflammatory cytokines and chemokines by p31–43 observed by *in vitro* tests using human cell lines, isolated intestinal human cells, or mouse models highlights the relevance of the role of p31–43 for the generation of a strong proinflammatory setting, necessary for the full development of a specific adaptive response against gluten-derived peptides.

5.7 Consequences of vesicular trafficking delay on signaling pathways

In Caco-2 cells (Barone et al., 2007, 2010; Maiuri et al., 2010) and enterocytes isolated from biopsies of the small intestine (Barone et al., 2007; Zimmer et al., 2010). Both p31–43 and p57–68 gliadin peptides have been observed to enter the cells by endocytosis and interact with the vesicular system. However, while the p57–68 passes from early to late endosomes in a short time, p31–43 remains localized in early vesicles, slowing down their motility and maturation toward late endosomes (Barone et al., 2007; Zimmer et al., 2010). In fact, using two very different experimental approaches, namely, pulse-chase immunofluorescence experiments (Barone et al., 2007) and electron microscopy (Zimmer et al., 2010), it has been demonstrated that in intestinal biopsies of CD patients, p31–43 trafficking is delayed at the stage of early vesicles (Barone and Zimmer, 2016). Therefore, p31–43, but not p57–68, slows the maturation of early vesicles carrying EGFR (Epidermal Growth factor receptor) and, presumably, other receptors, which can then remain active for longer periods of time. This is a well-known effect by which vesicular trafficking regulates the signaling of extracellular receptors and *vice versa* (Ceresa and Schmid, 2000).

Vesicular trafficking is regulated by complex mechanisms involving several molecules, such as Hepatocyte growth factor-regulated tyrosine kinase (HRS), which is a key protein for endocytic maturation because it recruits clathrin to early endosomes (Raiborg et al., 2001). HRS is located on early vesicles and the multivesicular body, which participates in the ubiquitination of proteins destined for lysosomes. HRS can be phosphorylated in cells treated with growth factors or cytokines, and can also become ubiquitinated (Komada and Kitamura, 1995). These post-translational modifications are necessary for the efficient HRS-mediated sorting of various proteins for degradation. Interestingly, p31–43 shares a strong amino acid sequence similarity with a small domain of HRS rich in proline/glutamine, which is necessary for the localization of HRS on the membrane of vesicles. Although gliadins are known to contain domains rich in glutamine and proline, the high homology of p31–43 with the HRS domain is restricted to this peptide (Barone et al., 2010). Experiments carried out in HRS-EGFP transfected Caco-2 cells, showed that p31–43, but not p57–68, colocalizes with HRS on the membrane of endocytic vesicles, mediating the translocation of HRS from the vesicular membrane into the cytosol, suggesting that p31–43 interferes with HRS-mediated early endosome maturation (Barone et al., 2010).

Growth factor receptors come to the scene when the consequences of the vesicular trafficking delay need to be considered. The activity of growth factor receptors is influenced by feedback mechanisms that regulate their degradation (Burke et al., 2001; Wilde et al., 1999). In cells in which HRS has been silenced, mutated, or dislocated, EGFR (Epidermal Growth Factor Receptor), which mediates the most potent mitogen signal in epithelial cells (Marshall, 1995), and other tyrosine kinase receptors, can be recycled on the cell surface extending their lifespan (Raiborg and Stenmark, 2002; Raiborg et al., 2008).

A delayed transition from early to late endocytosis, with a consequent delayed inactivation of EGFR, has been observed in enterocytes of intestinal biopsies from CD patients treated with p31–43 (Barone and Zimmer, 2016; Barone et al., 2007). In this context, p31–43 increased the proliferation of crypt enterocytes in CD patients, but not in healthy controls, an effect that was prevented by EGFR inhibitors. In fact, p31–43 slows down the trafficking of early vesicles that carrying EGFR, which therefore, remains active for longer time (Barone et al., 2007). Moreover, pulse and chase experiments with EGF-Alexa in intestinal biopsies of untreated patients, showed that p31–43 treatment delayed the trafficking of vesicles carrying the EGF/EGFR complex (Barone and Zimmer, 2016; Barone et al., 2007) (Fig. 3).

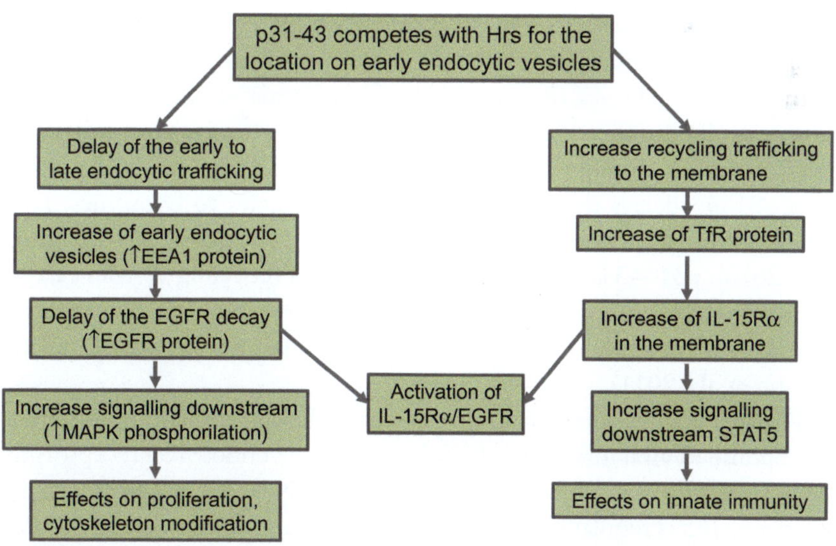

Fig. 3 Effects of p31–43 on endocytic trafficking. Cascade of events affected by p31–43 on early and recycling endocytic trafficking and biological consequences.

The biological effects of trafficking delay depend on the array of receptors present on the cell. The interplay between different receptors is a known phenomenon in biology, and EGFR is especially prone to form complex network with other receptors, including cytokine receptors (Sanchez-Guerrero et al., 2013). One such complex having multiple functions in the intestinal mucosa is the one formed between EGFR and IL-15Rα. The EGFR/IL-15Rα complex is crucial for the proliferation of crypt enterocytes. In Caco-2 cells, the presence of p31–43 increased complex formation between EGFR and IL-15Rα (Nanayakkara et al., 2013). Each of the ligands (EGF of IL-15) can initiate signaling by the EGFR/IL-15Rα complex. Furthermore, EGF can activate the transcription of IL-15, and IL-15 the transcription of EGF (Nanayakkara et al., 2013). The proliferation of cells induced by p31–43 can be prevented by treatment with IL-15 and EGFR specific antibodies in duodenal biopsies from CD patients, or by specific antibodies or siRNA in Caco-2 cells (Barone et al., 2011; Nanayakkara et al., 2013). These findings demonstrate the interplay between p31–43 and EGF/IL-15, occurring at different levels (Figs. 3 and 4A).

The delay to transition from early to late vesicles due to HRS silencing or deletion can interfere with the degradation of tyrosine kinase receptors through recycling. Interestingly, cytokine receptors are recycled to the cell surface, whereas tyrosine kinase receptors are mainly degraded in late vesicles. Therefore, the trafficking blockade induced by the presence of p31–43 or an HRS malfunction can also interfere also with the function of cytokine receptors. As a consequence of this process, an increasing amount of transferrin receptor (TfR) and IL-15Rα accumulates on the cell surface upon treatment with p31–43 (Barone et al., 2011; Lania et al., 2019). IL-15 is known to be localized in the recycling endocytic vesicles and in the Golgi complex (Fehniger and Caligiuri, 2001; Gaggero et al., 1999). Since p31–43 delays the maturation of these vesicles to lysosomes, IL-15 accumulation has also been observed (Barone et al., 2007, 2010; Zimmer et al., 2010). p31–43 also increases the number of recycling vesicles carrying IL-15, as well as the amount of IL-15 and its receptor on the surface of Caco-2 cells, but not the amount of the cytoplasmic or secreted protein (Barone et al., 2011).

The mechanisms that regulate vesicular trafficking are also important for the response to viral infections. TLR7 is one of the endosomal receptors that interact with the viral genome (specifically, ssRNA, occurring in rotavirus infection), and is regulated by endosomal trafficking (Heil et al., 2003; Pane et al., 2014). HRS is also a key factor in the trafficking of endosomal

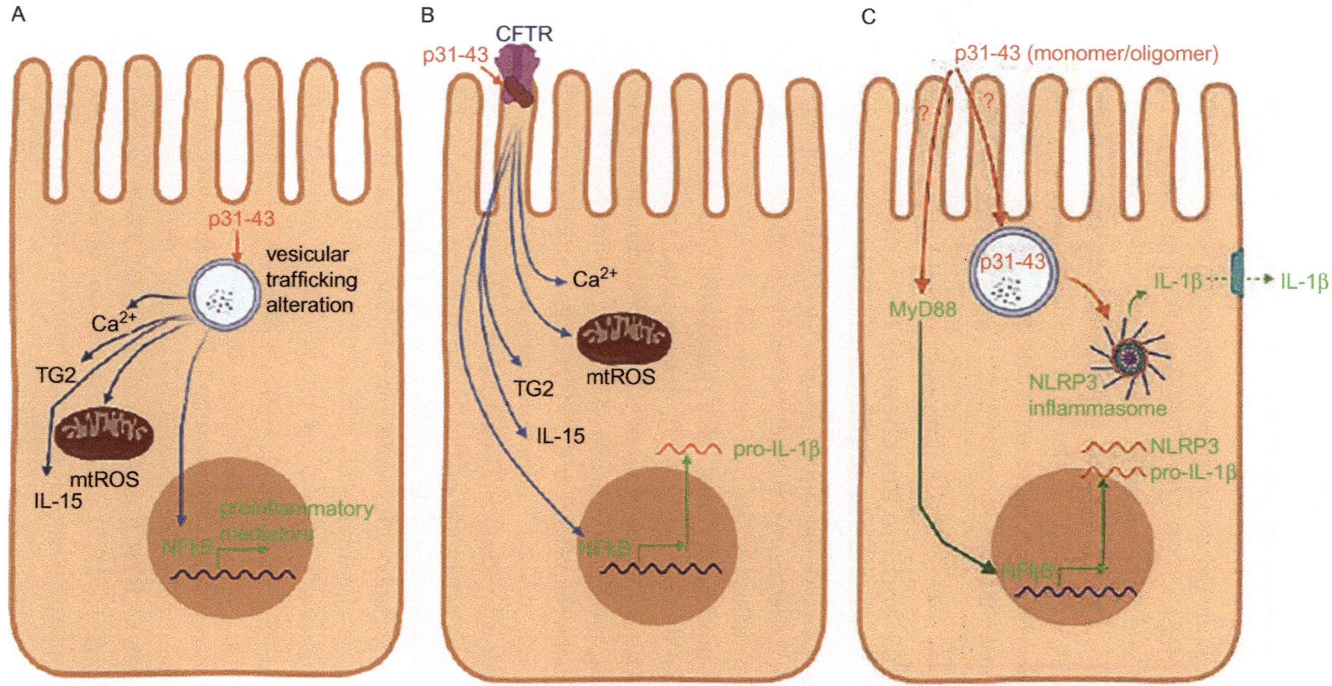

Fig. 4 Schematic representation of multiple proinflammatory effects driven by gliadin p31–43 peptide in enterocytes. (A) Gliadin p31–43 peptide causes alteration on vesicular trafficking leading to vesicular stress associated to multiple effects such as increased intracellular Ca^{2+} concentration, activation of TG2, alteration of mitochondria function and release of ROS, with a parallel NFκB activation and upregulation of IL-15 and proinflammatory mediators. (B) Gliadin p31–43 peptide binds to the NBD1 domain of CFTR, keeping it in inactive conformation, thus blocking CFTR function as an anion channel. As a consequence, TG2 is activated and forms a tripartite complex that stabilizes p31–43 binding to CFTR. CFTR inhibition leads to impaired endosomal trafficking, cytoskeleton disassembly, NFκB activation, IL-15 production, and inflammasome activation resulting in IL-1β secretion. (C) Mucosal damage induced by p31–43 requires MyD88. The activation of this pathway leads to upregulation of proinflammatory genes, among them NLRP3 and pro-IL-1β. In addition, p31–43 causes vesicular stress, which may release signals leading to assembly and activation of the NLRP3 inflammasome. Active caspase-1 produces mature IL-1β and N term GSDMD which oligomerizes and forms membrane pores. These pores allow the release of IL-1β and alarmins, and can cause cell death by pyroptosis. Mechanisms of p31–43 cell entry, accumulation in vesicles and activation of MyD88 pathway are still unknown.

receptors such as TLR7, and is necessary for the ubiquitin-dependent targeting to the lysosomes (Chiang et al., 2012). In fact, the early/late delay of the maturation of the vesicular system, observed in Caco-2 cells upon treatment with p31–43 or HSR silencing, produces effects relevant for the TLR7 pathway, such as the delay in trafficking of the TLR7 in the early vesicles, the increase in complex TLR7/MyD88 formation, the increase in the levels of type I IFN and MxA proteins, and the activation of NFκB (Nanayakkara et al., 2018). In addition, the viral ligand loxrubine, which is specific for TLR7, and p31–43 activate the same signaling pathways that depend on TLR7 activation, and cooperate in inducing the activation of the type I IFN pathway (Nanayakkara et al., 2018).

5.8 A new link: p31–43 and the ion channel cystic fibrosis transmembrane conductance regulator (CFTR)

Cystic fibrosis transmembrane conductance regulator (CFTR) is a cAMP-activated Cl^- and HCO_3^- conducting channel expressed in epithelial cells. The structure of CFTR presents two transmembrane domains (TMD1 and TMD2), two nucleotide-binding domains (NBD1 and NBD2) and a regulatory domain. Studies by Villella et al. (2019), demonstrated that p31–43, but not p57–68, binds to the NBD1 subunit of CFTR, reducing its ATPase activity. *In silico* modeling predicted that p31–43 interacts with NBD1 by binding at the boundary between NBD1 and TMD1, stabilizing the inactive conformation of CFTR. Loss of CFTR function causes an increase in ROS generation and a persistent activation of TG2, leading to increased nuclear translocation of NFκB. The activation of the NFκB pathway induces the transcription of proinflammatory cytokines, such as IL-17A, IL-21, and IL-15, which are closely related with different effector mechanisms in CD pathogenesis, as well as of components of the NLRP3 inflammasome pathway, which, upon activation, leads to the production of active IL-1β. These results were obtained *in vitro*, in assays using Caco-2 cells or PBMC from CD patients, as well as *in vivo*, in mouse models (Villella et al., 2019). The use of a compound maintaining CFTR in the active conformation prevented these p31–43 effects. Collectively, these findings showed that p31–43 can alter CFTR function, with multiple biological consequences for the pathogenesis of CD and, importantly, for other processes associated with the inflammatory response that may have an impact in other tissues as well (Fig. 4B).

6. Use of experimental mice models to unveil the *in vivo* effects of p31–43

Experimental mouse models have been developed to increase our understanding of the pathogenic mechanisms of gliadin peptides. Initial work from Troncone and Ferguson (1991) showed that mouse models were useful to investigate the immune response in the small intestine and to assess the pathogenic mechanisms underlying CD. Further work, using more complex models based on transgenic mice, were aimed at dissecting the mechanisms of mucosal damage involving the whole immune response (Ryan, 2015). Some of these models investigated the role of HLA molecules (Black et al., 2002; Marietta et al., 2004), IL-15 and T cell activation (Korneychuk et al., 2015; Yokoyama et al., 2011), the role of TG2 (Dafik et al., 2012), regulatory T cells (Du Pré et al., 2011; Galipeau et al., 2011; Van Leeuwen et al., 2017), the effect of microbiota (Caminero et al., 2019), and viral infections (Bouziat et al., 2017). All this information helped building the most complex mouse system, which allows drawing a more complete picture of CD pathology, incorporating most of the aspects related to IL-15, T cell activation, and IELs cytotoxicity (Abadie et al., 2020). Since the model developed by Bana Jabri's group recapitulates innate and adaptive mechanisms playing a role in CD pathogenesis, it could be also a very useful tool for the preclinical validation of therapeutic interventions.

The main goal of the various therapeutic strategies is the restoration of gluten tolerance in CD patients. Freitag et al. (2020) demonstrated the efficacy of nanoparticles encapsulating gliadin proteins to reduce gliadin-specific T cell proliferation, the production of inflammatory cytokines and circulating gliadin-specific antibodies, in three different mouse models (C57BL/6, RAG-deficient and huDQ8-transgenic). In addition, FoxP3 levels and the expression of genes associated with the regulatory response were enhanced. These findings in mice suggest an interesting potential strategy for therapeutic intervention in patients.

These mouse models were developed to assess the pathogenic mechanisms underlying CD, but they were mainly focused on the analysis of the gluten-specific response, and as a consequence, the role of innate immunity has been given less attention.

To evaluate whether p31–43 is able to drive innate mechanisms in the intestinal mucosa *in vivo*, a mouse model of enteropathy based on the intraluminal delivery of p31–43 by a surgical procedure was developed (Araya et al., 2016). This study showed that p31–43 reduces the villus height/crypt depth (V/C) ratio and increases the number of intraepithelial lymphocytes (IELs) (the most common morphological parameters used to evaluate changes in the mucosa of the small intestine). Experiments using deficient mice showed that these changes were dependent on MyD88 and type I IFNs, but not on TLR4. Inflammatory mediators (IFNγ, IFNβ, and CXCL10) were upregulated in the intestinal mucosa 4h after p31–43 treatment. Later, to avoid the inflammation caused by the surgical procedure, we demonstrated the toxic effects of p31–43 by intragastric delivery (Gómez Castro et al., 2019). In this model, the pathology in the upper small intestine, characterized by a reduced V/C ratio and a higher number of IELs, was observed 16h after intragastric administration of 10μg of p31–43. Treated mice also showed increased levels of IFNβ mRNA in the mucosa 4h after treatment. Importantly, the effects of p31–43 on the intestinal mucosa were sequence specific, since synthetic peptides with an inverted (YPQQPFPPQQQGL) or scrambled sequence (YQPLFQPQGPQPQ) did not show any effect even at a 10 times higher dose. The structure of p31–43 described in Section 4.1, markedly different to that of the scrambled or inverted peptides, may be responsible for such biological effects. Together, these results indicate that p31–43 activates *in vivo* innate immune pathways, such as type I IFNs dependent inflammation, and particularly, the inflammasome pathways (see Section 7).

7. Inflammasome activation by p31–43

Inflammasomes are cytosolic macromolecular complexes that assemble upon sensing components from pathogens or host damage-associated molecular patterns (DAMPs). The assembly of the inflammasome triggers the activation of caspase-1, which subsequently, induces the cleavage and maturation of the proinflammatory cytokines IL-1β and IL-18. Caspase-1 also cleaves gasdermin D (GSDMD), which oligomerizes and forms pores in the cellular membrane, leading to a programmed cell death called pyroptosis.

Evidence of inflammasome activation by gliadins came first from studies by Harris et al. (2010), which showed that incubation of PBMC from CD patients with PT-gliadins led to the release of IL-1β. Then, Palová-Jelínková et al. (2013) demonstrated the activation of the NLRP3 inflammasome by PT-gliadins. In this study, treatment of PBMC from CD patients with

PT-gliadins induced IL-1β and IL-1α production, while treatment with the caspase-1 inhibitor Z-YVAD-fmk reduced the release of IL-1β but not of IL-1α. Blocking K^+ efflux by incubating with a high extracellular K^+ concentration or by inhibiting K^+ channels, using quinidine or glybenclamide, suppressed IL-1β production. Experiments using cells from deficient mice showed that maximal production of IL-1β induced by PT-gliadins required TLR4/MyD88 signaling and was NLRP3- and ASC-dependent. Since these studies highlight the induction of inflammatory mediators by PT-gliadins through the inflammasome, later work aimed at determining directly whether p31–43 could activate the inflammasome platform.

In vitro assays, using THP1-ASC-GFP reporter cells, showed that p31–43 promotes ASC speck formation and the assembly of the inflammasome multimolecular platform (Gómez Castro et al., 2019). Importantly, intragastric administration of p31–43 to wild-type mice induced activation of NLRP3, caspase-1, IL-1β, and GSDMD. *In vivo* inhibition of caspase-1 precluded the activation of IL-1β and GSDMD upon p31–43 treatment, demonstrating the rapid induction of the NLRP3 inflammasome (Gómez Castro et al., 2019; Ruera et al., 2020). Our work demonstrated that activation of the inflammasome by p31–43 depends on NLRP3, ASC, and caspase-1, but the underlying molecular mechanism is unknown. As described above, p31–43 has a number of different toxic effects, known to produce various types of cellular stress, such as ER-stress and alterations of the mitochondria, which may promote NLRP3 activation. In contrast with other rather selective inflammasomes, NLRP3 is activated by a range of stimuli, including bacterial toxins (e.g., pore-forming toxins), host metabolites (e.g., uric acid and cholesterol crystals), and environmental irritants (e.g., asbestos and silica) (Swanson et al., 2019), which make difficult to identify the molecular connection between p31–43 and NLRP3 inflammasome assembly and activation. However, the particular conformation of p31–43 and its tendency to form oligomers are likely to be responsible for NLRP3 inflammasome activation (Fig. 4C).

8. A modern view of the role of p31–43 in CD pathogenesis

Here, we have reviewed literature of more than 20 years, describing the multiple proinflammatory effects of gliadin in general, and the gliadin p31–43 peptide in particular, in several cellular systems, mouse models, and intestinal tissues from CD patients. Subsequently, the question to be answered after having collected evidence from different experimental approaches is: why p31–43 affects only CD patients?

The answer may reside in the constitutive alterations present in celiac subjects. These involve several biological pathways, such as signaling/proliferation, stress/innate immune response and inflammation (Barone et al., 2014; Dieterich et al., 2020; Lania et al., 2019). A notable example is the NFκB pathway, constitutively altered in CD, with more than 20 components of the pathway increased in biopsies from patients on a GFD (Fernandez-jimenez et al., 2014; Lania et al., 2019). The constitutive inflammatory condition in CD cells is confirmed both in skin and intestinal fibroblasts from CD patients on a GFD with an increase of the activation of NFκB and MAPK1 (Lania et al., 2019). One possible candidate as responsible for this condition of inflammatory environment is the alteration of vesicular trafficking. Interestingly in fact, inducing a delay of the endocytic trafficking by silencing the HRS protein, produces at cellular level the same alterations that p31–43 does. Moreover, many non-HLA loci and e-qtl effects found related to CD are involved in the regulation of various stages of the vesicular trafficking (Lania et al., 2019). Finally, recent studies on intestinal organoids, derived from CD patients at the remission phase of the disease, indicate that constitutive alterations of the intestinal epithelial cells are present also in absence of gluten (Dieterich et al., 2020). Taken all together these data indicate that in CD constitutive alterations exist that renders the cells more sensitive to the proinflammatory activity of p31–43.

The variety of signaling pathways that can be activated by p31–43, such as the NFκB and MAPK pathways, Ca^{2+} mobilization, TG2 activation, inhibition of CFTR, and NLRP3 inflammasome activation, that have been reviewed here, suggest that p31–43 is able to alter cell biology in different ways. The particular biophysical characteristics of p31–43 likely play a role in the toxic effects described, and particularly in the activation of the inflammasome. Inflammation represents the common ground for several multifactorial diseases, such as rheumatic disease, type 2 diabetes, obesity, inflammatory bowel diseases, cardiovascular and neurodegenerative diseases, asthma, and cancer (Furman et al., 2019; King et al., 2020; Singh et al., 2018). Environmental factors, such as infections, western diet, but also the constituents of wheat proteins, amylase trypsin inhibitors (ATIs), and p31–43 can be considered important items in the list of drivers of the inflammatory reaction (Barone et al., 2014; Spisni et al., 2019). Although p31–43 may induce inflammation in normal tissues, CD patients seem to be particularly susceptible and react even to lower dose (Lania et al., 2019). Interestingly, combinations of proinflammatory factors may have synergistic effects. Thus, gliadins and viruses may cooperate to induce the activation of innate immunity and inflammatory pathways, which create a microenvironment with a lower threshold for the emergence and the amplification of gliadin-specific T cells (Lindfors et al., 2019; Nanayakkara et al., 2018).

To examine the long-lasting effects of inflammation even when it has apparently subsided, information gained from genomic studies must be considered. In addition to the HLA locus, 43 other loci are known to contribute to 40% of CD susceptibility. The non-HLA loci are mainly SNPs, which have also been associated with other autoimmune diseases. However, the most relevant finding is that the vast majority of CD-associated SNPs correspond to non-coding regions, thus, a simple connection with a protein function cannot be established (Withoff et al., 2016). For this reason, to link the genetic with the disease mechanism, additional information is required on the role of the particular SNPs on gene expression and how they are regulated, as well as on the affected pathways.

One example demonstrating such complexity was found by Castellanos-Rubio et al. (2016) and Castellanos-Rubio and Ghosh (2019) who studied long non-coding (lnc)RNAs. Since SNPs may localize in non-coding regions and affect the expression of genes by different mechanisms, the role of lncRNAs has received significant attention. lncRNA13 binds to a nuclear ribonucleoprotein, hnRNPD, and controls the expression of proinflammatory genes. Appropriate stimulation reduced the levels of lncRNA13, allowing the expression of the repressed genes. Interestingly, the lncRNA13 levels are significantly decreased in biopsy samples from the small intestine of CD patients, and a SNP variant of lncRNA13, found in CD patients, has less affinity to hnRNPD, leading to poor control of the repressed proinflammatory genes, and possibly contributing to the underlying chronic inflammation.

To increase the complexity, epigenetic changes, such as DNA methylation, may also contribute to the mechanisms that lead to the disease or, once the disease has been initiated, establish a chronic disorder. As an example, the evaluation of the methylome of the epithelial and immune cell populations of duodenal biopsies in CD patients and healthy individuals, showed that these cell populations present different methylation signatures, which may have a broad effect on the phenotype in a cell type-specific manner (Fernandez-Jimenez et al., 2019).

In addition, the analysis of the effects of SNPs requires taking into account the microenvironment and cell-cell contacts in the tissue. Thus, genetic and epigenetic studies have shown that not only T cells are mandatory to execute the tissue damage in CD but neutrophils, monocytes, enterocytes, and B cells also contribute to the whole picture of CD pathogenesis (Withoff et al., 2016). The use of new technologies and *in silico* analysis has begun to tackle the problem of the underlying complexity, and SNPs and epigenetic changes may be used in the future as biomarkers of the disease. Studies assessing whether p31–43 affects in the context of some particular SNPs or drives epigenetic changes on duodenal tissue may provide further insight on the understanding of the effects of this peptide on CD patients.

Since many of the biological effects reported for p31–43 are not completely understood, further investigations are required to determine the mechanisms of action of this peptide at the molecular and subcellular levels. We just begin to understand the relative contribution of genetic predisposition and environmental factors, such as viral infections (Fig. 5).

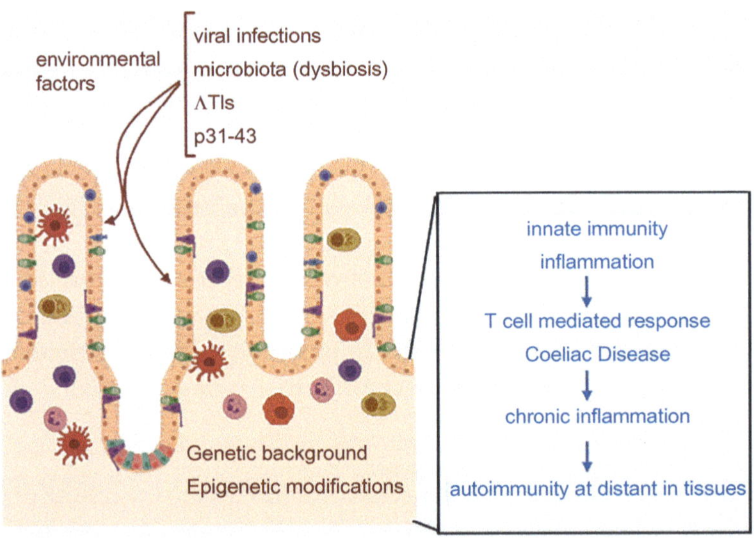

Fig. 5 Drivers of innate response in small intestine and its long term consequences. Environmental factors such as viral infections, changes in microbiota (dysbiosis), amylase trypsin inhibitors (ATIs) and p31–43, as the proposed stimuli driving the inflammatory response that promote the development of adaptive response in genetically predisposed individuals leading to Coeliac disease. Further chronic inflammation may be responsible to initiate or potentiate autoimmunity in different tissues.

Altogether, knowledge of the molecular mechanisms underlying the p31–43 action will have important implications in terms of both prevention and therapy, and will help us to understand the triggering events in CD-associated autoimmune conditions.

9. Conclusions

CD is the prototype of a chronic inflammatory disorder induced by dietary components. Cells in tissues from CD patients show increased sensitivity to the inflammatory insult from wheat-derived peptides. This is probably due to several layers of susceptibility factors: genes, non-coding regions, and epigenetic modifications. In this context, p31–43 comes into the spotlight as an important player in CD pathogenesis. Its particular conformation and its ability to induce different forms of cellular stress drive multiple inflammatory pathways, which, in the presence of appropriate susceptibility and environmental factors, may act together to drive the disease.

References

Abadie, V., et al., 2020. IL-15, gluten and HLA-DQ8 drive tissue destruction in coeliac disease. Nature 578 (7796), 600–604. https://doi.org/10.1038/s41586-020-2003-8.

Allegretti, Y.L., et al., 2013. Broad MICA/B expression in the small bowel mucosa: a link between cellular stress and celiac disease. PLoS One 8 (9), e73658. https://doi.org/10.1371/journal.pone.0073658.

Araya, R.E., et al., 2016. Mechanisms of innate immune activation by gluten peptide p31-43 in mice. Am. J. Physiol. Gastrointest. Liver Physiol. 311 (1), G40–G49. https://doi.org/10.1152/ajpgi.00435.2015.

Barone, M.V., Zimmer, K.P., 2016. Endocytosis and transcytosis of gliadin peptides. Mol. Cell. Pediatr. 3 (1), 4–8. https://doi.org/10.1186/s40348-015-0029-z.

Barone, M.V., et al., 2007. Growth factor-like activity of gliadin, an alimentary protein: implications for coeliac disease. Gut 56 (4), 480–488. https://doi.org/10.1136/gut.2005.086637.

Barone, M.V., et al., 2010. Gliadin peptide P31-43 localises to endocytic vesicles and interferes with their maturation. PLoS One 5 (8), e17039. https://doi.org/10.1371/journal.pone.0012246.

Barone, M.V., et al., 2011. Gliadin-mediated proliferation and innate immune activation in celiac disease are due to alterations in vesicular trafficking. PLoS One 6 (2), e17039. https://doi.org/10.1371/journal.pone.0017039.

Barone, M.V., Troncone, R., Auricchio, S., 2014. Gliadin peptides as triggers of the proliferative and stress/innate immune response of the celiac small intestinal mucosa. Int. J. Mol. Sci. 15 (11), 20518–20537. https://doi.org/10.3390/ijms151120518.

Barrera, E., Chirdo, F., Pantano, S., 2019. P31-43 gliadin peptide forms oligomers and induces NLRP3 inflammasome/caspase 1-dependent mucosal damage in small intestine. Commentary. Front. Immunol. https://doi.org/10.3389/fimmu.2019.02792.

Bayardo, M., et al., 2012. Transglutaminase 2 expression is enhanced synergistically by interferon-γ and tumour necrosis factor-α in human small intestine. Clin. Exp. Immunol. 168 (1), 95–104. https://doi.org/10.1111/j.1365-2249.2011.04545.x.

Bechara, C., Sagan, S., 2013. Cell-penetrating peptides: 20 years later, where do we stand? FEBS Lett. 587 (12), 1693–1702. https://doi.org/10.1016/j.febslet.2013.04.031.

Bernardin, J.E., Kasarda, D.D., Mecham, D.K., 1967. Preparation and characterization of alpha-gliadin. J. Biol. Chem. 242 (3), 445–450.

Bernardo, D., et al., 2008. Higher constitutive IL15Rα expression and lower IL-15 response threshold in coeliac disease patients. Clin. Exp. Immunol. 154 (1), 64–73.

Black, K.E., Murray, J.A., David, C.S., 2002. HLA-DQ determines the response to exogenous wheat proteins: a model of gluten sensitivity in transgenic knockout mice. J. Immunol. 169 (10), 5595–5600. https://doi.org/10.4049/jimmunol.169.10.5595.

Bouziat, R., et al., 2017. Reovirus infection triggers inflammatory responses to dietary antigens and development of celiac disease. Science 356 (6333), 44–50. https://doi.org/10.1126/science.aah5298.

Browning, T.H., Trier, J.S., 1969. Organ culture of mucosal biopsies of human small intestine. J. Clin. Invest. 48 (8), 1423–1432. https://doi.org/10.1172/JCI106108.

Budagian, V., et al., 2006. IL-15/IL-15 receptor biology: a guided tour through an expanding universe. Cytokine Growth Factor Rev. 17 (4), 259–280. https://doi.org/10.1016/j.cytogfr.2006.05.001.

Burke, P., Schooler, K., Wiley, H.S., 2001. Regulation of epidermal growth factor receptor signaling by endocytosis and intracellular trafficking. Mol. Biol. Cell 12 (6), 1897–1910. https://doi.org/10.1091/mbc.12.6.1897.

Calvanese, L., et al., 2019. Structural insights on P31-43, a gliadin peptide able to promote an innate but not an adaptive response in celiac disease. J. Pept. Sci. 25 (5), e3161. https://doi.org/10.1002/psc.3161.

Caminero, A., et al., 2019. Mechanisms by which gut microorganisms influence food sensitivities. Nat. Rev. Gastroenterol. Hepatol. 16 (1), 7–18. https://doi.org/10.1038/s41575-018-0064-z. Springer US.

Capozzi, A., et al., 2013. Modulatory effect of gliadin peptide 10-mer on epithelial intestinal CACO-2 cell inflammatory response. PLoS One 8 (6), e66561. https://doi.org/10.1371/journal.pone.0066561.

Caputo, I., et al., 2012. Gliadin peptides induce tissue transglutaminase activation and ER-stress through Ca2+ mobilization in Caco-2 cells. PLoS One 7 (9), e45209. https://doi.org/10.1371/journal.pone.0045209.

Caputo, I., et al., 2013. Anti-tissue transglutaminase antibodies activate intracellular tissue transglutaminase by modulating cytosolic Ca2+ homeostasis. Amino Acids 44 (1), 251–260. https://doi.org/10.1007/s00726-011-1120-y.

Castellanos-Rubio, A., Bilbao, J.R., 2018. Profiling celiac disease-related transcriptional changes. 1st edn. In: International Review of Cell and Molecular Biology, first ed. Elsevier Inc, https://doi.org/10.1016/bs.ircmb.2017.07.003.

Castellanos-Rubio, A., Ghosh, S., 2019. Disease-associated SNPs in inflammation-related lncRNAs. Front. Immunol. 10 (Mar), 1–9. https://doi.org/10.3389/fimmu.2019.00420.

Castellanos-Rubio, A., et al., 2016. A long noncoding RNA associated with susceptibility to celiac disease. Science 352 (6281), 91–95. https://doi.org/10.1126/science.aad0467.

Ceresa, B.P., Schmid, S.L., 2000. Regulation of signal transduction by endocytosis. Curr. Opin. Cell Biol. 12 (2), 204–210. https://doi.org/10.1016/S0955-0674(99)00077-0.

Chiang, C.Y., et al., 2012. Cofactors required for TLR7- and TLR9-dependent innate immune responses. Cell Host Microbe 11 (3), 306–318. https://doi.org/10.1016/j.chom.2012.02.002. Elsevier Inc.

Ciclitira, P.J., Evans, D.J., Fagg, N.L., Lennox, E.S., Dowling, R., 1984. Clinical testing of gliadin fractions in coeliac patients. Clin. Sci. (Lond.) 66 (3), 357–364.

D'Eletto, M., et al., 2018. Transglutaminase type 2 regulates ER-mitochondria contact sites by interacting with GRP75. Cell Rep. 25 (13), 3573–3581.e4. https://doi.org/10.1016/j.celrep.2018.11.094. ElsevierCompany.

Dafik, L., et al., 2012. Activation and inhibition of transglutaminase 2 in mice. PLoS One 7 (2), 1–7. https://doi.org/10.1371/journal.pone.0030642.

de Ritis, G., et al., 1988. In vitro (organ culture) studies of the toxicity of specific A-gliadin peptides in celiac disease. Gastroenterology 94 (1), 41–49. https://doi.org/10.1016/0016-5085(88)90607-5. Elsevier Inc.

Dieterich, W., Neurath, M.F., Zopf, Y., 2020. Intestinal ex vivo organoid culture reveals altered programmed crypt stem cells in patients with celiac disease. Sci. Rep. 10 (1), 1–10. https://doi.org/10.1038/s41598-020-60521-5.

Du Pré, M.F., et al., 2011. Tolerance to ingested deamidated gliadin in mice is maintained by splenic, type 1 regulatory T cells. Gastroenterology 141 (2). https://doi.org/10.1053/j.gastro.2011.04.048. AGA Institute American Gastroenterological Association.

Duitman, E.H., et al., 2008. How a cytokine is chaperoned through the secretory pathway by complexing with its own receptor: lessons from interleukin-15 (IL-15)/IL-15 receptor α. Mol. Cell. Biol. 28 (15), 4851–4861. https://doi.org/10.1128/mcb.02178-07.

Esposito, C., 2003. Expression and enzymatic activity of small intestinal tissue transglutaminase in celiac disease. Am. J. Gastroenterol. 98 (8), 1813–1820. https://doi.org/10.1016/s0002-9270(03)00431-3.

Falchuk, Z.M., et al., 1974. An in vitro model of gluten sensitive enteropathy. Effect of gliadin on intestinal epithelial cells of patients with gluten sensitive enteropathy in organ culture. J. Clin. Investig. 53 (2), 487–500. https://doi.org/10.1172/JCI107582.

Fehniger, T.A., Caligiuri, M.A., 2001. Interleukin 15: biology and relevance to human disease. Blood 97 (1), 14–32. https://doi.org/10.1182/blood.V97.1.14.

Fernandez-jimenez, N., et al., 2014. Coregulation and modulation of NFκB-related genes in celiac disease: uncovered aspects of gut mucosal inflammation. Hum. Mol. Genet. 23 (5), 1298–1310. https://doi.org/10.1093/hmg/ddt520.

Fernandez-Jimenez, N., et al., 2019. The methylome of the celiac intestinal epithelium harbours genotype-independent alterations in the HLA region. Sci. Rep. 9 (1), 1–13. https://doi.org/10.1038/s41598-018-37746-6.

Festenstein, G.N., Hay, F.C., Miflin, B.J., Shewry, P., 1984. Immunochemical studies on barley seed storage proteins: the specificity of an antibody to "C" hordein and its reaction with prolamins from other cereals. Planta 162 (6), 524–531.

Field, J.M., et al., 1982. The purification and characterization of homologous high molecular weight storage proteins from grain of wheat, rye and barley. Theor. Appl. Genet. 62 (4), 329–336. https://doi.org/10.1007/BF00275097.

Franz, J., et al., 2016. SAP(E)—a cell-penetrating polyproline helix at lipid interfaces. Biochim. Biophys. Acta, Biomembr. 1858 (9), 2028–2034. https://doi.org/10.1016/j.bbamem.2016.05.021. Elsevier B.V.

Freitag, T.L., et al., 2020. Gliadin nanoparticles induce immune tolerance to gliadin in mouse models of celiac disease. Gastroenterology 158 (6), 1667–1681.e12. https://doi.org/10.1053/j.gastro.2020.01.045. The American Gastroenterological Association.

Frossi, B., et al., 2017. Mast cells are associated with the onset and progression of celiac disease. J. Allergy Clin. Immunol. 139 (4), 1266–1274.e1. https://doi.org/10.1016/j.jaci.2016.08.011. Elsevier Ltd.

Furman, D., et al., 2019. Chronic inflammation in the etiology of disease across the life span. Nat. Med. 25 (12), 1822–1832. https://doi.org/10.1038/s41591-019-0675-0. Springer US.

Gaggero, A., et al., 1999. Differential intracellular trafficking, secretion and endosomal localization of two IL-15 isoforms. Eur. J. Immunol. 29 (4), 1265–1274. https://doi.org/10.1002/(sici)1521-4141(199904)29:04<1265::aid-immu1265>3.0.co;2-v.

Galipeau, H.J., et al., 2011. Sensitization to gliadin induces moderate enteropathy and insulitis in nonobese diabetic-DQ8 mice. J. Immunol. 187 (8), 4338–4346. https://doi.org/10.4049/jimmunol.1100854.

Gianfrani, C., et al., 2015. Extensive in vitro gastrointestinal digestion markedly reduces the immune-toxicity of Triticum monococcum wheat: implication for celiac disease. Mol. Nutr. Food Res. 59 (9), 1844–1854. https://doi.org/10.1002/mnfr.201500126.

Gómez Castro, M.F., et al., 2019. P31-43 gliadin peptide forms oligomers and induces NLRP3 inflammasome/caspase 1-dependent mucosal damage in small intestine. Front. Immunol. 10 (Jan), 1–11. https://doi.org/10.3389/fimmu.2019.00031.

Groh, V., et al., 1996. Cell stress-regulated human major histocompatibility complex class I gene expressed in gastrointestinal epithelium. Proc. Natl. Acad. Sci. U. S. A. 93 (22), 12445–12450. https://doi.org/10.1073/pnas.93.22.12445.

Hansson, T., et al., 2002. Transforming growth factor-β (TGF-β) and tissue transglutaminase expression in the small intestine in children with coeliac disease. Scand. J. Immunol. 56 (5), 530–537. https://doi.org/10.1046/j.1365-3083.2002.01157.x.

Harris, K.M., Fasano, A., Mann, D.L., 2010. Monocytes differentiated with IL-15 support Th17 and Th1 responses to wheat gliadin: implications for celiac disease. Clin. Immunol. 135 (3), 430–439. https://doi.org/10.1016/j.clim.2010.01.003. Elsevier Inc.

Heil, F., et al., 2003. The Toll-like receptor 7 (TLR7)-specific stimulus loxoribine uncovers a strong relationship within the TLR7, 8 and 9 subfamily. Eur. J. Immunol. 33 (11), 2987–2997. https://doi.org/10.1002/eji.200324238.

Herrera, M.G., et al., 2020. Structural conformation and self-assembly process of p31-43 gliadin peptide in aqueous solution. Implications for celiac disease. FEBS J. 287 (10), 2134–2149. https://doi.org/10.1111/febs.15109.

Hüe, S., et al., 2004. A direct role for NKG2D/MICA interaction in villous atrophy during celiac disease. Immunity 21 (3), 367–377. https://doi.org/10.1016/j.immuni.2004.06.018.

Husby, S., et al., 2020. European society paediatric gastroenterology, hepatology and nutrition guidelines for diagnosing coeliac disease 2020. J. Pediatr. Gastroenterol. Nutr. 70 (1), 141–156. https://doi.org/10.1097/MPG.0000000000002497.

Iversen, R., et al., 2020. Evidence that pathogenic transglutaminase 2 in celiac disease derives from enterocytes. Gastroenterology 159 (2), 788–790. https://doi.org/10.1053/j.gastro.2020.04.018. The American Gastroenterological Association.

Iversen, R., Sollid, L.M., 2020. Autoimmunity provoked by foreign antigens: in celiac disease, exogenous gluten drives T cell-B cell interactions that cause autoimmunity. Science 368 (6487), 132–133. https://doi.org/10.1126/science.aay3037.

Jabri, B., Abadie, V., 2015. IL-15 functions as a danger signal to regulate tissue-resident T cells and tissue destruction. Nat. Rev. Immunol. 15 (12), 771–783. https://doi.org/10.1038/nri3919. Nature Publishing Group.

Jabri, B., Sollid, L.M., 2017. T cells in celiac disease. J. Immunol. 198 (8), 3005–3014. https://doi.org/10.4049/jimmunol.1601693.

Junker, Y., et al., 2012. Wheat amylase trypsin inhibitors drive intestinal inflammation via activation of toll-like receptor 4. J. Exp. Med. 209 (13), 2395–2408. https://doi.org/10.1084/jem.20102660.

Kasarda, D.D., Bernardin, J.E., Thomas, R.S., 1967. Reversible aggregation of α-gliadin to fibrils. Science 155 (3759), 203–205. https://doi.org/10.1126/science.155.3759.203.

Kasarda, D.D., Bernardin, J.E., Gaffield, W., 1968. Circular dichroism and optical rotatory dispersion of alpha-gliadin. Biochemistry 7 (11), 3950–3957.

Kasarda, D.D., et al., 1984. Nucleic acid (cDNA) and amino acid sequences of alpha-type gliadins from wheat (Triticum aestivum). Proc. Natl. Acad. Sci. U. S. A. 81 (15), 4712–4716. https://doi.org/10.1073/pnas.81.15.4712.

Kim, S.M., Mayassi, T., Jabri, B., 2015. Innate immunity: actuating the gears of celiac disease pathogenesis. Best Pract. Res. Clin. Gastroenterol. 29 (3), 425–435. https://doi.org/10.1016/j.bpg.2015.05.001. Elsevier Ltd.

King, J.A., et al., 2020. Incidence of celiac disease is increasing over time: a systematic review and meta-analysis. Am. J. Gastroenterol. 115 (4), 507–525. https://doi.org/10.14309/ajg.0000000000000523.

Komada, M., Kitamura, N., 1995. Growth factor-induced tyrosine phosphorylation of Hrs, a novel 115-kilodalton protein with a structurally conserved putative zinc finger domain. Mol. Cell. Biol. 15 (11), 6213–6221. https://doi.org/10.1128/mcb.15.11.6213.

Korneychuk, N., Meresse, B., Cerf-Bensussan, N., 2015. Lessons from rodent models in celiac disease. Mucosal Immunol. 8 (1), 18–28. https://doi.org/10.1038/mi.2014.102. Nature Publishing Group.

Kreis, M., et al., 1985. Molecular evolution of the seed storage proteins of barley, rye and wheat. J. Mol. Biol. 183 (3), 499–502. https://doi.org/10.1016/0022-2836(85)90017-8.

Lania, G., et al., 2019. Constitutive alterations in vesicular trafficking increase the sensitivity of cells from celiac disease patients to gliadin. Commun. Biol. 2 (1), 1–13. https://doi.org/10.1038/s42003-019-0443-1. Springer US.

Lee, J., et al., 2004. Transglutaminase 2 induces nuclear factor-κB activation via a novel pathway in BV-2 microglia. J. Biol. Chem. 279 (51), 53725–53735. https://doi.org/10.1074/jbc.M407627200.

Lim, S.D., et al., 2018. Gliadin intake induces oxidative-stress responses in Caenorhabditis elegans. Biochem. Biophys. Res. Commun. 503 (3), 2139–2145. https://doi.org/10.1016/j.bbrc.2018.08.002. Elsevier Ltd.

Lindfors, K., et al., 2019. Coeliac disease. Nat. Rev. Dis. Primers. 5, 1. https://doi.org/10.1038/s41572-018-0054-z. Springer US.

Maiuri, L., Picarelli, A., et al., 1996a. Definition of the initial immunologic modifications upon in vitro gliadin challenge in the small intestine of celiac patients. Gastroenterology 110 (5), 1368–1378. https://doi.org/10.1053/gast.1996.v110.pm8613040.

Maiuri, L., Troncone, R., et al., 1996b. In vitro activities of A-gliadin-related synthetic peptides: damaging effect on the atrophic coeliac mucosa and activation of mucosal immune response in the treated coeliac mucosa. Scand. J. Gastroenterol. 31 (3), 247–253. https://doi.org/10.3109/00365529609004874.

Maiuri, L., et al., 2000. Interleukin 15 mediates epithelial changes in celiac disease. Gastroenterology 119 (4), 996–1006. https://doi.org/10.1053/gast.2000.18149.

Maiuri, L., et al., 2003. Association between innate response to gliadin and activation of pathogenic T cells in coeliac disease. Lancet 362 (9377), 30–37. https://doi.org/10.1016/S0140-6736(03)13803-2.

Maiuri, L., et al., 2010. Lysosomal accumulation of gliadin p31-43 peptide induces oxidative stress and tissue transglutaminase-mediated PPARγ downregulation in intestinal epithelial cells and coeliac mucosa. Gut 59 (3), 311–319. https://doi.org/10.1136/gut.2009.183608.

Mamone, G., et al., 2007. Identification of a peptide from α-gliadin resistant to digestive enzymes: implications for celiac disease. J. Chromatogr. B Anal. Technol. Biomed. Life Sci. 855 (2 Spec. Iss), 236–241. https://doi.org/10.1016/j.jchromb.2007.05.009.

Marietta, E., et al., 2004. A new model for dermatitis herpetiformis that uses HLA-DQ8 transgenic NOD mice. J. Clin. Investig. 114 (8), 1090–1097. https://doi.org/10.1172/JCI200421055.

Marshall, M.S., 1995. Ras target proteins in eukaryotic cells. FASEB J. 9 (13), 1311–1318. https://doi.org/10.1096/fasebj.9.13.7557021.

Martucciello, S., et al., 2020. Interplay between type 2 transglutaminase (Tg2), gliadin peptide 31-43 and anti-tg2 antibodies in celiac disease. Int. J. Mol. Sci. 21 (10). https://doi.org/10.3390/ijms21103673.

Matysiak-Budnik, T., et al., 2003. Alterations of the intestinal transport and processing of gliadin peptides in celiac disease. Gastroenterology 125 (3), 696–707. https://doi.org/10.1016/S0016-5085(03)01049-7.

Matysiak-Budnik, T., et al., 2008. Secretory IgA mediates retrotranscytosis of intact gliadin peptides via the transferrin receptor in celiac disease. Exp. Med. 205 (1), 143–154. https://doi.org/10.1084/jem.20071204.

Mayassi, T., et al., 2019. Chronic inflammation permanently reshapes tissue-resident immunity in celiac disease. Cell 176 (5), 967–981.e19. https://doi.org/10.1016/j.cell.2018.12.039. Elsevier Inc.

Mehta, K., Kumar, A., Kim, H.I., 2010. Transglutaminase 2: a multi-tasking protein in the complex circuitry of inflammation and cancer. Biochem. Pharmacol. 80 (12), 1921–1929. https://doi.org/10.1016/j.bcp.2010.06.029. Elsevier Inc.

Mention, J.J., et al., 2003. Interleukin 15: a key to disrupted intraepithelial lymphocyte homeostasis and lymphomagenesis in celiac disease. Gastroenterology 125 (3), 730–745. https://doi.org/10.1016/S0016-5085(03)01047-3.

Meresse, B., et al., 2004. Coordinated induction by IL15 of a TCR-independent NKG2D signaling pathway converts CTL into lymphokine-activated killer cells in celiac disease. Immunity 21 (3), 357–366. https://doi.org/10.1016/j.immuni.2004.06.020.

Mnard, S., et al., 2012. Paracellular versus transcellular intestinal permeability to gliadin peptides in active celiac disease. Am. J. Pathol. 180 (2), 608–615. https://doi.org/10.1016/j.ajpath.2011.10.019. Elsevier Inc.

Molberg, O., Mcadam, S.N., Körner, R., Quarsten, H., Kristiansen, C., Madsen, L., Fugger, L., Scott, H., Norén, O., Roepstorff, P., Lundin, K.E., Sjöström, H., Sollid, L., 1998. Tissue transglutaminase selectively modifies gliadin peptides that are recognized by gut-derived T cells in celiac disease. Nat. Med. 4 (6), 713–717.

Monguzzi, E., et al., 2019. Gliadin effect on the oxidative balance and DNA damage: an in-vitro, ex-vivo study. Dig. Liver Dis. 51 (1), 47–54. https://doi.org/10.1016/j.dld.2018.06.020. Editrice Gastroenterologica Italiana.

Monteleone, G., et al., 2001. Role of interferon α in promoting T helper cell type 1 responses in the small intestine in coeliac disease. Gut 48 (3), 425–429. https://doi.org/10.1136/gut.48.3.425.

Mowat, A.M.I., 2018. To respond or not to respond—a personal perspective of intestinal tolerance. Nat. Rev. Immunol. 18 (6), 405–415. https://doi.org/10.1038/s41577-018-0002-x. Springer US.

Mowat, A.M., Agace, W.W., 2014. Regional specialization within the intestinal immune system. Nat. Rev. Immunol. 14 (10), 667–685. https://doi.org/10.1038/nri3738.

Nanayakkara, M., et al., 2013. An undigested gliadin peptide activates innate immunity and proliferative signaling in enterocytes: the role in celiac disease. Am. J. Clin. Nutr. 98 (4), 1123–1135. https://doi.org/10.3945/ajcn.112.054544.

Nanayakkara, M., et al., 2018. P31-43, an undigested gliadin peptide, mimics and enhances the innate immune response to viruses and interferes with endocytic trafficking: a role in celiac disease. Sci. Rep. 8 (1), 1–12. https://doi.org/10.1038/s41598-018-28830-y. Springer US.

Oba, M., et al., 2019. Secondary structures and cell-penetrating abilities of arginine-rich peptide foldamers. Sci. Rep. 9 (1), 1–9. https://doi.org/10.1038/s41598-018-38063-8. Springer US.

Orlando, A., et al., 2019. An in vitro study on mitochondrial compensatory response induced by gliadin peptides in caco-2 cells. Int. J. Mol. Sci. 20 (8), 1862. https://doi.org/10.3390/ijms20081862.

Osborne, T., 1908. Our present knowledge of plant proteins. Science 28, 417–427.

Palová-Jelínková, L., et al., 2013. Pepsin digest of wheat gliadin fraction increases production of IL-1β via TLR4/MyD88/TRIF/MAPK/NF-κB signaling pathway and an NLRP3 inflammasome activation. PLoS One 8 (4), e62426. https://doi.org/10.1371/journal.pone.0062426.

Pane, J.A., Webster, N.L., Coulson, B.S., 2014. Rotavirus activates lymphocytes from non-obese diabetic mice by triggering toll-like receptor 7 signaling and interferon production in plasmacytoid dendritic cells. PLoS Pathog. 10 (3), 1003998. https://doi.org/10.1371/journal.ppat.1003998.

Paolella, G., et al., 2018. The toxic alpha-gliadin peptide 31–43 enters cells without a surface membrane receptor. Cell Biol. Int. 42 (1), 112–120. https://doi.org/10.1002/cbin.10874.

Perera, L., et al., 2007. Expression of nonclassical class I molecules by intestinal epithelial cells. Inflamm. Bowel Dis. 13 (3), 298–307. https://doi.org/10.1002/ibd.20026.

Przemioslo, R., et al., 1995. Analysis of crypt cell proliferation in coeliac disease using MI-B1 antibody shows an increase in growth fraction. Gut 36 (1), 22–27. https://doi.org/10.1136/gut.36.1.22.

Qiao, S.W., et al., 2014. Biased usage and preferred pairing of α- and β-chains of TCRs specific for an immunodominant gluten epitope in coeliac disease. Int. Immunol. 26 (1), 13–19. https://doi.org/10.1093/intimm/dxt037.

Raiborg, C., Stenmark, H., 2002. Hrs and endocytic sorting of ubiquitinated membrane proteins. Cell Struct. Funct. 27 (6), 403–408. https://doi.org/10.1247/csf.27.403.

Raiborg, C., et al., 2001. Hrs recruits clathrin to early endosomes. EMBO J. 20 (17), 5008–5021. https://doi.org/10.1093/emboj/20.17.5008.

Raiborg, C., et al., 2008. Differential functions of Hrs and ESCRT proteins in endocytic membrane trafficking. Exp. Cell Res. 314 (4), 801–813. https://doi.org/10.1016/j.yexcr.2007.10.014.

Raulet, D.H., 2003. Roles of the NKG2D immunoreceptor and its ligands. Nat. Rev. Immunol. 3 (10), 781–790. https://doi.org/10.1038/nri1199.

Reinecker, H.C., et al., 1996. Intestinal epithelial cells both express and respond to interleukin 15. Gastroenterology 111 (6), 1706–1713. https://doi.org/10.1016/S0016-5085(96)70036-7.

Rivabene, R., Mancini, E., De Vincenzi, M., 1999. In vitro cytotoxic effect of wheat gliadin-derived peptides on the Caco-2 intestinal cell line is associated with intracellular oxidative imbalance: implications for coeliac disease. Biochim. Biophys. Acta Mol. basis Dis. 1453 (1), 152–160. https://doi.org/10.1016/S0925-4439(98)00095-7.

Ruera, C.N., Miculán, E., Pérez, F., Ducca, G., Carasi, P., Chirdo, F.G., 2020. Sterile Inflammation drives multiple programmed cell death pathways in the gut. J. Leukoc. Biol. https://doi.org/10.1002/JLB.3MA0820-660R.

Ryan, A.W., 2015. Celiac disease. Methods Mol. Biol. 1326. https://doi.org/10.1007/978-1-4939-2839-2.

Sanchez-Guerrero, E., Jo, S.R., Chong, B.H., Khachigian, L., 2013. EGFR and the complexity of receptor crosstalk in the cardiovascular system. Curr. Mol. Med. 13 (1), 3–12.

Shan, L., Molberg, Ø., Parrot, I., Hausch, F., Filiz, F., Gray, G.M., Sollid, L.M., Khosla, C., 2002. Structural basis for gluten intolerance in celiac sprue. Science 297 (5590), 2275–2279. https://doi.org/10.1126/science.1074129.

Shewry, P., 2019. What is gluten—why is it special? Front. Nutr. 6 (July), 1–10. https://doi.org/10.3389/fnut.2019.00101.

Shewry, P.R., Halford, N.G., 2002. Cereal seed storage proteins: structures, properties and role in grain utilization. J. Exp. Bot. 53 (370), 947–958. https://doi.org/10.1093/jexbot/53.370.947.

Shidrawi, R.G., et al., 1995. In vitro toxicity of gluten peptides in coeliac disease assessed by organ culture. Scand. J. Gastroenterol. 30 (8), 758–763. https://doi.org/10.3109/00365529509096324.

Siegel, M., et al., 2008. Extracellular transglutaminase 2 is catalytically inactive, but is transiently activated upon tissue injury. PLoS One 3 (3), e1861. https://doi.org/10.1371/journal.pone.0001861.

Singh, P., et al., 2018. Global prevalence of celiac disease: systematic review and meta-analysis. Clin. Gastroenterol. Hepatol. 16 (6), 823–836.e2. https://doi.org/10.1016/j.cgh.2017.06.037.

Sollid, L.M., Jabri, B., 2013. Triggers and drivers of autoimmunity: lessons from coeliac disease. Nat. Rev. Immunol. 13 (4), 294–302. https://doi.org/10.1038/nri3407. Nature Publishing Group.

Sollid, L.M., et al., 2012. Nomenclature and listing of celiac disease relevant gluten T-cell epitopes restricted by HLA-DQ molecules. Immunogenetics 64 (6), 455–460. https://doi.org/10.1007/s00251-012-0599-z.

Sollid, L.M., et al., 2020. Update 2020: nomenclature and listing of celiac disease–relevant gluten epitopes recognized by CD4+ T cells. Immunogenetics 72 (1–2), 85–88. https://doi.org/10.1007/s00251-019-01141-w.

Spisni, E., et al., 2019. Differential physiological responses elicited by ancient and heritage wheat cultivars compared to modern ones. Nutrients 11 (12), 1–20. https://doi.org/10.3390/nu11122879.

Stene, L.C., et al., 2006. Rotavirus infection frequency and risk of celiac disease autoimmunity in early childhood: a longitudinal study. Am. J. Gastroenterol. 101 (10), 2333–2340. https://doi.org/10.1111/j.1572-0241.2006.00741.x.

Sturgess, R., Day, P., Ellis, H.J., Lundin, K.E., Gjertsen, H.A., Kontakou, M., Ciclitira, P., 1994. Wheat peptide challenge in coeliac disease. Lancet 343 (8900), 758–761.

Swanson, K.V., Deng, M., Ting, J.P.Y., 2019. The NLRP3 inflammasome: molecular activation and regulation to therapeutics. Nat. Rev. Immunol. 19 (8), 477–489. https://doi.org/10.1038/s41577-019-0165-0. Springer US.

Szondy, Z., et al., 2017. Transglutaminase 2 in human diseases. BioMedicine (France) 7 (3), 1–13. https://doi.org/10.1051/bmdcn/2017070315.

Tatham, A.S., Drake, A.F., Shewry, P.R., 1985. A conformational study of a glutamine- and proline-rich cereal seed protein, C hordein. Biochem. J. 226 (2), 557–562. https://doi.org/10.1042/bj2260557.

Tatham, A.S., et al., 1990. Conformational studies of peptides corresponding to the coeliac-activating regions of wheat α-gliadin. Biochem. J. 270 (2), 313–318. https://doi.org/10.1042/bj2700313.

Thomas, K.E., et al., 2006. Gliadin stimulation of murine macrophage inflammatory gene expression and intestinal permeability are MyD88-dependent: role of the innate immune response in celiac disease. J. Immunol. 176 (4), 2512–2521. https://doi.org/10.4049/jimmunol.176.4.2512.

Troncone, R., Ferguson, A., 1991. Animal model of gluten induced enteropathy in mice. Gut 32 (8), 871–875.

Van Leeuwen, M.A., et al., 2017. Macrophage-mediated gliadin degradation and concomitant IL-27 production drive IL-10- and IFN-γ 3-secreting Tr1-like-cell differentiation in a murine model for gluten tolerance. Mucosal Immunol. 10 (3), 635–649. https://doi.org/10.1038/mi.2016.76.

Vilasi, S., et al., 2010. Interaction of "toxic" and "immunogenic" A-gliadin peptides with a membrane-mimetic environment. J. Mol. Recognit. 23 (3), 322–328. https://doi.org/10.1002/jmr.987.

Villella, V.R., et al., 2019. A pathogenic role for cystic fibrosis transmembrane conductance regulator in celiac disease. EMBO J. 38 (2), 1–22. https://doi.org/10.15252/embj.2018100101.

Vincentini, O., et al., 2015. Gliadin-dependent cytokine production in a bidimensional cellular model of celiac intestinal mucosa. Clin. Exp. Med. 15 (4), 447–454. https://doi.org/10.1007/s10238-014-0325-2. Springer Milan.

Wieser, H., Belitz, H.D., Ashkenazi, A., 1984. Amino-acid sequence of the coeliac active gliadin peptide B 3142. Z. Lebensm. Unters. Forsch. 179 (5), 371–376. https://doi.org/10.1007/BF01043432.

Wieser, H., et al., 1986. Coeliac activity of the gliadin peptides CT-1 and CT-2. Z. Lebensm. Unters. Forsch. 182 (2), 115–117. https://doi.org/10.1007/BF01454241.

Wilde, A., et al., 1999. EGF receptor signaling stimulates SRC kinase phosphorylation of clathrin, influencing clathrin redistribution and EGF uptake. Cell 96 (5), 677–687. https://doi.org/10.1016/S0092-8674(00)80578-4.

Withoff, S., et al., 2016. Understanding celiac disease by genomics. Trends Genet. 32 (5), 295–308. https://doi.org/10.1016/j.tig.2016.02.003. Elsevier Ltd.

Yokoyama, S., et al., 2011. Transgenic mice that overexpress human IL-15 in enterocytes recapitulate both B and T cell-mediated pathologic manifestations of celiac disease. J. Clin. Immunol. 31 (6), 1038–1044. https://doi.org/10.1007/s10875-011-9586-7.

Zemskov, E.A., et al., 2011. Unconventional secretion of tissue transglutaminase involves phospholipid-dependent delivery into recycling endosomes. PLoS One 6 (4), e19414. https://doi.org/10.1371/journal.pone.0019414.

Zimmer, K.P., et al., 2010. Endocytotic segregation of gliadin peptide 31-49 in enterocytes. Gut 59 (3), 300–310. https://doi.org/10.1136/gut.2008.169656.

Zimmermann, C., et al., 2014. Epithelial transport of immunogenic and toxic gliadin peptides in vitro. PLoS One 9 (11), 1–17. https://doi.org/10.1371/journal.pone.0113932.

CHAPTER SEVEN

Cellular and molecular bases of refractory celiac disease

Craig R. Soderquist* and Govind Bhagat*
Department of Pathology and Cell Biology, Columbia University Irving Medical Center, New York, NY, United States
*Corresponding authors: e-mail address: crs2130@cumc.columbia.edu; gb96@cumc.columbia.edu

Contents

Abstract

Refractory celiac disease (RCD) encompasses biologically heterogeneous disorders that develop in a small proportion (0.3%) of individuals with celiac disease that are associated with high morbidity. Two broad categories are currently recognized, type I (RCD I) and type II (RCD II), based on immunophenotypic and molecular features of the intraepithelial lymphocytes (IELs). RCD I is characterized by a polyclonal expansion of IELs displaying a normal immunophenotype, while RCD II represents a clonal proliferation of immunophenotypically "aberrant" IELs, and is considered a low-grade

International Review of Cell and Molecular Biology, Volume 358
ISSN 1937-6448
https://doi.org/10.1016/bs.ircmb.2020.12.001

lymphoproliferative disorder. The pathogenesis of RCD I has not been clarified, but limited studies suggest multifactorial etiology. On the other hand, recent immunologic, molecular and immunophenotypic analyses have proposed lineage-negative innate IELs to be the cell of origin of a proportion of RCD II cases. Furthermore, sequencing studies have identified frequent, recurrent, activating mutations in members of the JAK-STAT pathway in RCD II. This finding, in conjunction with prior in vitro experimental observations, suggests roles of deregulated cytokine signaling in disease pathogenesis. In this review, we describe current understanding of environmental, immune and genetic factors associated with the development of RCD and briefly discuss diagnostic and therapeutic considerations.

1. Introduction

Celiac disease (CD) is a common (prevalence ~1%) (Singh et al., 2018) chronic immune-mediated disorder, which develops in genetically susceptible individuals in response to gluten ingestion (Green and Cellier, 2007). Although CD patients universally have small bowel mucosal alterations and often present with gastrointestinal (GI) symptoms, studies over the past couple of decades have revealed a high frequency of extra-intestinal manifestations and autoimmune diseases in those afflicted with CD (Hadjivassiliou et al., 2015; Leffler et al., 2015). The diagnosis of CD is based on a constellation of clinical findings, abnormal small bowel histopathology, and detection of circulating celiac-associated autoantibodies. A gluten-free diet (GFD) is effective treatment for the majority of patients, leading to mucosal healing and resolution of symptoms, but a subset (7–30% Leffler et al., 2007; O'Mahony et al., 1996) experience ongoing or recurrent GI symptoms and have villous atrophy, despite adhering to a strict GFD, and are considered to have non-responsive celiac disease (NRCD) (Penny et al., 2020). Slow or incomplete mucosal healing can be attributed to surreptitious gluten ingestion in a significant proportion of individuals with NRCD. However, if gluten ingestion and other CD-associated etiologies for persistent disease symptoms lasting >12 months have been excluded, the patient is deemed to have refractory celiac disease (RCD) (Ludvigsson et al., 2013; Penny et al., 2020).

Ever since its initial description, RCD has been recognized to be biologically and clinically heterogeneous (Ashton-Key et al., 1997; Murray et al., 1995; Wright et al., 1991), with multiple investigators demonstrating clonal expansions of phenotypically abnormal intraepithelial lymphocytes (IELs) in small intestinal biopsies from certain RCD patients

(Brousse et al., 1999; Carbonnel et al., 1998; Cellier et al., 1998). Subsequently a clonal relationship between RCD and enteropathy-associated T-cell lymphoma (EATL) was established (Bagdi et al., 1999; Cellier et al., 2000; Daum et al., 2001), affirming that some cases of RCD are precursors to EATL. In 2005, RCD was subclassified into two subtypes, type I (RCD I) and type II (RCD II), based on immunophenotypic and molecular features of the IELs (Daum et al., 2005). RCD I is characterized by a polyclonal expansion of immunophenotypically normal IELs, similar to that observed in uncomplicated CD. Conversely, RCD II constitutes a clonal expansion of immunophenotypically "aberrant" IELs, and is considered to represent a low grade lymphoma of intraepithelial T-lymphocytes or "cryptic" EATL (Alfsen et al., 1989; Daum et al., 2005; van Wanrooij et al., 2014; Verbeek et al., 2008a).

2. Epidemiology

Studies from specialized centers have described the prevalence of RCD to be 1.5–10% among celiacs (Arguelles-Grande et al., 2013; Leffler et al., 2007; O'Mahony et al., 1996; Roshan et al., 2011; Rubio-Tapia et al., 2009) and 8–23% among those with NRCD (Abdulkarim et al., 2002; Dewar et al., 2012; Roshan et al., 2011). However, population based studies report a much lower prevalence of RCD — 0.31–0.38% among celiacs (Biagi et al., 2014; Ilus et al., 2014) and 0.002% in the general population (Ilus et al., 2014).

RCD is most often diagnosed in older individuals, usually after the age of 50, though younger patients have been documented (Ilus et al., 2014; Malamut et al., 2009; Rubio-Tapia et al., 2009). RCD I is more common than RCD II, accounting for ~75% of all RCD cases, and women are more commonly affected in both types (~80% in RCD I, 60% of RCD II) (Al-Toma et al., 2007a; Daum et al., 2009; Ilus et al., 2014; Malamut et al., 2009; Roshan et al., 2011; Rubio-Tapia et al., 2009). Progression from RCD I to RCD II is considered uncommon (Daum et al., 2009).

3. Genetic and epigenetic risk factors

3.1 Genetic susceptibility

CD has a strong hereditary component, as demonstrated by familial clustering and high concordance in monozygotic twins (Fasano et al., 2003; Kuja-Halkola et al., 2016; Singh et al., 2015). The majority of CD patients carry the HLA-DQ2.5 (DQA1*0501, DQB1*0201) or HLA-DQ8 (DQA1*0301, DQB1*0302) risk alleles (Jabri and Sollid, 2017; Meresse et al., 2012), the remaining harboring either HLA-DQ2.2 (DQA1*0201, DQB1*0201) or HLA-DQ7.5 (DQA1*0505, DQB1*0301) (Karell et al., 2003). These alleles are necessary but insufficient for CD pathogenesis, as only approximately 2% of individuals possessing them develop CD. It is estimated that HLA-DQ2.5 and HLA-DQ8 account for ~35% of the heritable risk of CD (Dubois et al., 2010; Hunt et al., 2008; Trynka et al., 2011; Van Heel et al., 2007). In addition, genome-wide association studies (GWAS) have identified >40 non-HLA loci that together contribute to 15–20% of CD heritability (Dubois et al., 2010; Hunt et al., 2008; Kumar et al., 2012; Trynka et al., 2011; Van Heel et al., 2007). Many of these loci harbor single nucleotide polymorphisms (SNPs) adjacent genes involved in cytokine signaling (e.g., *IL2/IL21*) and T-cell receptor activation (e.g., *ICOS/CTLA4/CD28/CD80*), or in genes modulating downstream signal transduction cascades (e.g., *SOCS1/SH2B3/TNFAIP3*). Some SNPs are also associated with changes in the expression of noncoding RNAs, which can alter expression of inflammatory genes, among others (Castellanos-Rubio et al., 2016; Castellanos-Rubio and Bilbao, 2018; Ricaño-Ponce et al., 2015).

To date, only a few genetic factors have been specifically linked with RCD I or II. Patients with RCD I and RCD II are more likely to be homozygous for HLA-DQ2.5 (26–40% and 44–67% of cases, respectively), compared to those with uncomplicated celiac disease (21% of cases) (Al-Toma et al., 2006a; Malamut et al., 2009). Homozygosity of this allele has been shown to lead to an increase in the number and diversity of gliadin epitopes presented by APCs (Fallang et al., 2009; Sollid et al., 2012; Vader et al., 2003). In the only RCD GWAS study performed to date, a SNP (rs2041570) on chromosome 7p14.3 was associated with progression of CD to RCD II (odds ratio: 2.36) (Hrdlickova et al., 2018). Transcriptome analysis showed increased expression of *C20orf114* (*BPIFB1*) and differential expression of many epithelial innate immune genes and antimicrobial defensins, suggesting roles of Paneth cells and the gut microbiome in disease pathogenesis (Hrdlickova et al., 2018).

3.2 Environmental factors

The cumulative dose of gluten exposure has been associated with the development of RCD I and II (Al-Toma et al., 2006a; Biagi and Corazza, 2010;

Holmes et al., 1989; Ilus et al., 2014; Malamut et al., 2009, 2012; Rubio-Tapia et al., 2009; Verbeek et al., 2008b). Initial support for this idea came from studies showing a higher risk of developing lymphoma in GFD non-compliant patients (Askling et al., 2002; Catassi et al., 2002, 2005; Corrao et al., 2001; Gao et al., 2009; Green et al., 2003; Holmes et al., 1989; Silano et al., 2008). Advanced age at RCD I and II diagnosis also corroborates this notion (Ilus et al., 2014; Malamut et al., 2009; Rubio-Tapia et al., 2009). A substantial fraction of RCD I and II patients have chronic and/or recurrent gastrointestinal or hepatic viral infections (Malamut et al., 2009, 2012; Meresse et al., 2012), which has led to speculation that components of the antiviral immune responses, particularly IFN-α, which can induce IL-15 expression (Colpitts et al., 2012), may not only trigger an autoimmune disorder, but also bolster chronic inflammation. Reovirus infection has been shown to induce expression of type-1 IFNs, resulting in loss of gluten tolerance in a mouse model of CD (Bouziat et al., 2017) and some individuals have been reported to develop CD after IFN-α therapy (Cammarota et al., 2000; Kim et al., 2015). However, mucosal expression of this cytokine has not been investigated in RCD I or II.

4. Pathogenesis

4.1 Mucosal lymphoid populations in the small intestine

Normal small intestinal mucosa hosts a variety of lymphocytes within the epithelium and lamina propria, including distinct subtypes of lymphocytes and innate lymphoid cells (Calleja et al., 2011; Eberl et al., 2015; Eiras et al., 2000; Jarry et al., 1990; Leon et al., 2003; Schmitz et al., 2016; van Wijk and Cheroutre, 2009). These cells orchestrate myriad functions, including regulation of mucosal immunity and inflammation, antimicrobial defense and maintenance of oral tolerance to dietary proteins and commensal bacteria (Kim et al., 2015). The lymphoid populations are relatively stable in healthy mucosa, but changes in their number and function due to age and microenvironment alterations have been documented (Camarero et al., 2007; Jabri and Sollid, 2017; Schmitz et al., 2016; Verbeek et al., 2008c).

Different types of IELs and lamina propria lymphocytes are recognized to play important roles in the pathogenesis of CD and RCD (Fig. 1A). Thymus derived CD8+ TCRαβ+ T-cells comprise the largest population of IELs, accounting for 75–80% of lymphoid cells in normal small bowel epithelium (Guy-Grand et al., 1991; Jarry et al., 1990). These cells resemble conventional antigen-experienced T-cells found in the blood and express

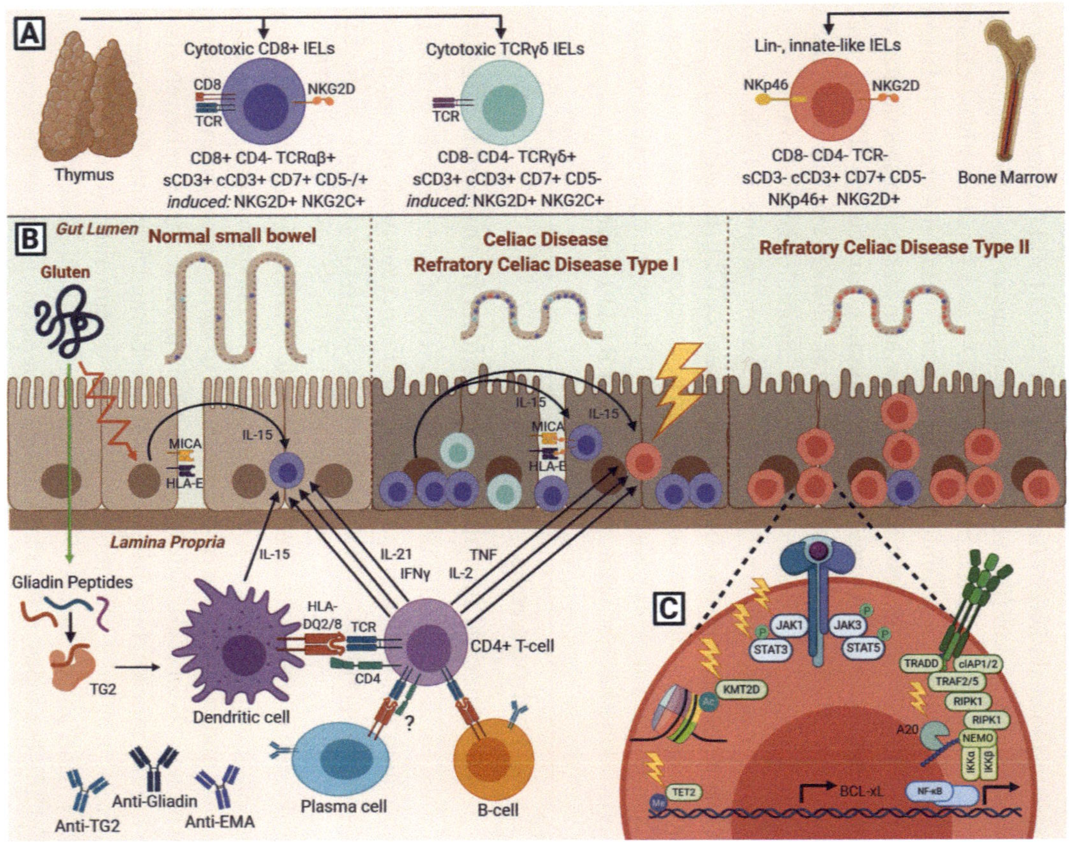

Fig. 1 See figure legend on opposite page.

surface (s) CD3, cytoplasmic (c) CD3, CD7, and CD2, but frequently show variable CD5 expression. CD4+ TCRαβ+ T-cells only represent a minor fraction of IELs (5–10%) (Guy-Grand et al., 1991; Jarry et al., 1990). TCRγδ T-cells are another thymus derived T-cell subset, which constitute 5–10% of IELs, with the majority lacking expression of the TCR co-receptors CD4 and CD8 (Guy-Grand et al., 1991; Jarry et al., 1990). Similar to their TCRαβ+ counterparts, these T-cells also express sCD3, cCD3, CD7, and CD2, and lack CD5 (Spencer et al., 1991; Verbeek et al., 2008c). Innate lymphoid cells (ILCs), which are lineage negative (Lin-), innate immune cells derived from bone marrow progenitors, comprise up to 10% of IELs (Ettersperger et al., 2016; van Wanrooij et al., 2014). In contrast to the TCRαβ and γδ T-cells, these innate IELs do not express surface TCR, CD3 or CD5, most show bright CD7 expression, and the majority express activating NK-receptors NKG2D and NKp46 (Eiras et al., 1998, 2000; Ettersperger et al., 2016; Kooy-Winkelaar et al., 2017; Schmitz et al., 2013, 2016; Tack et al., 2012b; Verbeek et al., 2008a).

Apart from diverse types of chronic inflammatory cells, the lamina propria harbors different types of T-cells and innate lymphoid cells, B-cells, usually arranged in follicles (constitutive GALT–solitary intestinal lymphoid tissue [SILT] and Peyer's patches Buettner and Lochner, 2016; Gibbons and Spencer, 2011) and plasma cells. The latter represent the predominant cell type in the lamina propria (Farstad et al., 2000; Høydahl et al.,

Fig. 1 Immune and molecular mechanisms of CD and RCD pathogenesis. (A) Different IEL subsets play roles in pathogenesis of CD and RCD: cytotoxic CD8+ TCRαβ+ T-cells (blue), CD8− TCRγδ+ T-cells (green), and CD8− innate ILC-like IELs (red). (B) Partially digested gliadin peptides cross the epithelial barrier into the lamina propria, are deamidated by TG2, and presented by dendritic cells to gluten-reactive CD4+ T-cells, which upon activation secrete Th1 cytokines (IFNγ, TNFα, IL-21, and IL-2), stimulating IELs and triggering a humoral immune response leading to the production of celiac autoantibodies. Certain gliadin peptides (p31–43) damage enterocytes and elicit an epithelial "stress" response, resulting in upregulation of MICA and HLA-E and expression of IL-15. In uncomplicated CD (and RCD I), IL-15 and cytokines derived from gluten-specific T-cells induce proliferation and NK-like reprogramming of intraepithelial CD8+ TCRαβ+ T-cells and these lymphokine activated killer-like cells cause cytolysis of epithelial cells. TCRγδ T-cells are also expanded in CD. In RCD II, epithelial IL-15, and gluten-specific CD4 + T-cell derived cytokines stimulate the proliferation of innate ILC-like IELs. (C) RCD II IELs subsequently acquire activating *JAK1* and *STAT3* mutations that enhance cytokine signaling via the JAK-STAT pathway. Additional mutations (Soderquist et al., 2021) occur in genes modulating other biologic processes, such as epigenetic modifications (*TET2, KMT2D*), NF-κB (*TNFAIP3*), DNA damage response and repair (*POT1*, not pictured) signaling and immune evasion (*CD58*, not pictured). *Figure created with BioRender.com.*

2019). Most plasma cells produce secretory IgA directed toward luminal commensal and pathogenic bacteria, which prevents translocation of bacteria across the epithelial barrier (Benckert et al., 2011; Sutherland et al., 2016). CD4+ T-cells make up the largest lymphocytic subset in the lamina propria (~70%) (Jahnsen et al., 1998) and comprise a variety, including type 1 and 2 T-helper (Th1, Th2) cells and regulatory T-cells (Tregs) (Gibbons and Spencer, 2011).

4.2 Pathogenesis of CD

Details of the cellular and molecular mechanisms underlying uncomplicated CD have been described extensively (Abadie et al., 2011; Jabri and Sollid, 2017) and are briefly summarized here. Gluten, the generic term for the family of prolamin proteins (rich in proline and glutamine), found in wheat, barley, and rye, is digested by enzymes located in the brush border of small intestinal enterocytes (Dupont et al., 2011; Shan et al., 2002; Shewry et al., 2002) (Fig. 1B). The undigested peptides, including α-gliadin, traverse the epithelium and are deamidated by the enzyme tissue transglutaminase type 2 (TG2) in the lamina propria. Deamidation results in increased numbers of negatively charged amino acids, particularly at key anchor residues that contact HLA molecules. The modified peptides are phagocytized by dendritic cells, loaded onto HLA-DQ2 or HLA-DQ8 molecules and presented to CD4+ Th cells (Dahal-Koirala et al., 2019; Jabri and Sollid, 2017; Meresse et al., 2012; Sollid, 2017; Sollid et al., 2012). The activated gliadin-specific CD4+ Th1 cells in the lamina propria secrete a variety of effector cytokines, including interferon gamma (IFNγ), tumor necrosis factor alpha (TNFα), IL-21, and IL-2, which actuate the B-cell mediated humoral immune response (Jabri and Sollid, 2017; Meresse et al., 2012) resulting in the production of anti-TG2, anti-gliadin, and anti-endomysial antibodies by plasma cells. Recent studies indicate that the plasma cells also bear surface HLA-DQ2.5-gluten complexes in CD and, hence, may contribute to the activation of lamina propria gliadin-responsive CD4+ T-cells (Di Niro et al., 2012; Dørum et al., 2016; Høydahl et al., 2019).

Additionally, certain peptides, such as α-gliadin p31-43, have been shown to directly damage small intestinal enterocytes, eliciting an epithelial "stress" response (Hüe et al., 2004; Luciani et al., 2010; Meresse et al., 2004), which leads to the upregulation of non-classical MHC class I molecules (MICA and HLA-E) by epithelial cells as well as IL-15 expression by the enterocytes and lamina propria mononuclear cells (Hüe et al., 2004; Maiuri et al., 2003; Meresse et al., 2004, 2006). IL-15 and the cytokines secreted by lamina propria Th1 T-cells synergistically stimulate the proliferation of IELs, including intraepithelial cytotoxic CD8+ TCRαβ+ T-cells and CD8− TCRαβ− ILCs (Hüe et al., 2004; Kooy-Winkelaar et al., 2017; Jabri and Sollid, 2017; Meresse et al., 2006, 2012). Inhibition of the Smad3-TGF-β signaling axis by IL-15 activates intraepithelial cytotoxic CD8+ TCRαβ+ T-cells (Benahmed et al., 2007), which undergo reprogramming to lymphokine-activated killer (LAK) cells and upregulate the activating NK-receptors NKG2D and CD94/NKG2C (Hüe et al., 2004; Korneychuk et al., 2014; Meresse et al., 2004, 2006). NKG2D binding to MICA and HLA-E on the surface of enterocytes leads to epithelial cell killing, which results in villous atrophy and malabsorption (Hüe et al., 2004; Jabri and Sollid, 2017; Meresse et al., 2006, 2012). IL-15 and cytokines elaborated by gliadin-specific CD4+ T-cells also trigger JAK3-STAT5 signaling, which upregulates expression of the antiapoptotic protein Bcl-xL, prolonging the survival of innate IELs, and activation of PI-3 kinase signaling increases their resistance to FOXP3+ Tregs (Ben Ahmed et al., 2009; Bodd et al., 2010; Ettersperger et al., 2016; Fina et al., 2008; Hmida et al., 2012; Malamut et al., 2010; Mention et al., 2003; Sarra et al., 2013; Schmitz et al., 2013, 2016).

Intraepithelial CD4− CD8− TCRγδ T-cells are also expanded in CD (15–40%) (Dunne et al., 2013; Malamut et al., 2009; Nijeboer et al., 2019), likely by similar cytokines that stimulate proliferation of TCRαβ+ IELs, but the molecular signals are not known. In normal small intestinal mucosa, butyrophilin-like molecules 3 (BTNL3) and 8 (BTNL8) expressed by epithelial cells appear to regulate different subsets of TCRγδ T-cells (Di Marco Barros et al., 2016). In CD, BTNL8 expression is down-regulated, leading to depletion of naturally occurring Vγ4+/Vδ1+ IELs and expansion of IFNγ expressing Vγ1+ IELs (Mayassi et al., 2019). Murine studies have suggested roles in epithelial repair and a regulatory function of TCRγδ T-cells has been proposed in CD (Bhagat et al., 2008; Boismenu and Havran, 1994; Chen et al., 2002; Ogata and Itoh, 2016). However, triggering of cytotoxic activity of TCRγδ T-cells by proinflammatory cytokines is well known in a variety of diseases (Hayday, 2019) and likely also occurs in untreated CD.

In contrast to TCRαβ and TCRγδ IELs, intraepithelial ILCs, which constitute distinct subsets, do not appear to proliferate to a significant extent in CD, and their numbers are relatively reduced (<1% of CD45+ cells)

(Nijeboer et al., 2019; Schmitz et al., 2016) in patients with active or untreated disease. The role of innate IELs in CD pathogenesis is uncertain. Ex vivo culture with IL-15 has been shown to induce proliferation of a subset (Schmitz et al., 2016) and enhanced IFN-γ and cytotoxic granule protein expression by NKp44− CD127− ILCs has been proposed to contribute to inflammation and epithelial damage (Uhde et al., 2020).

4.3 Pathogenesis of RCD I

The immunologic and molecular mechanisms underlying RCD I are not well characterized, but are considered similar to uncomplicated CD. Despite adherence to a strict GFD, trace gluten exposure is thought to sustain low-level mucosal inflammation, demonstrated by persistent anti-TG2 antibodies (~30% of cases) (Arguelles-Grande et al., 2013; Rubio-Tapia et al., 2009). Certain GFD-adherent, seronegative individuals are believed to develop an IL-15-dependent autoimmune disorder, possibly resulting in altered responses to dietary and microbial antigens (Hüe et al., 2004) that sustain mucosal inflammation (Malamut et al., 2010; Meresse et al., 2012). It is conceivable that deregulated IL-15 expression may render the autoreactive T-cells resistant to the inhibitory control of Tregs (Ben Ahmed et al., 2009).

4.4 Genetic alterations in RCD I

No somatic genetic abnormalities have been reported in RCD I.

4.5 Pathogenesis of RCD II

Until recently, RCD II was considered to represent a clonal, neoplastic expansion of immunophenotypically "aberrant" cytotoxic TCRαβ IELs, which had downregulated or lost sCD3, sTCR, and invariably CD8 expression (Meresse et al., 2006; Tjon et al., 2008). Using flow cytometric, transcriptional, and functional analyses, Ettersperger et al. and Schmitz et al. have suggested the cell of origin of aberrant RCD II IELs to be Lin-, CD103+, sCD3−, cCD3+, CD7+, sTCR−, CD127−, CD56−, NKp46+, CD122 (IL-2/15Rβ)+ innate IELs that resemble ILCs (Ettersperger et al., 2016; Schmitz et al., 2013, 2016). The proliferation of innate IELs in RCD II has been shown to be stimulated by similar cytokines that propagate CD (and RCD I) (Kooy-Winkelaar et al., 2017; Meresse et al., 2012). IL-15 expression by damaged enterocytes activates IL-15Rβ bearing innate IELs, triggering downstream JAK3-STAT5 signaling, which upregulates expression of the antiapoptotic protein Bcl-xL, resulting in prolonged survival of innate IELs (Ettersperger et al., 2016; Malamut et al., 2010; Mention et al., 2003; Schmitz et al., 2013, 2016). IL-21, secreted by gliadin-specific CD4+ T-cells (Bodd et al., 2010; Fina et al., 2008; Sarra et al., 2013), as well as IL-2 and TNFα, also activate similar downstream signaling cascades as IL-15 (Kooy-Winkelaar et al., 2017). In addition, contact with dendritic cells has been demonstrated to increase proliferation of RCD II IELs independently of IL-15 in vitro; the precise location of these cells in human small intestinal mucosa and the molecular mechanisms underlying contact mediated IEL stimulation are unclear (Schmitz et al., 2014). The aberrant RCD II IELs acquire the capability to lyse epithelial cells by upregulating the activating receptor DNAM-1(Tjon et al., 2011).

In healthy individuals, the innate ILC-like IELs are believed to be capable of differentiating into either NK-cells or T-cells, however, RCD II IELs lack this ability and arrest maturation at an early NK/T precursor stage. It was shown that, in the presence of IL-15, granzyme B cleaves NOTCH1, which abrogates the T-cell differentiation program (Ettersperger et al., 2016). RCD II IELs therefore lack markers of mature T-cells, but express NK-associated markers (KIR2DL4/CD158d, NKp46, NKG2D) (Ettersperger et al., 2016; Schmitz et al., 2013, 2016; Tack et al., 2012b). Additionally, ~70% of RCD II cases show absent or non-productive T-cell receptor gene (*TR*) rearrangements, supporting the view that RCD II IELs have commenced but not completed T-cell development (Ettersperger et al., 2016; Tack et al., 2012b). Interestingly, Tack et al. described evolution of RCD II to EATL in all cases with complete *TRB* rearrangements and hypothesized that the maturational stage of RCD II IELs may predict the risk for transformation to EATL (Tack et al., 2012b). In support of this concept, Garcia-Hoz et al. showed that increased cCD3 intensity predicts risk of RCD II progression to EATL (García-Hoz et al., 2020).

4.6 Microenvironmental changes in RCD II

Changes in the mucosal microenvironment have not been well studied in RCD II. It has been shown that Paneth cells are markedly reduced and α-defensin expression is decreased in RCD II (Di Sabatino et al., 2008). The implications of these findings for RCD II pathogenesis are not clear at present. In contrast to the increase in intraepithelial TCRγδ T-cells in CD and RCD I, TCRγδ T-cells are decreased in RCD II, but normalize upon elimination of aberrant IELs by chemotherapy or autologous stem cell transplantation (Verbeek et al., 2008c), potentially due to therapy-induced microenvironmental alterations, normalization of IL-15 levels, or other unknown mechanisms. Further investigations are warranted to determine alterations in the number, phenotype and function of other intraepithelial and lamina propria immune and non-immune cells in RCD II.

4.7 Genetic alterations in RCD II

The augmented proliferation and survival of innate IELs in RCD II makes them increasingly vulnerable to genotoxic stressors. Aberrant nuclear p53 expression, indicative of P53 dysregulation (Obermann et al., 2004), and decreased PCNA expression (Tack et al., 2012b) has been observed in RCD II IELs, suggesting impaired DNA damage response and repair mechanisms. Recurrent gain of chromosome 1q22-q44 has been detected in RCD II (Malamut et al., 2009; Verkarre et al., 2003), but the biological significance of this aberration is unknown. In 2016, Ettersperger et al. reported activating point mutations in *JAK1* and *STAT3* in all RCD II cases examined, providing the first evidence of specific genetic alterations in promoting the clonal expansion of RCD II IELs (Fig. 1C) (Ettersperger et al., 2016). RCD II cell lines harboring these mutations demonstrated increased STAT3 phosphorylation in the presence of IL-15, indicating increased sensitivity of RCD II IELs to cytokine stimulation (Ettersperger et al., 2016). These findings support earlier observations of dysregulated JAK-STAT signaling in RCD II (Abadie and Jabri, 2014).

We have confirmed the high prevalence of *JAK1* and *STAT3* mutations and also identified recurrent mutations in epigenetic modifiers (*TET2*, *KMT2D*) and NF-κB (*TNFAIP3*), DNA damage repair (*POT1*), and immune evasion (*CD58*) pathway genes in RCD II (Soderquist et al., 2021). Some of these mutations, particularly those affecting *TET2* and *KMT2D*, which are ubiquitous in hematologic neoplasms (Ntziachristos et al., 2016), have also been reported in intestinal T-cell lymphomas, including EATL (Moffitt et al., 2017), while others (*TNFAIP3*, *POT1*, *CD58*) are less frequently described in T-cell lymphomas, and have not yet been reported in EATL. Our observations indicate early deregulation of diverse cellular processes and signaling pathways fostering the neoplastic transformation of IELs. Larger studies are awaited to assess the functional implications of these and other genetic alterations and determine if any are predictive of EATL development.

5. Diagnostic considerations

5.1 Clinical features

By definition, patients with RCD I and RCD II experience persistent or recurrent gastrointestinal symptoms, including diarrhea, abdominal pain, and weight loss (Ilus et al., 2014; Malamut et al., 2009; Rubio-Tapia et al., 2009), which tend to be more severe in RCD II. Laboratory studies often reveal evidence of malabsorption, including vitamin deficiencies and hypoalbuminemia. Primary refractoriness (lack of response to GFD at diagnosis) is noted in approximately 30% of RCD I and 50% of RCD II patients, while the remainder develop symptoms after variable periods of response to a GFD (secondary RCD) (Malamut et al., 2009).

Details of the clinical workup for evaluation of RCD have been reviewed elsewhere (Al-Toma et al., 2019; Hujoel and Murray, 2020; Lebwohl et al., 2018; Penny et al., 2020), and are only briefly mentioned here. The critical first step is to determine whether the patient has CD, another disorder such as food allergy, autoimmune enteropathy, irritable bowel syndrome (IBS) or lymphoma, or medication-associated small intestinal injury. Sorting through the differential diagnoses requires comprehensive review of clinical and pathology data, including review of serologies and biopsy findings (see proposed diagnostic flowchart in Fig. 2) (Hujoel and Murray, 2020; Lebwohl et al., 2018; Leffler et al., 2007; Penny et al., 2020). Once the diagnosis of CD has been confirmed, repeat endoscopy/biopsies and additional laboratory studies need to be performed to investigate the etiology of persistent symptoms, e.g., bacterial overgrowth, pancreatic insufficiency, and microscopic colitis (Hujoel and Murray, 2020; Lebwohl et al., 2018; Leffler et al., 2007; Penny et al., 2020). Since NRCD commonly results from dietary indiscretion (~30% of cases), adherence to GFD should be assessed by an expert dietician (Abdulkarim et al., 2002; Dewar et al., 2012; Malamut et al., 2009). Newly available urine and stool antigen tests can assist in evaluating dietary compliance (Comino et al., 2016; Moreno et al., 2017).

5.2 Endoscopy and imaging

Diagnosis and management of RCD has improved with advances in endoscopic and imaging techniques. Capsule endoscopy (CE) and magnetic resonance imaging (MRI) are recommended screening tests (Branchi et al., 2016).

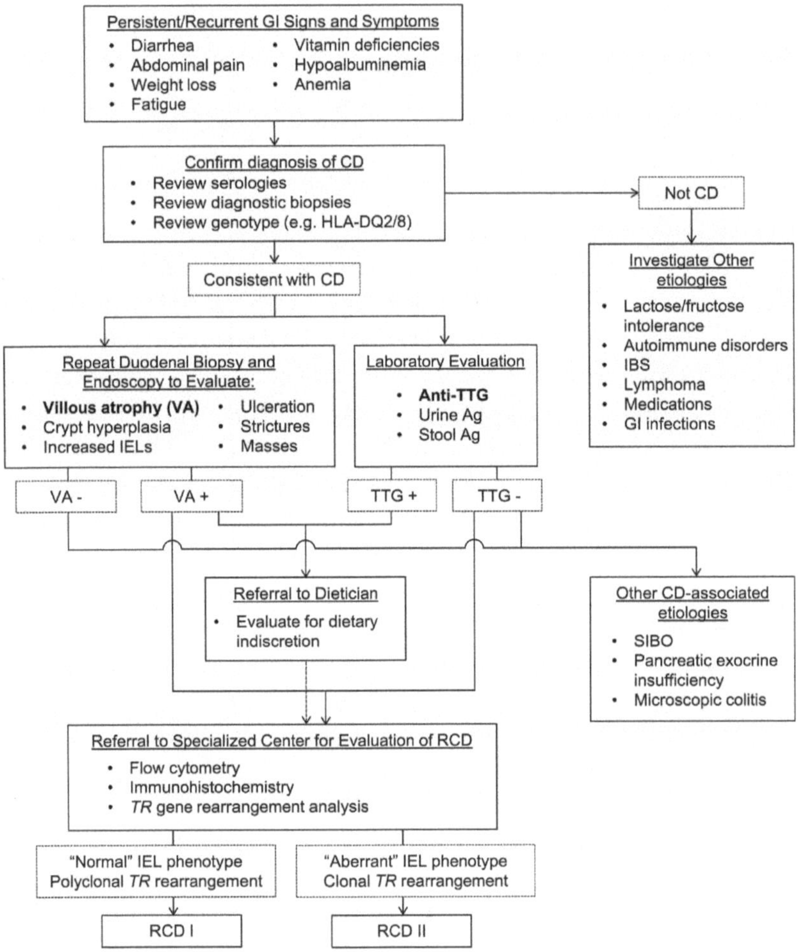

Fig. 2 Proposed algorithm for diagnosing RCD in patients with persistent or recurrent gastrointestinal symptoms. Ag: antigen; CD: celiac disease; GI: gastrointestinal; IBS: irritable bowel syndrome; IEL: intraepithelial lymphocyte; RCD: refractory celiac disease; SIBO: small intestinal bacterial overgrowth; *TR*: T-cell receptor gene; TTG: tissue transglutaminase; VA: villous atrophy.

On endoscopy, mucosal damage extending beyond the proximal small bowel and multiple ulcers are commonly observed in RCD; the presence of erosions, strictures, and large ulcers (>1 cm) is more common in RCD II (Ashton-Key et al., 1997; Bagdi et al., 1999; Cellier et al., 2000; Malamut et al., 2009). If ulcerations, strictures, or masses are observed, device-assisted enteroscopy (DAE) is recommend for further exploration (Branchi et al., 2016; Elli et al., 2017). Up to 70% of RCD patients show MRI abnormalities, which include jejunoileal fold reversal, small bowel wall thickening or dilation, and splenic atrophy (volume $<100\,cm^3$) (Al-Bawardy et al., 2017; Radmard et al., 2017; van Gils et al., 2017). Computed tomography (CT) and positron-emission tomography (PET) have both been shown to help identify early progression of RCD II to EATL, and ^{18}F-FDG PET has been shown to be more sensitive than CT in differentiating RCD II from EATL (Hadithi et al., 2006; Hoffmann et al., 2003).

5.3 Histopathology and immunophenotype

The morphologic features of small intestinal biopsies from patients with RCD I (Fig. 3A) and RCD II (Fig. 4A) are similar to those seen in uncomplicated CD, though cases of RCD II more commonly exhibit severe villous atrophy. The IELs are small and do not show overt cytologic atypia, and frequently extend into the crypts (Fig. 4B) (Al-Toma et al., 2007a; Bagdi et al., 1999; Cellier et al., 2000). Infiltration of IELs into the lamina propria, which occurs in up to 50% of RCD II cases (Fig. 4C), is not observed in RCD I (Verbeek et al., 2009). The IELs frequently spread to other GI organs (e.g., stomach, colon) in RCD II and can also traffic to extra-GI sites (e.g., skin, lungs, CNS) (Malamut et al., 2009, 2013). Epithelial ulcers with acute inflammatory exudate and granulation tissue containing lymphocytes, plasma cells, and macrophages ("ulcerative jejunitis" or "chronic jejunoileitis"), can be seen in both RCD I and RCD II, but are more common in RCD II (Ashton-Key et al., 1997; Bagdi et al., 1999; Cellier et al., 2000; Isaacson and Wright, 1978; Jewell, 1983).

As previously described, the immunophenotype of the IELs in RCD I and RCD II is distinct. In RCD I, the majority of IELs display an immunophenotype of thymus-derived, mature, cytotoxic TCRαβ+ T-cells that are s/cCD3+ (Fig. 3B, D), CD8+ (Fig. 3C, E), CD7+ (Fig. 3F), CD5(variable)+ (Fig. 3F), and sTCRαβ+ (Fig. 3G). An expansion of TCRγδ T-cells is usually also observed (Fig. 3G). In contrast, RCD II IELs show an immunophenotype similar to bone marrow derived ILCs: cCD3+ (Fig. 4D, F), sCD3− (Fig. 4F), CD8− (Fig. 4E, G), CD4− (Fig. 4G), CD7(bright)+ (Fig. 4H), CD5− (Fig. 4H), and sTCRαβ- (Fig. 4I) (Cellier et al., 2000; Roshan et al., 2011; Rubio-Tapia et al., 2009; Verbeek et al., 2008a; Verkarre et al., 2003). Cytotoxic granule proteins (granzyme B and perforin) are frequently expressed (Meresse et al., 2006; Setty et al., 2015) while CD56 and CD30 are negative. It should be noted that variant RCD II phenotypes have been documented,

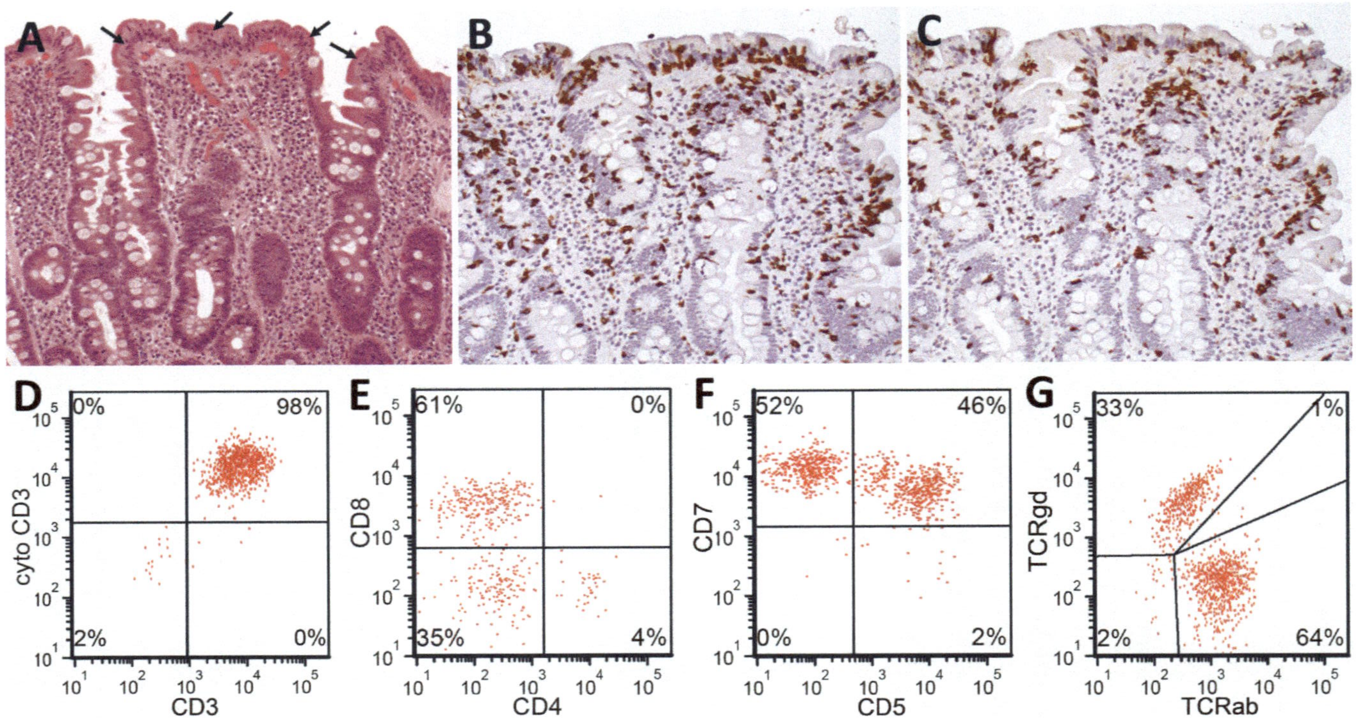

Fig. 3 Histopathologic and immunophenotypic features of RCD I. (A) Duodenal biopsy from a patient with RCD I demonstrates subtotal villous atrophy, crypt hyperplasia and increased IELs (arrows). (B) The IELs are CD3+ and (C) most, but not all, are CD8+; the number of CD8+ IELs is usually lower than CD3+ IELs due to an increase in intraepithelial CD8− TCRγδ T-cells. (D) Flow cytometry shows that the IELs express surface and cytoplasmic CD3. (E) The majority of IELs express CD8 and the CD8− cells largely represent TCRγδ T-cells. (F) Virtually all IELs express CD7, while CD5 expression is variable. (G) The majority of IELs express TCRαβ, but an increased population of TCRγδ IELs is also present.

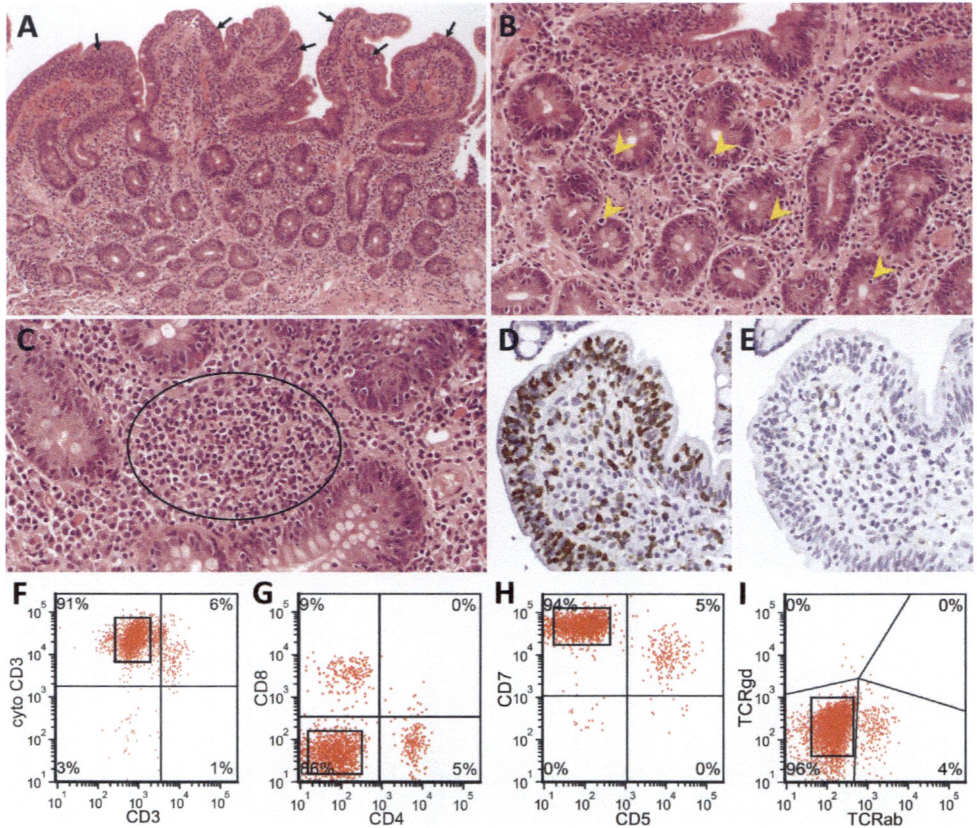

Fig. 4 Histopathologic and immunophenotypic features of RCD II. (A) Duodenal biopsy from a patient with RCD II shows partial villous atrophy, crypt hyperplasia, and increased IELs (arrows). (B) The IELs are small and lack cytologic atypia, but are seen infiltrating the crypts (arrowheads). (C) Infiltration of the lamina propria by IELs is noted, which form ill-defined clusters (circle). The IELs show immunohistochemical expression of (D) CD3, but are negative for (E) CD8. Flow cytometry shows that the aberrant IELs are: (F) sCD3−/cCD3+, (G) CD4−/CD8−, (H) CD5−/CD7(bright)+ and (I) TCRαβ−/TCRγδ−.

with CD8+ cases being the most frequent (De Mascarel et al., 2008; Nasr et al., 2015). Uncommon variant immunophenotypes include sCD3+/CD4+, sCD3+/TCRγδ+, and sCD3−/cCD3− (Cellier et al., 2019; Soderquist et al., 2020). Recently, NKp46 has been reported to be a useful marker for distinguishing RCD I from RCD II, since the IELs in RCD I are NKp46 negative (Cheminant et al., 2019; Ettersperger et al., 2016).

The percentage of aberrant IELs can be assessed by either flow cytometry or immunohistochemistry (IHC) and the threshold for RCD II diagnosis varies according to the technique used. Cut-offs of 20% and 50% aberrant IELs have been proposed for flow cytometry and IHC, respectively (Patey-Mariaud De Serre et al., 2000; van Wanrooij et al., 2010; Verbeek et al., 2008a, 2009). The lower threshold for flow cytometry reflects its superior discriminatory ability. Specifically, flow cytometry can differentiate between surface and cytoplasmic CD3 expression and can more precisely distinguish and enumerate different mucosal CD8− lymphoid populations, such as ILCs, aberrant RCDII IELs and TCRγδ T-cells, the latter being expanded in CD and RCD I (van Wanrooij et al., 2010; Verbeek et al., 2008a, 2009).

5.4 Molecular analysis

TR rearrangement analysis by polymerase chain reaction (PCR) is important in the diagnostic workup of RCD. Cases of RCD I often demonstrate polyclonal products while RCD II characteristically displays clonal *TR* rearrangements (Perfetti et al., 2012; van Wanrooij et al., 2014). However, one should be aware that clonal *TR* products can be observed in 6–16% of CD and 17% of RCD I cases and the clonal products may be transient or persist over time without any clinical evidence of progression to RCD II (Hussein et al., 2018; Liu et al., 2010). The location (epithelium or lamina propria) of these clonal expansions has not been determined. Conversely, up to 25% of RCD II cases lack clonal rearrangements at all *TR* loci (Ettersperger et al., 2016; Tack et al., 2012b). Hence, *TR* rearrangement results, in isolation, should not be used to diagnose RCD II.

Although results of cytogenetic and NGS analyses are not incorporated in the current diagnostic algorithm of RCD, abnormalities detected by these assays can be useful in establishing clonality and diagnosing RCD II.

6. Limitations of the current RCD classification

The current RCD classification does not fully integrate the spectrum of clinical, histopathologic and immunophenotypic features of RCD. Gastrointestinal symptoms are the only RCD clinical criterion. Extra-intestinal manifestations, including neurologic, dermatologic, arthritic, hepatic, and osteologic disorders, are well-documented in CD (Hadjivassiliou et al., 2014; Jericho et al., 2017; Sarrigiannis et al., 2014; Stier et al., 2019) and some RCD patients, including those with RCD II (Soderquist et al., 2020) can present with non-GI symptoms. Hence, a wider spectrum of clinical symptoms needs to be incorporated in future RCD diagnostic algorithms. It's also worth noting that while most RCD patients exhibit severe villous atrophy on small intestinal biopsy (Marsh 3b or 3c) (Rubio-Tapia et al., 2016), a subset may show mild or absent villous atrophy (Soderquist et al., 2020, 2021). Cases with variant immunophenotypes also need to be recognized as bona fide RCD II as they harbor similar genetic alterations as cases displaying the "typical" immunophenotype (Soderquist et al., 2021), suggesting similar underlying biology.

7. Management

Contemporary therapeutic management of RCD patients is discussed in more detail elsewhere (Al-Toma et al., 2019; Baggus et al., 2020; Hujoel and Murray, 2020; Penny et al., 2020), and the principles are briefly summarized here. While RCD patients show persistent symptoms despite adherence to a GFD by definition, compliance with a strict GFD is strongly recommended to dampen mucosal inflammation. There is currently no cure for RCD II and the goals of therapy are to mitigate malnutrition and limit progression to EATL. Therapies utilizing immunomodulatory agents (budesonide or systemic corticosteroids) have traditionally been the mainstay of initial treatment of both RCD I and II (Al-Toma et al., 2007a, 2019; Brar et al., 2007; Goerres et al., 2003; Malamut et al., 2009; Rowinski and Christensen, 2016; Rubio-Tapia et al., 2016). Corticosteroids have been shown to induce clinical and histologic responses in 90% and 40% of RCD I patients and 77% and 33% of RCD II patients, respectively (Malamut et al., 2009). Moderate improvement in symptoms has been reported for ~75% of RCD I and II patients treated with oral budesonide, although without significant histologic response (Brar et al., 2007). Better outcomes were observed with an "open-capsule" budesonide regimen (clinical response 92%, histologic response 89%, aberrant IEL diminution 53%), which is now considered to be first-line therapy for all RCD I and clinically stable RCD II patients (Al-Toma et al., 2019; Mukewar et al., 2017). The addition of azathioprine to prednisone may improve response in RCD I patients; however, concern for increased risk of EATL development has been raised in RCD II patients treated with this drug (Goerres et al., 2003). In a series of 12 RCD I patients, treatment with thioguanine resulted in clinical and histologic response in ~80% of cases (Tack et al., 2012a) and among 10 RCD I patients treated with small intestinal release mesalamine (SIRM), some of whom also received budesonide, 50% showed complete clinical response (Jamma et al., 2011). Case reports have also described clinical and histologic response in RCD I patients treated with infliximab (Costantino et al., 2008; Gillett et al., 2002). A pilot open-label study of recombinant human IL-10 did not show efficacy in a small group of RCD I and RCD II patients (Mulder et al., 2001). In RCD II patients treated with cladribine, clinical (36–81%) and histologic (47–59%) improvement, as well as a decrease in neoplastic IEL burden (35–41% of cases) has been documented, but this therapy does not appear to prevent development of EATL (Al-Toma et al., 2006b; Tack et al., 2011a). Other agents, such as alemtuzumab (anti-CD52) (Malamut et al., 2009; Vivas et al., 2006), methotrexate (Malamut et al., 2009), and cyclosporin (Malamut et al., 2009; Wahab et al., 2000), which have been utilized in a small number of RCD II patients, have not shown durable responses.

Relapsed or intractable RCD II has been treated with high-dose chemotherapy using fludarabine and melphalan as conditioning regimens, followed by autologous stem cell transplantation (Al-Toma et al., 2007b; Nijeboer et al., 2017; Tack et al., 2011b). However, long-term outcomes are not substantially improved and these interventions have significant side-effects. A recent phase 2 clinical trial of an anti-IL-15 monoclonal antibody (AMG-714) demonstrated safety and improvement in symptoms, though a statistically significant decrease in aberrant IELs or improvement in histologic parameters was not observed (Cellier et al., 2019). JAK-STAT pathway inhibitors (tofacitinib) (Yokoyama et al., 2013) and other biologic agents targeting NKp46 and BCL-xL are currently under investigation.

8. Unanswered questions and future directions

Advances in the clinical, pathologic, immunologic and molecular characterization of RCD have furthered our understanding of disease biology, however, many questions remain unanswered. It is unclear whether systematic evaluation of cytokine alterations will resolve biologic subsets of RCD I or II patients that require different therapies. The constellation of non-HLA germline genetic alterations that predispose to RCD I or II are not known, as are microenvironmental factors influencing the development of these conditions. Ongoing genomic profiling studies can help inform early events in RCD II pathogenesis, which might uncover biomarkers of disease progression, and potentially also elucidate similarities and differences between "de novo" and "secondary" EATLs. While there is currently no curative therapy for RCD II, several agents currently under investigation show promise. Future genetic and epigenetic analyses of RCD II could guide the development of additional novel and effective therapies.

References

Abadie, V., Jabri, B., 2014. IL-15: a central regulator of celiac disease immunopathology. Immunol. Rev. 260, 221–234.

Abadie, V., Sollid, L.M., Barreiro, L.B., Jabri, B., 2011. Integration of genetic and immunological insights into a model of celiac disease pathogenesis. Annu. Rev. Immunol. 29, 493–525.

Abdulkarim, A.S., Burgart, L.J., See, J., Murray, J.A., 2002. Etiology of nonresponsive celiac disease: results of a systematic approach. Am. J. Gastroenterol. 97, 2016–2021.

Al-Bawardy, B., Barlow, J.M., Vasconcelos, R.N., Kim, S.T., Bruining, D.H., Hansel, S.L., Sheedy, S.P., Murray, J.A., Rubio-Tapia, A., Rajan, E., Fidler, J.L., Fletcher, J.G., 2017. Cross-sectional imaging in refractory celiac disease. Abdom. Radiol. 42, 389–395.

Alfsen, C.G., Beiske, K., Bell, H., Marton, P.F., 1989. Low-grade intestinal lymphoma of intraepithelial T lymphocyties with concomitant enteropathy-associated T cell lymphoma: case report suggesting a possible histogenetic relationship. Hum. Pathol. 20, 909–913.

Al-Toma, A., Goerres, M.S., Meijer, J.W., Peña, A.S., Crusius, J.B.A., Mulder, C.J., 2006a. Human leukocyte antigen–DQ2 homozygosity and the development of refractory celiac disease and enteropathy-associated T-cell lymphoma. Clin. Gastroenterol. Hepatol. 4, 315–319.

Al-Toma, A., Goerres, M.S., Meijer, J.W.R., von Blomberg, B.M.E., Wahab, P.J., Kerckhaert, J.A.M., Mulder, C.J.J., 2006b. Cladribine therapy in refractory celiac disease with aberrant T cells. Clin. Gastroenterol. Hepatol. 4, 1322–1327.

Al-Toma, A., Verbeek, W.H.M., Hadithi, M., Von Blomberg, B.M.E., Mulder, C.J.J., 2007a. Survival in refractory coeliac disease and enteropathy-associated T-cell lymphoma: retrospective evaluation of single-centre experience. Gut 56, 1373–1378.

Al-Toma, A., Visser, O.J., Van Roessel, H.M., Von Blomberg, B.M.E., Verbeek, W.H.M., Scholten, P.E.T., Ossenkoppele, G.J., Huijgens, P.C., Mulder, C.J.J., 2007b. Autologous hematopoietic stem cell transplantation in refractory celiac disease with aberrant T cells. Blood 109, 2243–2249.

Al-Toma, A., Volta, U., Auricchio, R., Castillejo, G., Sanders, D.S., Cellier, C., Mulder, C.J., Lundin, K.E.A., 2019. European Society for the Study of Coeliac Disease (ESsCD) guideline for coeliac disease and other gluten-related disorders. United European Gastroenterol J 7, 583–613.

Arguelles-Grande, C., Brar, P., Green, P.H.R., Bhagat, G., 2013. Immunohistochemical and t-cell receptor gene rearrangement analyses as predictors of morbidity and mortality in refractory celiac disease. J. Clin. Gastroenterol. 47, 593–601.

Ashton-Key, M., Diss, T.C., Pan, L., Du, M.Q., Isaacson, P.G., 1997. Molecular analysis of T-cell clonality in ulcerative jejunitis and enteropathy-associated T-cell lymphoma. Am. J. Pathol. 151, 493–498.

Askling, J., Linet, M., Gridley, G., Halstensen, T.S., Ekström, K., Ekbom, A., 2002. Cancer incidence in a population-based cohort of individuals hospitalized with celiac disease or dermatitis herpetiformis. Gastroenterology 123, 1428–1435.

Bagdi, E., Diss, T.C., Munson, P., Isaacson, P.G., 1999. Mucosal intra-epithelial lymphocytes in enteropathy-associated T-cell lymphoma, ulcerative jejunitis, and refractory celiac disease constitute a neoplastic population. Blood 94, 260–264.

Baggus, E.M.R., Hadjivassiliou, M., Cross, S., Penny, H., Urwin, H., Watson, S., Woodward, J.M., Sanders, D.S., 2020. How to manage adult coeliac disease: perspective from the NHS England rare diseases collaborative network for non-responsive and refractory coeliac disease. Frontline Gastroenterol 11, 235–242.

Ben Ahmed, M., Belhadj Hmida, N., Moes, N., Buyse, S., Abdeladhim, M., Louzir, H., Cerf-Bensussan, N., 2009. IL-15 renders conventional lymphocytes resistant to suppressive functions of regulatory T cells through activation of the phosphatidylinositol 3-kinase pathway. J. Immunol. 182, 6763–6770.

Benahmed, M., Meresse, B., Arnulf, B., Barbe, U., Mention, J.J., Verkarre, V., Allez, M., Cellier, C., Hermine, O., Cerf-Bensussan, N., 2007. Inhibition of TGF-β signaling by IL-15: a new role for IL-15 in the loss of immune homeostasis in celiac disease. Gastroenterology 132, 994–1008.

Benckert, J., Schmolka, N., Kreschel, C., Zoller, M.J., Sturm, A., Wiedenmann, B., Wardemann, H., 2011. The majority of intestinal IgA+ and IgG+ plasmablasts in the human gut are antigen-specific. J. Clin. Invest. 121, 1946–1955.

Bhagat, G., Naiyer, A.J., Shah, J.G., Harper, J., Jabri, B., Wang, T.C., Green, P.H.R., Manavalan, J.S., 2008. Small intestinal CD8+TCRγδ+NKG2A+ intraepithelial lymphocytes have attributes of regulatory cells in patients with celiac disease. J. Clin. Invest. 118, 281–293.

Biagi, F., Corazza, G.R., 2010. Mortality in celiac disease. Nat. Rev. Gastroenterol. Hepatol. 7, 158–162.

Biagi, F., Gobbi, P., Marchese, A., Borsotti, E., Zingone, F., Ciacci, C., Volta, U., Caio, G., Carroccio, A., Ambrosiano, G., Mansueto, P., Corazza, G.R., 2014. Low incidence but poor prognosis of complicated coeliac disease: a retrospective multicentre study. Dig. Liver Dis. 46, 227–230.

Bodd, M., Ráki, M., Tollefsen, S., Fallang, L.E., Bergseng, E., Lundin, K.E.A., Sollid, L.M., 2010. HLA-DQ2-restricted gluten-reactive T cells produce IL-21 but not IL-17 or IL-22. Mucosal Immunol. 3, 594–601.

Boismenu, R., Havran, W.L., 1994. Modulation of epithelial cell growth by intraepithelial γδT cells. Science 266, 1253–1255.

Bouziat, R., Hinterleitner, R., Brown, J.J., Stencel-Baerenwald, J.E., Ikizler, M., Mayassi, T., Meisel, M., Kim, S.M., Discepolo, V., Pruijssers, A.J., Ernest, J.D., Iskarpatyoti, J.A., Costes, L.M.M., Lawrence, I., Palanski, B.A., Varma, M.,

Zurenski, M.A., Khomandiak, S., McAllister, N., Aravamudhan, P., Boehme, K.W., Hu, F., Samsom, J.N., Reinecker, H.C., Kupfer, S.S., Guandalini, S., Semrad, C.E., Abadie, V., Khosla, C., Barreiro, L.B., Xavier, R.J., Ng, A., Dermody, T.S., Jabri, B., 2017. Reovirus infection triggers inflammatory responses to dietary antigens and development of celiac disease. Science 356, 44–50.

Branchi, F., Locatelli, M., Tomba, C., Conte, D., Ferretti, F., Elli, L., 2016. Enteroscopy and radiology for the management of celiac disease complications: time for a pragmatic roadmap. Dig. Liver Dis. 48, 578–586.

Brar, P., Lee, S., Lewis, S., Egbuna, I., Bhagat, G., Green, P.H.R., 2007. Budesonide in the treatment of refractory celiac disease. Am. J. Gastroenterol. 102, 2265–2269.

Brousse, N., Verkarre, V., De Serre, N.P.M., Cellier, C., Cerf-Bensussan, N., Delabesse, E., Macintyre, E., 1999. Is complicated celiac disease or refractory sprue an intestinal intra-epithelial cryptic T-cell lymphoma? Blood 93, 3154–3155.

Buettner, M., Lochner, M., 2016. Development and function of secondary and tertiary lymphoid organs in the small intestine and the colon. Front. Immunol 7, 342.

Calleja, S., Vivas, S., Santiuste, M., Arias, L., Hernando, M., Nistal, E., Casqueiro, J., Ruiz De Morales, J.G., 2011. Dynamics of non-conventional intraepithelial lymphocytes-NK, NKT, and γδ T-in celiac disease: relationship with age, diet, and histopathology. Dig. Dis. Sci. 56, 2042–2049.

Camarero, C., Leon, F., Sanchez, L., Asensio, A., Roy, G., 2007. Age-related variation of intraepithelial lymphocytes subsets in normal human duodenal mucosa. Dig. Dis. Sci. 52, 685–691.

Cammarota, G., Cuoco, L., Cianci, R., Pandolfi, F., Gasbarrini, G., 2000. Onset of coeliac disease during treatment with interferon for chronic hepatitis C. Lancet 356, 1494–1495.

Carbonnel, F., Grollet-Bioul, L., Brouet, J.C., Teilhac, M.F., Cosnes, J., Angonin, R., Deschaseaux, M., Châtelet, F.P., Gendre, J.P., Sigaux, F., 1998. Are complicated forms of celiac disease cryptic T-cell lymphomas? Blood 92, 3879–3886.

Castellanos-Rubio, A., Bilbao, J.R., 2018. Profiling Celiac Disease-Related Transcriptional Changes, in: International Review of Cell and Molecular Biology. Elsevier Inc., pp. 149–174.

Castellanos-Rubio, A., Fernandez-Jimenez, N., Kratchmarov, R., Luo, X., Bhagat, G., Green, P.H.R., Schneider, R., Kiledjian, M., Bilbao, J.R., Ghosh, S., 2016. A long noncoding RNA associated with susceptibility to celiac disease. Science 352, 91–95.

Catassi, C., Fabiani, E., Corrao, G., Barbato, M., De Renzo, A., Carella, A.M., Gabrielli, A., Leoni, P., Carroccio, A., Baldassarre, M., Bertolani, P., Caramaschi, P., Sozzi, M., Guariso, G., Volta, U., Corazza, G.R., 2002. Risk of non-Hodgkin lymphoma in celiac disease. JAMA 287, 1413–1419.

Catassi, C., Bearzi, I., Holmes, G.K.T., 2005. Association of celiac disease and intestinal lymphomas and other cancers. Gastroenterology 128, S79–S86.

Cellier, C., Patey, N., Mauvieux, L., Jabri, B., Delabesse, E., Cervoni, J.P., Burtin, M.L.L., Guy-Grand, D., Bouhnik, Y., Modigliani, R., Barbier, J.P., Cervoni, P., Burtin, M.L.L., Delphine, G.G., Bouhnik, Y., Modigliani, R., Barbier, P., Macintyre, E., Brousse, N., Cerf-Bensussan, N., 1998. Abnormal intestinal intraepithelial lymphocytes in refractory sprue. Gastroenterology 114, 471–481.

Cellier, C., Delabesse, E., Helmer, C., Patey, N., Matuchansky, C., Jabri, B., MacIntyre, E., Cerf-Bensussan, N., Brousse, N., 2000. Refractory sprue, coeliac disease, and enteropathy-associated T-cell lymphoma. Lancet 356, 203–208.

Cellier, C., Bouma, G., van Gils, T., Khater, S., Malamut, G., Crespo, L., Collin, P., Green, P.H.R., Crowe, S.E., Tsuji, W., Butz, E., Cerf-Bensussan, N., Macintyre, E., Parnes, J.R., Leon, F., Hermine, O., Mulder, C.J., 2019. Safety and efficacy of AMG 714 in patients with type 2 refractory coeliac disease: a phase 2a, randomised, double-blind, placebo-controlled, parallel-group study. Lancet Gastroenterol. Hepatol. 4, 960–970.

Cheminant, M., Bruneau, J., Malamut, G., Sibon, D., Guegan, N., Van Gils, T., Cording, S., Trinquand, A., Verkarre, V., Lhermitte, L., Brousse, N., Anne-S, J., Khater, S., Frenzel, L., Delarue, R., Suarez, F., Marçais, A., Mulder, C.J.J.J., Macintyre, E., Asnafi, V., Pouyet, L., Bonnafous, C., Lhospice, F., Molina, T.J., Meresse, B., Cellier, C., Cerf-Bensussan, N., Hermine, O., Jo Molina, T., Meresse, B., Cellier, C., Cerf-Bensussan, N., Hermine, O., Olivier Hermine, P., Molina, T.J., Meresse, B., Cellier, C., Cerf-Bensussan, N., Hermine, O., 2019. NKp46 is a diagnostic biomarker and may be a therapeutic target in gastrointestinal T-cell lymphoproliferative diseases: a CELAC study. Gut 68, 1396–1405.

Chen, Y., Chou, K., Fuchs, E., Havran, W.L., Boismenu, R., 2002. Protection of the intestinal mucosa by intraepithelial γδ T cells. Proc. Natl. Acad. Sci. U. S. A. 99, 14338–14343.

Colpitts, S.L., Stoklasek, T.A., Plumlee, C.R., Obar, J.J., Guo, C., Lefrançois, L., 2012. Cutting edge: the role of IFN-α receptor and MyD88 signaling in induction of IL-15 expression in vivo. J. Immunol. 188, 2483–2487.

Comino, I., Fernández-Bañares, F., Esteve, M., Ortigosa, L., Castillejo, G., Fambuena, B., Ribes-Koninckx, C., Sierra, C., Rodríguez-Herrera, A., Salazar, J.C., Caunedo, Á., Marugán-Miguelsanz, J.M., Garrote, J.A., Vivas, S., Lo Iacono, O., Nuñez, A., Vaquero, L., Vegas, A.M., Crespo, L., Fernández-Salazar, L., Arranz, E., Jiménez-García, V.A., Antonio Montes-Cano, M., Espín, B., Galera, A., Valverde, J., Girón, F.J., Bolonio, M., Millán, A., Cerezo, F.M., Guajardo, C., Alberto, J.R., Rosinach, M., Segura, V., León, F., Marinich, J., Muñoz-Suano, A., Romero-Gómez, M., Cebolla, Á., Sousa, C., 2016. Fecal gluten peptides reveal limitations of serological tests and food questionnaires for monitoring gluten-free diet in celiac disease patients. Am. J. Gastroenterol. 111, 1456–1465.

Corrao, G., Corazza, G.R., Bagnardi, V., Brusco, G., Ciacci, C., Cottone, M., Sategna Guidetti, C., Usai, P., Cesari, P., Pelli, M.A., Loperfido, S., Volta, U., Calabró, A., Certo, M., 2001. Mortality in patients with coeliac disease and their relatives: a cohort study. Lancet 358, 356–361.

Costantino, G., Della Torre, A., Lo Presti, M.A., Caruso, R., Mazzon, E., Fries, W., 2008. Treatment of life-threatening type I refractory coeliac disease with long-term infliximab. Dig. Liver Dis. 40, 74–77.

Dahal-Koirala, S., Ciacchi, L., Petersen, J., Risnes, L.F., Neumann, R.S., Christophersen, A., Lundin, K.E.A., Reid, H.H., Qiao, S.W., Rossjohn, J., Sollid, L.M., 2019. Discriminative T-cell receptor recognition of highly homologous HLA-DQ2– bound gluten epitopes. J. Biol. Chem. 294, 941–952.

Daum, S., Weiss, D., Hummel, M., Ullrich, R., Heise, W., Stein, H., Riecken, E.O., Foss, H.D., 2001. Frequency of clonal intraepithelial T lymphocyte proliferations in enteropathy-type intestinal T cell lymphoma, coeliac disease, and refractory sprue. Gut 49, 804–812.

Daum, S., Cellier, C., Mulder, C.J.J., 2005. Refractory coeliac disease. Best Pract. Res. Clin. Gastroenterol. 19, 413–424.

Daum, S., Ipczynski, R., Schumann, M., Wahnschaffe, U., Zeitz, M., Ullrich, R., 2009. High rates of complications and substantial mortality in both types of refractory sprue. Eur. J. Gastroenterol. Hepatol. 21, 66–70.

De Mascarel, A., Belleannée, G., Stanislas, S., Merlio, C., Parrens, M., Laharie, D., Dubus, P., Merlio, J.-P.P., 2008. Mucosal intraepithelial T-lymphocytes in refractory celiac disease: a neoplastic population with a variable CD8 phenotype. Am. J. Surg. Pathol. 32, 744–751.

Dewar, D.H., Donnelly, S.C., McLaughlin, S.D., Johnson, M.W., Ellis, H.J., Ciclitira, P.J., 2012. Celiac disease: management of persistent symptoms in patients on a gluten-free diet. World J. Gastroenterol. 18, 1348–1356.

Di Marco Barros, R., Roberts, N.A., Dart, R.J., Vantourout, P., Jandke, A., Nussbaumer, O., Deban, L., Cipolat, S., Hart, R., Iannitto, M.L., Laing, A., Spencer-Dene, B., East, P., Gibbons, D., Irving, P.M., Pereira, P., Steinhoff, U., Hayday, A., 2016. Epithelia use butyrophilin-like molecules to shape organ-specific γδ T cell compartments. Cell 167, 203–218. e17.

Di Niro, R., Mesin, L., Zheng, N.Y., Stamnaes, J., Morrissey, M., Lee, J.H., Huang, M., Iversen, R., Du Pré, M.F., Qiao, S.W., Lundin, K.E.A., Wilson, P.C., Sollid, L.M., 2012. High abundance of plasma cells secreting transglutaminase 2-specific IgA autoantibodies with limited somatic hypermutation in celiac disease intestinal lesions. Nat. Med. 18, 441–445.

Di Sabatino, A., Miceli, E., Dhaliwal, W., Biancheri, P., Salerno, R., Cantoro, L., Vanoli, A., De Vincenzi, M., Blanco, C.D.V., MacDonald, T.T., Corazza, G.R., 2008. Distribution, proliferation, and function of paneth cells in uncomplicated and complicated adult celiac disease. Am. J. Clin. Pathol. 130, 34–42.

Dørum, S., Steinsbø, Ø., Bergseng, E., Arntzen, M., De Souza, G.A., Sollid, L.M., 2016. Gluten-specific antibodies of celiac disease gut plasma cells recognize long proteolytic fragments that typically harbor T-cell epitopes. Sci. Rep. 6, 25565.

Dubois, P.C.A., Trynka, G., Franke, L., Hunt, K.A., Romanos, J., Curtotti, A., Zhernakova, A., Heap, G.A.R., Ádány, R., Aromaa, A., Bardella, M.T., Van Den Berg, L.H., Bockett, N.A., De La Concha, E.G., Dema, B., Fehrmann, R.S.N., Fernández-Arquero, M., Fiatal, S., Grandone, E., Green, P.M., Groen, H.J.M., Gwilliam, R., Houwen, R.H.J., Hunt, S.E., Kaukinen, K., Kelleher, D., Korponay-Szabo, I., Kurppa, K., MacMathuna, P., Mäki, M., Mazzilli, M.C., McCann, O.T., Mearin, M.L., Mein, C.A., Mirza, M.M., Mistry, V., Mora, B., Morley, K.I., Mulder, C.J., Murray, J.A., Ñez, C., Oosterom, E., Ophoff, R.A., Polanco, I., Peltonen, L., Platteel, M., Rybak, A., Salomaa, V., Schweizer, J.J., Sperandeo, M.P., Tack, G.J., Turner, G., Veldink, J.H., Verbeek, W.H.M., Weersma, R.K., Wolters, V.M., Urcelay, E., Cukrowska, B., Greco, L., Neuhausen, S.L., McManus, R., Barisani, D., Deloukas, P., Barrett, J.C., Saavalainen, P., Wijmenga, C., Van Heel, D.A., 2010. Multiple common variants for celiac disease influencing immune gene expression. Nat. Genet. 42, 295–302.

Dunne, M.R., Elliott, L., Hussey, S., Mahmud, N., Kelly, J., Doherty, D.G., Feighery, C.F., 2013. Persistent changes in circulating and intestinal γδ T cell subsets, invariant natural killer T cells and mucosal-associated invariant T cells in children and adults with coeliac disease. PLoS One 8, e76008.

Dupont, F.M., Vensel, W.H., Tanaka, C.K., Hurkman, W.J., Altenbach, S.B., 2011. Deciphering the complexities of the wheat flour proteome using quantitative two-dimensional electrophoresis, three proteases and tandem mass spectrometry. Proteome Sci. 9, 1–29.

Eberl, G., Colonna, M., Santo, J.P.D., McKenzie, A.N.J., 2015. Innate lymphoid cells: a new paradigm in immunology. Science 348, aaa6566.

Eiras, P., Roldan, E., Camarero, C., Olivares, F., Bootello, A., Roy, G., 1998. Flow cytometry description of a novel CD3-/CD7+ intraepithelial lymphocyte subset in human duodenal biopsies: potential diagnostic value in coeliac disease. Cytometry 34, 95–102.

Eiras, P., Leon, F., Camarero, C., Lombardia, M., Roldan, E., Bootello, A., Roy, G., 2000. Intestinal intraepithelial lymphocytes contain a CD3- CD7+ subset expressing natural killer markers and a singular pattern of adhesion molecules. Scand. J. Immunol. 52, 1–6.

Elli, L., Casazza, G., Locatelli, M., Branchi, F., Ferretti, F., Conte, D., Fraquelli, M., 2017. Use of enteroscopy for the detection of malignant and premalignant lesions of the small bowel in complicated celiac disease: a meta-analysis. Gastrointest. Endosc. 86, 264–273. e1.

Ettersperger, J., Montcuquet, N., Malamut, G., Guegan, N., Lopez-Lastra, S., Gayraud, S., Reimann, C., Vidal, E., Cagnard, N., Villarese, P., Andre-Schmutz, I., Gomes

Domingues, R., Godinho-Silva, C., Veiga-Fernandes, H., Lhermitte, L., Asnafi, V., Macintyre, E., Cellier, C., Beldjord, K., Di Santo, J.P., Cerf-Bensussan, N., Meresse, B., Gomes Domingues, R., Godinho-Silva, C., Veiga-Fernandes, H., Lhermitte, L., Asnafi, V., Macintyre, E., Cellier, C., Beldjord, K., Di Santo, J.P., Cerf-Bensussan, N., Meresse, B., 2016. Interleukin-15-dependent T-cell-like innate intraepithelial lymphocytes develop in the intestine and transform into lymphomas in celiac disease. Immunity 45, 610–625.

Fallang, L.E., Bergseng, E., Hotta, K., Berg-Larsen, A., Kim, C.Y., Sollid, L.M., 2009. Differences in the risk of celiac disease associated with HLA-DQ2.5 or HLA-DQ2.2 are related to sustained gluten antigen presentation. Nat. Immunol. 10, 1096–1101.

Farstad, I.N., Carlsen, H., Morton, H.C., Brandtzaeg, P., 2000. Immunoglobulin A cell distribution in the human small intestine: phenotypic and functional characteristics. Immunology 101, 354–363.

Fasano, A., Berti, I., Gerarduzzi, T., Not, T., Colletti, R.B., Drago, S., Elitsur, Y., Green, P.H.R., Guandalini, S., Hill, I.D., Pietzak, M., Ventura, A., Thorpe, M., Kryszak, D., Fornaroli, F., Wasserman, S.S., Murray, J.A., Horvath, K., 2003. Prevalence of celiac disease in at-risk and not-at-risk groups in the United States: a large multicenter study. Arch. Intern. Med. 163, 286–292.

Fina, D., Sarra, M., Caruso, R., Del Vecchio Blanco, G., Pallone, F., MacDonald, T.T., Monteleone, G., 2008. Interleukin 21 contributes to the mucosal T helper cell type 1 response in coeliac disease. Gut 57, 887–892.

Gao, Y., Kristinsson, S.Y., Goldin, L.R., Björkholm, M., Caporaso, N.E., Landgren, O., 2009. Increased risk for non-Hodgkin lymphoma in individuals with celiac disease and a potential familial association. Gastroenterology 136, 91–98.

García-Hoz, C., Crespo, L., Lopez, N., De Andrés, A., Ríos León, R., Santón, A., Garriga, M., Butz, E., León, F., Roy, G., 2020. The intracellular intensity of CD3 on aberrant intraepithelial lymphocytes is a prognostic factor of the progression to overt lymphoma in refractory celiac disease type II (RCD-II/pre-enteropathy-associated T cell lymphoma). Dig. Dis. 38, 490–499.

Gibbons, D.L., Spencer, J., 2011. Mouse and human intestinal immunity: same ballpark, different players; different rules, same score. Mucosal Immunol. 4, 148–157.

Gillett, H.R., Arnott, I.D.R., McIntyre, M., Campbell, S., Dahele, A., Priest, M., Jackson, R., Ghosh, S., 2002. Successful infliximab treatment for steroid-refractory celiac disease: a case report. Gastroenterology 122, 800–805.

Goerres, M.S., Meijer, J.W.R., Wahab, P.J., Kerckhaert, J.A.M., Groenen, P.J.T.A., Van Krieken, J.H.J.M., Mulder, C.J.J., 2003. Azathioprine and prednisone combination therapy in refractory coeliac disease. Aliment. Pharmacol. Ther. 18, 487–494.

Green, P.H., Cellier, C., 2007. Celiac disease. N. Engl. J. Med. 357, 1731–1743.

Green, P.H.R., Fleischauer, A.T., Bhagat, G., Goyal, R., Jabri, B., Neugut, A.I., 2003. Risk of malignancy in patients with celiac disease. Am. J. Med. 115, 191–195.

Guy-Grand, D., Cerf-Bensussan, N., Malissen, B., Malassis-Seris, M., Briottet, C., Vassalli, P., 1991. Two gut intraepithelial CD8+ lymphocyte populations with different T cell receptors: a role for the gut epithelium in T cell differentiation. J. Exp. Med. 173, 471–481.

Hadithi, M., Mallant, M., Oudejans, J., Van Waesberghe, J.H.T.M., Mulder, C.J., Comans, E.F.I., 2006. 18F-FDG PET versus CT for the detection of enteropathy- associated T-cell lymphoma in refractory celiac disease. J. Nucl. Med. 47, 1622–1627.

Hadjivassiliou, M., Duker, A.P., Sanders, D.S., 2014. Gluten-Related Neurologic Dysfunction, Handbook of Clinical Neurology. Elsevier.

Hadjivassiliou, M., Sanders, D.D., Aeschlimann, D.P., 2015. Gluten-Related Disorders: Gluten Ataxia. Dig. Dis. 33, 264–268.

Hayday, A.C., 2019. γδ T cell update: adaptate orchestrators of immune surveillance. J. Immunol. 203, 311–320.

Hmida, N.B., Ahmed, M.B., Moussa, A., Rejeb, M.B., Said, Y., Kourda, N., Meresse, B., Abdeladhim, M., Louzir, H., Cerf-Bensussan, N., 2012. Impaired control of effector T cells by regulatory T cells: a clue to loss of oral tolerance and autoimmunity in celiac disease? Am. J. Gastroenterol. 107, 604–611.

Hoffmann, M., Vogelsang, H., Kletter, K., Zettinig, G., Chott, A., Raderer, M., 2003. 18F-fluoro-deoxy-glucose positron emission tomography (18F-FDG-PET) for assessment of enteropathy-type T cell lymphoma. Gut 52, 347–351.

Holmes, G.K.T., Prior, P., Lane, M.R., Pope, D., Allan, R.N., 1989. Malignancy in coeliac disease—effect of a gluten free diet. Gut 30, 333–338.

Høydahl, L.S., Richter, L., Frick, R., Snir, O., Gunnarsen, K.S., Landsverk, O.J.B., Iversen, R., Jeliazkov, J.R., Gray, J.J., Bergseng, E., Foss, S., Qiao, S.W., Lundin, K.E.A., Jahnsen, J., Jahnsen, F.L., Sandlie, I., Sollid, L.M., Løset, G.Å., 2019. Plasma cells are the Most abundant gluten peptide MHC-expressing cells in inflamed intestinal tissues from patients with celiac disease. Gastroenterology 156, 1428–1439. e10.

Hrdlickova, B., Mulder, C.J., Malamut, G., Meresse, B., Platteel, M., Kamatani, Y., Ricaño-Ponce, I., van Wanrooij, R.L.J.J., Zorro, M.M., Jan Bonder, M., Gutierrez-Achury, J., Cellier, C., Zhernakova, A., Nijeboer, P., Galan, P., Withoff, S., Lathrop, M., Bouma, G., Xavier, R.J., Jabri, B., Bensussan, N.C., Wijmenga, C., Kumar, V., Bonder, M.J., Gutierrez-Achury, J., Cellier, C., Zhernakova, A., Nijeboer, P., Galan, P., Withoff, S., Lathrop, M., Bouma, G., Xavier, R.J., Jabri, B., Bensussan, N.C., Wijmenga, C., Kumar, V., 2018. A locus at 7p14.3 predisposes to refractory celiac disease progression from celiac disease. Eur. J. Gastroenterol. Hepatol. 30, 828–837.

Hüe, S., Mention, J.J., Monteiro, R.C., Zhang, S.L., Cellier, C., Schmitz, J., Verkarre, V., Fodil, N., Bahram, S., Cerf-Bensussan, N., Caillat-Zucman, S., 2004. A direct role for NKG2D/MICA interaction in villous atrophy during celiac disease. Immunity 21, 367–377.

Hujoel, I.A., Murray, J.A., 2020. Refractory celiac disease. Curr. Gastroenterol. Rep. 22, 1–8.

Hunt, K.A., Zhernakova, A., Turner, G., Heap, G.A.R., Franke, L., Bruinenberg, M., Romanos, J., Dinesen, L.C., Ryan, A.W., Panesar, D., Gwilliam, R., Takeuchi, F., McLaren, W.M., Holmes, G.K.T., Howdle, P.D., Walters, J.R.F., Sanders, D.S., Playford, R.J., Trynka, G., Mulder, C.J.J., Mearin, M.L., Verbeek, W.H.M., Trimble, V., Stevens, F.M., O'Morain, C., Kennedy, N.P., Kelleher, D., Pennington, D.J., Strachan, D.P., McArdle, W.L., Mein, C.A., Wapenaar, M.C., Deloukas, P., McGinnis, R., McManus, R., Wijmenga, C., Van Heel, D.A., 2008. Newly identified genetic risk variants for celiac disease related to the immune response. Nat. Genet. 40, 395–402.

Hussein, S., Gindin, T., Lagana, S.M., Arguelles-Grande, C., Krishnareddy, S., Alobeid, B., Lewis, S.K., Mansukhani, M.M., Green, P.H.R., Bhagat, G., 2018. Clonal T cell receptor gene rearrangements in coeliac disease: implications for diagnosing refractory coeliac disease. J. Clin. Pathol. 71, 825–831.

Ilus, T., Kaukinen, K., Virta, L.J., Huhtala, H., Mäki, M., Kurppa, K., Heikkinen, M., Heikura, M., Hirsi, E., Jantunen, K., Moilanen, V., Nielsen, C., Puhto, M., Pölkki, H., Vihriälä, I., Collin, P., 2014. Refractory coeliac disease in a country with a high prevalence of clinically-diagnosed coeliac disease. Aliment. Pharmacol. Ther. 39, 418–425.

Isaacson, P., Wright, D., 1978. Intestinal lymphoma associated with malabsorption. Lancet 1, 67–70.

Jabri, B., Sollid, L.M., 2017. T cells in celiac disease. J. Immunol. 198, 3005–3014.

Jahnsen, F.L., Farstad, I.N., Aanesen, J.P., Brandtzaeg, P., 1998. Phenotypic distribution of T cells in human nasal mucosa differs from that in the gut. Am. J. Respir. Cell Mol. Biol. 18, 392–401.

Jamma, S., Leffler, D.A., Dennis, M., Najarian, R.M., Schuppan, D.B., Sheth, S., Kelly, C.P., 2011. Small intestinal release mesalamine for the treatment of refractory celiac disease type I. J. Clin. Gastroenterol. 45, 30–33.

Jarry, A., Cerf-Bensussan, N., Brousse, N., Selz, F., Guy-Grand, D., Cerf-Bensussan, N., Brousse, N., Selz, F., Guy-grand, D., 1990. Subsets of CD3+ (T cell receptor α/β or γ/δ) and CD3− lymphocytes isolated from normal human gut epithelium display phenotypical features different from their counterparts in peripheral blood. Eur. J. Immunol. 20, 1097–1103.

Jericho, H., Sansotta, N., Guandalini, S., 2017. Extraintestinal manifestations of celiac disease: effectiveness of the gluten-free diet. J. Pediatr. Gastroenterol. Nutr. 65, 75–79.

Jewell, D.P., 1983. Ulcerative enteritis. Br. Med. J. (Clin. Res. Ed) 287, 1740–1741.

Karell, K., Louka, A.S., Moodie, S.J., Ascher, H., Clot, F., Greco, L., Ciclitira, P.J., Sollid, L.M., Partanen, J., 2003. HLA types in celiac disease patients not carrying the DQA1 *05-DQB1 *02 (DQ2) heterodimer: results from the European genetics cluster on celiac disease. Hum. Immunol. 64, 469–477.

Kim, S.M., Mayassi, T., Jabri, B., 2015. Innate immunity: actuating the gears of celiac disease pathogenesis. Best Pract. Res. Clin. Gastroenterol. 29, 425–435.

Kooy-Winkelaar, Y.M.C.C., Bouwer, D., Janssen, G.M.C.C., Thompson, A., Brugman, M.H., Schmitz, F., De Ru, A.H., Van Gils, T., Bouma, G., Van Rood, J.J., Van Veelen, P.A., Mearin, M.L., Mulder, C.J., Koning, F., Van Bergen, J., 2017. CD4 T-cell cytokines synergize to induce proliferation of malignant and nonmalignant innate intraepithelial lymphocytes. Proc. Natl. Acad. Sci. U. S. A. 114, E980–E989.

Korneychuk, N., Ramiro-Puig, E., Ettersperger, J., Schulthess, J., Montcuquet, N., Kiyono, H., Meresse, B., Cerf-Bensussan, N., 2014. Interleukin 15 and CD4+ T cells cooperate to promote small intestinal enteropathy in response to dietary antigen. Gastroenterology 146, 1017–1027.

Kuja-Halkola, R., Lebwohl, B., Halfvarson, J., Wijmenga, C., Magnusson, P.K.E., Ludvigsson, J.F., 2016. Heritability of non-HLA genetics in coeliac disease: a population-based study in 107 000 twins. Gut 65, 1793–1798.

Kumar, V., Wijmenga, C., Withoff, S., 2012. From genome-wide association studies to disease mechanisms: celiac disease as a model for autoimmune diseases. Semin. Immunopathol. 34, 567–580.

Lebwohl, B., Sanders, D.S., Green, P.H.R., 2018. Coeliac disease. Lancet 391, 70–81.

Leffler, D.A., Dennis, M., Hyett, B., Kelly, E., Schuppan, D., Kelly, C.P., 2007. Etiologies and predictors of diagnosis in nonresponsive celiac disease. Clin. Gastroenterol. Hepatol. 5, 445–450.

Leffler, D.A., Green, P.H.R., Fasano, A., 2015. Extraintestinal manifestations of coeliac disease. Nat. Rev. Gastroenterol. Hepatol. 12, 561–571.

Leon, F., Roldan, E., Sanchez, L., Camarero, C., Bootello, A., Roy, G.G., 2003. Human small-intestinal epithelium contains functional natural killer lymphocytes. Gastroenterology 125, 345–356.

Liu, H., Brais, R., Lavergne-Slove, A., Jeng, Q., Payne, K., Ye, H., Liu, Z., Carreras, J., Huang, Y., Bacon, C.M., Hamoudi, R.A., Save, V., Venkatraman, L., Isaacson, P.G., Woodward, J., Du, M.Q., 2010. Continual monitoring of intraepithelial lymphocyte immunophenotype and clonality is more important than snapshot analysis in the surveillance of refractory coeliac disease. Gut 59, 452–460.

Luciani, A., Villella, V.R., Vasaturo, A., Giardino, I., Pettoello-Mantovani, M., Guido, S., Cexus, O.N., Peake, N., Londei, M., Quaratino, S., Maiuri, L., 2010. Lysosomal accumulation of gliadin p31-43 peptide induces oxidative stress and tissue transglutaminase-mediated PPARγ downregulation in intestinal epithelial cells and coeliac mucosa. Gut 59, 311–319.

Ludvigsson, J.F., Leffler, D.A., Bai, J.C., Biagi, F., Fasano, A., Green, P.H.R., Hadjivassiliou, M., Kaukinen, K., Kelly, C.P., Leonard, J.N., Lundin, K.E.A., Murray, J.A., Sanders, D.S., Walker, M.M., Zingone, F., Ciacci, C., 2013. The Oslo definitions for coeliac disease and related terms. Gut 62, 43–52.

Maiuri, L., Ciacci, C., Ricciardelli, I., Vacca, L., Raia, V., Auricchio, S., Picard, J., Osman, M., Quaratino, S., Londei, M., 2003. Association between innate response to gliadin and activation of pathogenic T cells in coeliac disease. Lancet 362, 30–37.

Malamut, G., Afchain, P., Verkarre, V., Lecomte, T., Amiot, A., Damotte, D., Bouhnik, Y., Colombel, J.F., Delchier, J.C., Allez, M., Cosnes, J., Lavergne-Slove, A., Meresse, B., Trinquart, L., Macintyre, E., Radford-Weiss, I., Hermine, O., Brousse, N., Cerf-Bensussan, N., Cellier, C., 2009. Presentation and long-term follow-up of refractory celiac disease: comparison of type I with type II. Gastroenterology 136, 81–90.

Malamut, G., El Machhour, R., Montcuquet, N., Martin-Lannerée, S., Dusanter-Fourt, I., Verkarre, V., Mention, J.-J., Rahmi, G., Kiyono, H., Butz, E.A., Brousse, N., Cellier, C., Cerf-Bensussan, N., Meresse, B., 2010. IL-15 triggers an antiapoptotic pathway in human intraepithelial lymphocytes that is a potential new target in celiac disease-associated inflammation and lymphomagenesis. J. Clin. Invest. 120, 2131–2143.

Malamut, G., Meresse, B., Cellier, C., Cerf-Bensussan, N., 2012. Refractory celiac disease: from bench to bedside. Semin. Immunopathol. 34, 601–613.

Malamut, G., Chandesris, O., Verkarre, V., Meresse, B., Callens, C., Macintyre, E., Bouhnik, Y., Gornet, J.M., Allez, M., Jian, R., Berger, A., Châtellier, G., Brousse, N., Hermine, O., Cerf-Bensussan, N., Cellier, C., 2013. Enteropathy associated T cell lymphoma in celiac disease: a large retrospective study. Dig. Liver Dis. 45, 377–384.

Mayassi, T., Ladell, K., Gudjonson, H., McLaren, J.E., Shaw, D.G., Tran, M.T., Rokicka, J.J., Lawrence, I., Grenier, J.C., van Unen, V., Ciszewski, C., Dimaano, M., Sayegh, H.E., Kumar, V., Wijmenga, C., Green, P.H.R., Gokhale, R., Jericho, H., Semrad, C.E., Guandalini, S., Dinner, A.R., Kupfer, S.S., Reid, H.H., Barreiro, L.B., Rossjohn, J., Price, D.A., Jabri, B., 2019. Chronic inflammation permanently reshapes tissue-resident immunity in celiac disease. Cell 176, 967–981. e19.

Mention, J.J., Ahmed, M.B., Bègue, B., Barbe, U., Verkarre, V., Asnafi, V., Colombel, J.F., Cugnenc, P.H., Ruemmele, F.M., McIntyre, E., Brousse, N., Cellier, C., Cerf-Bensussan, N., 2003. Interleukin 15: A key to disrupted intraepithelial lymphocyte homeostasis and lymphomagenesis in celiac disease. Gastroenterology 125, 730–745.

Meresse, B., Chen, Z., Ciszewski, C., Tretiakova, M., Bhagat, G., Krausz, T.N., Raulet, D.H., Lanier, L.L., Groh, V., Spies, T., Ebert, E.C., Green, P.H., Jabri, B., 2004. Coordinated induction by IL15 of a TCR-independent NKG2D signaling pathway converts CTL into lymphokine-activated killer cells in celiac disease. Immunity 21, 357–366.

Meresse, B., Curran, S.A., Ciszewski, C., Orbelyan, G., Setty, M., Bhagat, G., Lee, L., Tretiakova, M., Semrad, C., Kistner, E., Winchester, R.J., Braud, V., Lanier, L.L., Geraghty, D.E., Green, P.H., Guandalini, S., Jabri, B., 2006. Reprogramming of CTLs into natural killer-like cells in celiac disease. J. Exp. Med. 203, 1343–1355.

Meresse, B., Malamut, G., Cerf-Bensussan, N., 2012. Celiac disease: an immunological jigsaw. Immunity 36, 907–919.

Moffitt, A.B., Ondrejka, S.L., McKinney, M., Rempel, R.E., Goodlad, J.R., Teh, C.H., Leppa, S., Mannisto, S., Kovanen, P.E., Tse, E., Au-Yeung, R.K.H., Kwong, Y.-L., Srivastava, G., Iqbal, J., Yu, J., Naresh, K., Villa, D., Gascoyne, R.D., Said, J.,

Czader, M.B., Chadburn, A., Richards, K.L., Rajagopalan, D., Davis, N.S., Smith, E.C., Palus, B.C., Tzeng, T.J., Healy, J.A., Lugar, P.L., Datta, J., Love, C., Levy, S., Dunson, D.B., Zhuang, Y., Hsi, E.D., Dave, S.S., 2017. Enteropathy-associated T cell lymphoma subtypes are characterized by loss of function of SETD2. J. Exp. Med. 214, 1371–1386.

Moreno, M.D.L., Cebolla, Á., Munõz-Suano, A., Carrillo-Carrion, C., Comino, I., Pizarro, Á., León, F., Rodríguez-Herrera, A., Sousa, C., 2017. Detection of gluten immunogenic peptides in the urine of patients with coeliac disease reveals transgressions in the gluten-free diet and incomplete mucosal healing. Gut 66, 250–257.

Mukewar, S.S., Sharma, A., Rubio-Tapia, A., Wu, T.-T.T., Jabri, B., Murray, J.A., 2017. Open-capsule budesonide for refractory celiac disease. Am. J. Gastroenterol. 112, 959–967.

Mulder, C.J.J., Wahab, P.J., Meijer, J.W.R., Metselaar, E., 2001. A pilot study of recombinant human interleukin-10 in adults with refractory coeliac disease. Eur. J. Gastroenterol. Hepatol. 13, 1183–1188.

Murray, A., Cuevas, E.C., Jones, D.B., Wright, D.H., 1995. Study of the immunohistochemistry and T cell clonality of enteropathy-associated T cell lymphoma. Am. J. Pathol. 146, 509–519.

Nasr, I., Nasr, I., Beyers, C., Chang, F., Donnelly, S., Ciclitira, P., 2015. Recognising and managing refractory coeliac disease: a tertiary centre experience. Nutrients 7, 9896–9907.

Nijeboer, P., van Wanrooij, R., van Gils, T., Wierdsma, N.J., Tack, G.J., Witte, B.I., Bontkes, H.J., Visser, O., Mulder, C., Bouma, G., 2017. Lymphoma development and survival in refractory coeliac disease type II: histological response as prognostic factor. United European Gastroenterol J 5, 208–217.

Nijeboer, P., van Gils, T., Reijm, M., Ooijevaar, R., Lissenberg-Witte, B.I., Bontkes, H.J., Mulder, C.J.J., Bouma, G., 2019. Gamma-delta T lymphocytes in the diagnostic approach of coeliac disease. J. Clin. Gastroenterol. 53, e208–e213.

Ntziachristos, P., Abdel-Wahab, O., Aifantis, I., 2016. Emerging concepts of epigenetic dysregulation in hematological malignancies. Nat. Immunol. 17, 1016–1024.

Obermann, E.C., Diss, T.C., Hamoudi, R.A., Wilkins, B.S., Camozzi, M.L.P., Isaacson, P.G., Du, M.Q., Dogan, A., 2004. Loss of heterozygosity at chromosome 9p21 is a frequent finding in enteropathy-type T-cell lymphoma. J. Pathol. 202, 252–262.

Ogata, M., Itoh, T., 2016. Gamma/delta intraepithelial lymphocytes in the mouse small intestine. Anat. Sci. Int. 91, 301–312.

O'Mahony, S., Howdle, P.D., Losowsky, M.S., 1996. Management of patients with non-responsive coeliac disease. Aliment. Pharmacol. Ther. 10, 671–680.

Patey-Mariaud De Serre, N., Cellier, C., Jabri, B., Delabesse, E., Verkarre, V., Roche, B., Lavergne, A., Brière, J., Mauvieux, L., Leborgne, M., Barbier, J.P., Modigliani, R., Matuchansky, C., Macintyre, E., Cerf-Bensussan, N., Brousse, N., 2000. Distinction between coeliac disease and refractory sprue: a simple immunohistochemical method. Histopathology 37, 70–77.

Penny, H.A., Baggus, E.M.R., Rej, A., Snowden, J.A., Sanders, D.S., 2020. Non-responsive coeliac disease: a comprehensive review from the NHS England national Centre for refractory coeliac disease. Nutrients 12, 216.

Perfetti, V., Brunetti, L., Biagi, F., Ciccocioppo, R., Bianchi, P.I., Corazza, G.R., 2012. TCRβ Clonality improves diagnostic yield of TCRγ clonality in refractory celiac disease. J. Clin. Gastroenterol. 46, 675–679.

Radmard, A.R., Hashemi Taheri, A.P., Salehian Nik, E., Kooraki, S., Kolahdoozan, S., Mirminachi, B., Sotoudeh, M., Ekhlasi, G., Malekzadeh, R., Shahbazkhani, B., 2017. MR enterography in nonresponsive adult celiac disease: correlation with endoscopic, pathologic, serologic, and genetic features. J. Magn. Reson. Imaging 46, 1096–1106.

Ricaño-Ponce, I., Wijmenga, C., Gutierrez-Achury, J., 2015. Genetics of celiac disease. Best Pract. Res. Clin. Gastroenterol. 29, 399–412.
Roshan, B., Leffler, D.A., Jamma, S., Dennis, M., Sheth, S., Falchuk, K., Najarian, R., Goldsmith, J., Tariq, S., Schuppan, D., Kelly, C.P., 2011. The incidence and clinical spectrum of refractory celiac disease in a north american referral center. Am. J. Gastroenterol. 106, 923–928.
Rowinski, S.A., Christensen, E., 2016. Epidemiologic and therapeutic aspects of refractory coeliac disease—a systematic review. Dan. Med. J. 63, A5307.
Rubio-Tapia, A., Kelly, D.G., Lahr, B.D., Dogan, A., Wu, T.T., Murray, J.A., 2009. Clinical staging and survival in refractory celiac disease: a single center experience. Gastroenterology 136, 99–107.
Rubio-Tapia, A., Malamut, G., Verbeek, W.H.M., van Wanrooij, R.L.J., Leffler, D.A., Niveloni, S.I., Arguelles-Grande, C., Lahr, B.D., Zinsmeister, A.R., Murray, J.A., Kelly, C.P., Bai, J.C., Green, P.H., Daum, S., Mulder, C.J.J., Cellier, C., 2016. Creation of a model to predict survival in patients with refractory coeliac disease using a multinational registry. Aliment. Pharmacol. Ther. 44, 704–714.
Sarra, M., Cupi, M.L., Monteleone, I., Franzè, E., Ronchetti, G., Di Sabatino, A., Gentileschi, P., Franceschilli, L., Sileri, P., Sica, G., Del Vecchio Blanco, G., Cretella, M., Paoluzi, O.A., Corazza, G.R., Pallone, F., Monteleone, G., 2013. IL-15 positively regulates IL-21 production in celiac disease mucosa. Mucosal Immunol. 6, 244–255.
Sarrigiannis, P.G., Hoggard, N., Aeschlimann, D., Sanders, D.S., Grünewald, R.A., Unwin, Z.C., Hadjivassiliou, M., 2014. Myoclonus ataxia and refractory coeliac disease. Cerebellum & Ataxias 1, 11.
Schmitz, F., Tjon, J.M.L., Lai, Y., Thompson, A., Kooy-Winkelaar, Y., Lemmers, R.J.L.F., Verspaget, H.W., Mearin, M.L., Staal, F.J., Schreurs, M.W., Cupedo, T., Langerak, A.W., Mulder, C.J., Van Bergen, J., Koning, F., 2013. Identification of a potential physiological precursor of aberrant cells in refractory coeliac disease type II. Gut 62, 509–519.
Schmitz, F., Tjon, J.M.L., van Bergen, J., Koning, F., 2014. Dendritic cells promote expansion and survival of aberrant TCR-negative intraepithelial lymphocyte lines from refractory celiac disease type II patients. Mol. Immunol. 58, 10–16.
Schmitz, F., Kooy-Winkelaar, Y., Wiekmeijer, A.S., Brugman, M.H., Mearin, M.L., Mulder, C., Chuva De Sousa Lopes, S., Mummery, C.L., Staal, F.J., Van Bergen, J., Koning, F., 2016. The composition and differentiation potential of the duodenal intraepithelial innate lymphocyte compartment is altered in coeliac disease. Gut 65, 1269–1278.
Setty, M., Discepolo, V., Abadie, V., Kamhawi, S., Mayassi, T., Kent, A., Ciszewski, C., Maglio, M., Kistner, E., Bhagat, G., Semrad, C., Kupfer, S.S., Green, P.H., Guandalini, S., Troncone, R., Murray, J.A., Turner, J.R., Jabri, B., 2015. Distinct and synergistic contributions of epithelial stress and adaptive immunity to functions of intraepithelial killer cells and active celiac disease. Gastroenterology 149. 681e10–691.e10.
Shan, L., Molberg, Ø., Parrot, I., Hausch, F., Filiz, F., Gray, G.M., Sollid, L.M., Khosla, C., 2002. Structural basis for gluten intolerance in celiac sprue. Science 297, 2275–2279.
Shewry, P.R., Halford, N.G., Belton, P.S., Tatham, A.S., 2002. The structure and properties of gluten: an elastic protein from wheat grain. Philos. Trans. R Soc. Lond B Biol. Sci. 357, 133–142.
Silano, M., Volta, U., De Vincenzi, A., Dessì, M., De Vincenzi, M., Gasbarrini, G., De Vitis, V., Greco, L., Auricchio, S., Santini, D., Scaggiante, M.F., Vincenzi, M., Federici, Castellano, E., Calvi, A., Grosso, S., Campanella, J., Corazza, G.R., Sandri, G., Giorgetti, G., Volta, U., Parisi, C., Lo Perfido, S., Perri, F., Festa, V., Pelli, M.A., Cavalletti, M.L., Segato, S., Curzio, M., Pennazio, M., Rossini, F.P.,

Picarelli, A., Pera, A., Ercole, E., Passaleva, M.T., Barbato, M., Usai, P., Dore, M.F., Chilovi, F., Piazzi, L., Zancanella, L., Boarino, V., Ferrari, A., 2008. Effect of a gluten-free diet on the risk of enteropathy-associated T-cell lymphoma in celiac disease. Dig. Dis. Sci. 53, 972–976.

Singh, P., Arora, S., Lal, S., Strand, T.A., Makharia, G.K., 2015. Risk of celiac disease in the first- and second-degree relatives of patients with celiac disease: a systematic review and meta-analysis. Am. J. Gastroenterol. 110, 1539–1548.

Singh, P., Arora, A., Strand, T.A., Leffler, D.A., Catassi, C., Green, P.H., Kelly, C.P., Ahuja, V., Makharia, G.K., 2018. Global prevalence of celiac disease: systematic review and meta-analysis. Clin. Gastroenterol. Hepatol. 16, 823–836. e2.

Soderquist, C.R., Hsiao, S., Mansukhani, M.M., Alobeid, B., Green, P.H., Bhagat, G., 2020. Refractory celiac disease type II: an atypical case highlighting limitations of the current classification system. Hematol. Oncol. 38, 399–405.

Soderquist, C.R., Lewis, S.K., Gru, A.A., Vlad, G., Williams, E.S., Hsiao, S., Mansukhani, M.M., Park, D.C., Bacchi, C.E., Alobeid, B., Green, P.H., Bhagat, G., 2021. Immunophenotypic Spectrum and Genomic Landscape of Refractory Celiac Disease Type II. Am. J. Surg. Pathol. https://doi.org/10.1097/PAS.0000000000001658.

Sollid, L.M., 2017. The roles of MHC class II genes and post-translational modification in celiac disease. Immunogenetics 69, 605–616.

Sollid, L.M., Qiao, S.W., Anderson, R.P., Gianfrani, C., Koning, F., 2012. Nomenclature and listing of celiac disease relevant gluten T-cell epitopes restricted by HLA-DQ molecules. Immunogenetics 64, 455–460.

Spencer, J., Isaacson, P.G., MacDonald, T.T., Thomas, A.J., Walker-Smith, J.A., 1991. Gamma/delta T cells and the diagnosis of coeliac disease. Clin. Exp. Immunol. 85, 109–113.

Stier, K., Lewis, S.K., Bhagat, G., Green, P.H., 2019. Research publication trends regarding the extraintestinal manifestations of celiac disease. Informatics Med. Unlocked 17, 100242.

Sutherland, D.B., Suzuki, K., Fagarasan, S., 2016. Fostering of advanced mutualism with gut microbiota by immunoglobulin A. Immunol. Rev. 270, 20–31.

Tack, G.J., Verbeek, W.H.M., Al-Toma, A., Kuik, D.J., Schreurs, M.W.J., Visser, O., Mulder, C.J.J., 2011a. Evaluation of cladribine treatment in refractory celiac disease type II. World J. Gastroenterol. 17, 506–513.

Tack, G.J., Wondergem, M.J., Al-Toma, A., Verbeek, W.H.M., Schmittel, A., MacHado, M.V., Perri, F., Ossenkoppele, G.J., Huijgens, P.C., Schreurs, M.W.J., Mulder, C.J.J., Visser, O.J., 2011b. Auto-SCT in refractory celiac disease type II patients unresponsive to cladribine therapy. Bone Marrow Transplant. 46, 840–846.

Tack, G.J., Van Asseldonk, D.P., van Wanrooij, R.L.J., Van Bodegraven, A.A., Mulder, C.J., 2012a. Tioguanine in the treatment of refractory coeliac disease—a single Centre experience. Aliment. Pharmacol. Ther. 36, 274–281.

Tack, G.J., van Wanrooij, R.L.J., Langerak, A.W., Tjon, J.M.L., Von Blomberg, M.B.E., Heideman, D.A.M., Van Bergen, J., Koning, F., Bouma, G., Mulder, C.J.J., Schreurs, M.W.J., 2012b. Origin and immunophenotype of aberrant IEL in RCDII patients. Mol. Immunol. 50, 262–270.

Tjon, J.M., Verbeek, W.H., Kooy-Winkelaar, Y.M., Nguyen, B., van der Slik, A., Thompson, A., Heemskerk, M., Schreurs, M., Dekking, L., Mulder, C., van Bergen, J., Koning, F., 2008. Defective synthesis or association of T-cell receptor chains underlies loss of surface T-cell receptor-CD3 expression in enteropathy-associated T-cell lymphoma. Blood 112, 5103–5110.

Tjon, J.M.L., Kooy-Winkelaar, Y.M.C., Tack, G.J., Mommaas, A.M., Schreurs, M.W.J., Schilham, M.W., Mulder, C.J., van Bergen, J., Koning, F., 2011. DNAM-1 mediates epithelial cell-specific cytotoxicity of aberrant intraepithelial lymphocyte lines from refractory celiac disease type II patients. J. Immunol. 186, 6304–6312.

Trynka, G., Hunt, K.A., Bockett, N.A., Romanos, J., Mistry, V., Szperl, A., Bakker, S.F., Bardella, M.T., Bhaw-Rosun, L., Castillejo, G., De La Concha, E.G., De Almeida, R.C., Dias, K.R.M., Van Diemen, C.C., Dubois, P.C.A., Duerr, R.H., Edkins, S., Franke, L., Fransen, K., Gutierrez, J., Heap, G.A.R., Hrdlickova, B., Hunt, S., Izurieta, L.P., Izzo, V., Joosten, L.A.B., Langford, C., Mazzilli, M.C., Mein, C.A., Midah, V., Mitrovic, M., Mora, B., Morelli, M., Nutland, S., Núñez, C., Onengut-Gumuscu, S., Pearce, K., Platteel, M., Polanco, I., Potter, S., Ribes-Koninckx, C., Ricaño-Ponce, I., Rich, S.S., Rybak, A., Santiago, J.L., Senapati, S., Sood, A., Szajewska, H., Troncone, R., Varadé, J., Wallace, C., Wolters, V.M., Zhernakova, A., Thelma, B.K., Cukrowska, B., Urcelay, E., Bilbao, J.R., Mearin, M.L., Barisani, D., Barrett, J.C., Plagnol, V., Deloukas, P., Wijmenga, C., Van Heel, D.A., 2011. Dense genotyping identifies and localizes multiple common and rare variant association signals in celiac disease. Nat. Genet. 43, 1193–1201.

Uhde, M., Yu, X., Bunin, A., Brauner, C., Lewis, S.K., Lebwohl, B., Krishnareddy, S., Alaedini, A., Reizis, B., Ghosh, S., Green, P.H., Bhagat, G., 2020. Phenotypic shift of small intestinal intra-epithelial type 1 innate lymphoid cells in celiac disease is associated with enhanced cytotoxic potential. Clin. Exp. Immunol. 200, 163–175.

Vader, W., Stepniak, D., Kooy, Y., Mearin, L., Thompson, A., Van Rood, J.J., Spaenij, L., Koning, F., 2003. The HLA-DQ2 gene dose effect in celiac disease is directly related to the magnitude and breadth of gluten-specific T cell responses. Proc. Natl. Acad. Sci. U. S. A. 100, 12390–12395.

van Gils, T., Nijeboer, P., van Waesberghe, J.H.T., Coupé, V.M., Janssen, K., Zegers, J.A., Nurmohamed, S.A., Kraal, G., Jiskoot, S.C., Bouma, G., Mulder, C.J., 2017. Splenic volume differentiates complicated and non-complicated celiac disease. United European Gastroenterol J 5, 374–379.

Van Heel, D.A., Franke, L., Hunt, K.A., Gwilliam, R., Zhernakova, A., Inouye, M., Wapenaar, M.C., Barnardo, M.C.N.M., Bethel, G., Holmes, G.K.T., Feighery, C., Jewell, D., Kelleher, D., Kumar, P., Travis, S., Walters, J.R.F., Sanders, D.S., Howdle, P., Swift, J., Playford, R.J., McLaren, W.M., Mearin, M.L., Mulder, C.J., McManus, R., McGinnis, R., Cardon, L.R., Deloukas, P., Wijmenga, C., 2007. A genome-wide association study for celiac disease identifies risk variants in the region harboring IL2 and IL21. Nat. Genet. 39, 827–829.

van Wanrooij, R.L.J., Müller, D.M.J., Neefjes-Borst, E.A., Meijer, J., Koudstaal, L.G., Heideman, D.A.M., Bontkes, H.J., von Blomberg, B.M.E., Bouma, G., Mulder, C.J.J., 2014. Optimal strategies to identify aberrant intra-epithelial lymphocytes in refractory coeliac disease. J. Clin. Immunol. 34, 828–835.

van Wanrooij, R.L.J., Schreurs, M.W.J., Bouma, G., Von Blomberg, B.M.E., Tack, G.J., Verbeek, W.H.M., Mulder, C.J.J., 2010. Accurate classification of RCD requires flow cytometry. Gut 59, 1732.

van Wijk, F., Cheroutre, H., 2009. Intestinal T cells: facing the mucosal immune dilemma with synergy and diversity. Semin. Immunol. 21, 130–138.

Verbeek, W.H.M., Goerres, M.S., von Blomberg, B.M.E., Oudejans, J.J., Scholten, P.E.T., Hadithi, M., Al-Toma, A., Schreurs, M.W.J., Mulder, C.J.J., 2008a. Flow cytometric determination of aberrant intra-epithelial lymphocytes predicts T-cell lymphoma development more accurately than T-cell clonality analysis in refractory celiac disease. Clin. Immunol. 126, 48–56.

Verbeek, W.H.M., Van De Water, J.M.W., Al-Toma, A., Oudejans, J.J., Mulder, C.J.J., Coupé, V.M.H., 2008b. Incidence of enteropathy-associated T-cell lymphoma: a nation-wide study of a population-based registry in the Netherlands. Scand. J. Gastroenterol. 43, 1322–1328.

Verbeek, W.H.M., von Blomberg, B.M.E., Scholten, P.E., Mulder, C.J., Schreurs, M.W., 2008c. The presence of small intestinal intraepithelial gamma/delta T-lymphocytes is

inversely correlated with lymphoma development in refractory celiac disease. Am. J. Gastroenterol. 103, 3152.
Verbeek, W.H.M., von Blomberg, B.M.E., Oudejans, J., Mulder, C.J., Schreurs, M.W., 2009. Aberrant T-lymphocytes in refractory coeliac disease are not strictly confined to a small intestinal intraepithelial localization. Cytometry B Clin. Cytom. 76, 367–374.
Verkarre, V., Asnafi, V., Lecomte, T., Patey Mariaud-de Serre, N., Leborgne, M., Grosdidier, E., Le Bihan, C., Macintyre, E., Cellier, C., Cerf-Bensussan, N., Brousse, N., 2003. Refractory coeliac sprue is a diffuse gastrointestinal disease. Gut 52, 205–211.
Vivas, S., Ruiz De Morales, J.M., Ramos, F., Suárez-Vilela, D., 2006. Alemtuzumab for refractory celiac disease in a patient at risk for enteropathy-associated T-cell lymphoma. N. Engl. J. Med. 354, 2514–2515.
Wahab, Crusius, Meijer, Uil, Mulder, 2000. Cyclosporin in the treatment of adults with refractory coeliac disease—an open pilot study. Aliment. Pharmacol. Ther. 14, 767–774.
Wright, D.H.H., Jones, D.B.B., Clark, H., Mead, G.M.M., Hodges, E., Howell, W.M.M., Jones, D.B.B., Mead, G.M.M., Clark, H., Hodges, E., 1991. Is adult-onset coeliac disease due to a low-grade lymphoma of intraepithelial T lymphocytes? Lancet 337, 1373–1374.
Yokoyama, S., Perera, P.Y., Waldmann, T.A., Hiroi, T., Perera, L.P., 2013. Tofacitinib, a Janus kinase inhibitor demonstrates efficacy in an IL-15 transgenic mouse model that recapitulates pathologic manifestations of celiac disease. J. Clin. Immunol. 33, 586–594.

CHAPTER EIGHT

Involvement of lncRNAs in celiac disease pathogenesis

Ane Olazagoitia-Garmendia[a,b], Maialen Sebastian-delaCruz[a,b], and Ainara Castellanos-Rubio[a,b,c,d,*]

[a]Department of Genetics, Physical Anthropology and Animal Physiology, University of the Basque Country, UPV/EHU, Leioa, Spain
[b]Biocruces Bizkaia Health Research Institute, Barakaldo, Spain
[c]CIBER de Diabetes y Enfermedades Metabólicas Asociadas (CIBERDEM), Instituto de Salud Carlos III, Madrid, Spain
[d]Ikerbasque, Basque Foundation for Science, Bilbao, Spain
*Corresponding author: e-mail address: ainara.castellanos@ehu.eus

Contents

Abstract

Celiac disease (CD) is an immune-mediated disease that develops in genetically susceptible individuals upon gluten exposure. Human Leukocyte Antigen (HLA) genes in the Major Histocompatibility Complex (MHC) have been described to represent the 40% of the genetic risk to develop CD. Aiming to gain understanding of the genetic involvement in CD, high throughput studies have been performed, revealing that many CD-associated variants are located in non-coding regions, hindering the study of the functional implications of these single nucleotide polymorphisms (SNPs). In the last decade, long non-coding RNAs (lncRNAs) have been described to be influenced by disease-associated SNPs and to drive many important mechanisms involved in the

International Review of Cell and Molecular Biology, Volume 358
ISSN 1937-6448
https://doi.org/10.1016/bs.ircmb.2020.10.004

development of inflammatory diseases. Here we describe the lncRNAs identified and characterized in the context of celiac disease and highlight the importance of the study of these molecules in inflammatory and autoimmune disorders.

1. Introduction

Celiac disease (CD) is a complex, chronic immune-mediated enteropathy that develops in genetically susceptible individuals in response to ingested gluten and affects 1–2% of the population. Active CD patients present villous atrophy, crypt hyperplasia and lymphocytic infiltration of the intestinal epithelia. Nowadays, the only available treatment for CD is a strict, lifelong gluten-free diet (GFD) (Abadie et al., 2011; Husby et al., 2012).

While gluten proteins are known to be the environmental trigger in CD, the genetic risk factors are not totally described yet. It has been known for a long time that Human Leukocyte Antigen (HLA) genes in the Major Histocompatibility Complex (MHC) on chromosome 6p21 account for around 40% of the genetic risk to develop CD and the vast majority of CD patients carry the risk alleles HLA-DQ2 or HLA-DQ8 (de Haas et al., 2014; Gutierrez-Achury et al., 2015). However, other predisposing genetic variants, both within or outside the HLA region are also involved in the genetic risk to CD (Dieli-Crimi et al., 2015). Hence, identification of new susceptibility variants has been crucial for a better understanding of the molecular pathways that lead to the development of this disease (Plaza-Izurieta et al., 2015).

In this sense, genome-wide association studies (GWAS) have contributed to the identification of several associated single nucleotide polymorphisms (SNPs), while the development of high throughput sequencing technologies has pointed out the implication of rare variants in the development of the disease (Lee et al., 2018). GWAS together with the Immunochip genotyping project have described 57 genetic variants within 39 non-HLA loci associated with CD (Dubois et al., 2010; Plaza-Izurieta et al., 2015; Trynka et al., 2011; van Heel et al., 2007). These data together with pathway enrichment analysis have helped establish disease related molecular networks, the majority of them related with the immune response (Castellanos-Rubio and Bilbao, 2018; de Haas et al., 2014). However, the molecular mechanisms by which most of these variants contribute to the development of CD remain to be clarified.

More than 85% of the complex disease-associated SNPs are located in non-coding regions of the genome, which makes it difficult to describe their molecular mechanisms. It has been shown that SNPs located in non-coding genomic regions are able to regulate the expression of nearby genes as well as of genes located somewhere else in the genome (Ricaño-Ponce et al., 2016). Thus, elucidating the functional effects of non-coding disease-associated variants is a complicated but indispensable task that would help clarify the role of immune disease-associated SNPs in disease susceptibility.

In this line, RNA sequencing techniques, not limited to already described transcripts, have shown that almost 70% of the non-coding human genome has transcriptional activity and helped in the identification of novel transcripts that could be involved in the pathogenesis of disease (Castellanos-Rubio and Bilbao, 2018; Mowel et al., 2018). Long non-coding RNAs (lncRNAs) are the most abundant among these newly described elements, with up to thousands identified lncRNAs. LncRNAs are RNA molecules longer than 200 nucleotides that do not have coding potential and are able to regulate many biological processes. The study of these RNA molecules has recently gained much interest, but due to the challenges of their functional study, the molecular mechanisms of the majority of these non-coding transcripts are still unknown (Mowel et al., 2018; Robinson et al., 2020).

Recent studies have demonstrated that lncRNAs are involved in cell homeostasis, development, pathophysiologic changes or immune response through a wide range of molecular mechanisms (Castellanos-Rubio and Ghosh, 2019; de Haas et al., 2014). LncRNAs are expressed across different tissues and can bind to their target sites in the nucleus or cytoplasm. Moreover, these molecules can act rapidly and in a cell-specific manner in response to different environmental signals (Mowel et al., 2018; Robinson et al., 2020). LncRNAs are able to regulate gene expression at transcriptional or post-transcriptional levels; directly interacting with other RNAs, proteins or DNA molecules, or by acting as scaffolds or guides in the target genes (Chen et al., 2017; Mowel et al., 2018; Robinson et al., 2020). Some lncRNAs, as Xist, interact with chromatin and nuclear proteins to regulate gene expression (McHugh et al., 2015; Wutz, 2011) while others (i.e. lncRNA NORAD) are present in the cytoplasm and can affect mRNA stability or translation (Carlevaro-Fita et al., 2016; Lee et al., 2016) (Fig. 1). A high number of disease-associated SNPs are located within lncRNA

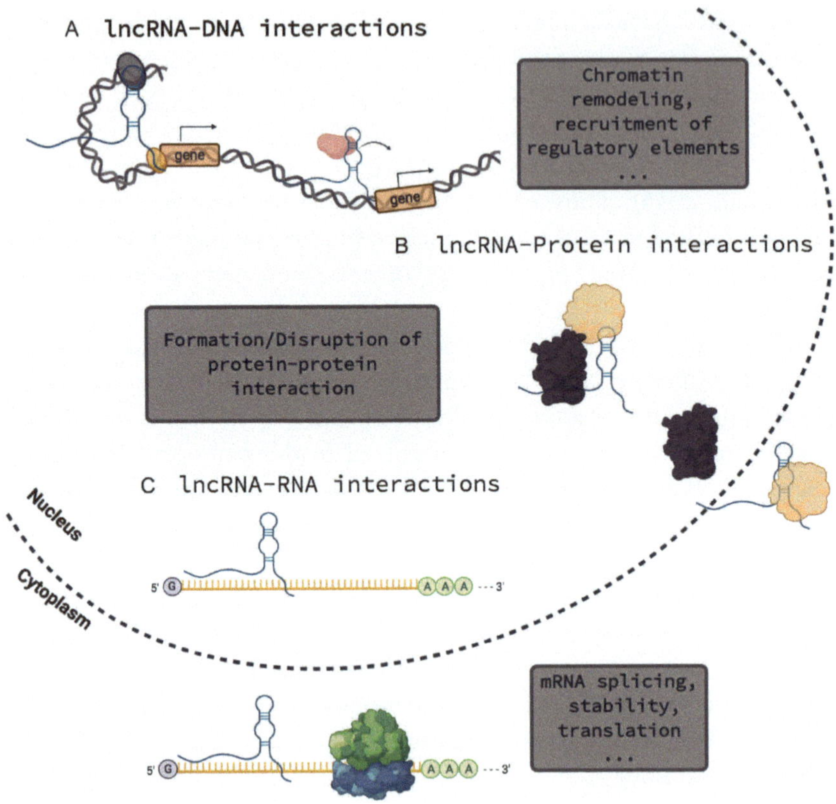

Fig. 1 LncRNAs have been shown to regulate protein coding genes transcriptionally and post-transcriptionally interacting with their targets in the nucleus or cytoplasm. (A) LncRNA-DNA interactions. LncRNAs can bind chromatin to change its structure or to recruit regulatory elements to specific loci to regulate gene expression. (B) LncRNA-protein interactions. LncRNAs can facilitate protein complex formation or disrupt protein-protein interactions in the nucleus or in the cytoplasm. (C) LncRNA-RNA interactions. LncRNAs can bind mRNAs in different ways to regulate diverse mRNA metabolism aspects as splicing, stability, degradation or translation.

sequences, altering lncRNA structure which affects their functions (Castellanos-Rubio and Ghosh, 2019; Chen et al., 2017). In addition, a recent study has demonstrated that disease-associated SNPs can also modulate RNA modifications, as m^6A methylation (Zhang et al., 2020), which have also been described in non-coding RNAs (Patil et al., 2018; Robinson et al., 2020). Thus, increasing evidence suggests that SNPs may be regulating lncRNA functionality due to changes in their secondary structures (Castellanos-Rubio and Bilbao, 2018; Castellanos-Rubio et al., 2016).

The wide diversity of lncRNAs together with the presence of associated SNPs within their sequences have arisen the interest on clarifying the involvement of these molecules in disease development, including intestinal inflammatory and immune diseases such as CD.

The role of the main coding genes involved in inflammation and immune response has been broadly studied while the functional implications of lncRNAs in immunity has just started emerging (Chen et al., 2017; Mowel et al., 2018; Robinson et al., 2020). The wide variety of mechanism of action presented by lncRNAs and the diverse origin of these molecules (intergenic lncRNAs, antisense lncRNAs, enhancer lncRNAs, ...) make their study challenging, but efforts from different groups are highlighting the implication of lncRNAs in the immune processes and disease development (Hadjicharalambous and Lindsay, 2019). In this sense, a growing number of lncRNAs are being involved in innate and adaptive immune pathways.

Increasing identification of disease-associated lncRNAs diverge from the small amount of functionally characterized lncRNAs, mainly due to the tedious work behind these studies. Deciphering the implication mechanisms of these newly identified lncRNAs is of great importance in order to gain understanding of the disease pathogenesis and to open the door to new therapeutic approaches. Regarding the involvement of lncRNAs in the development of CD, our group and others have identified and functionally characterized several CD-associated lncRNAs that are implicated in the disease pathogenesis (Castellanos-Rubio et al., 2016; Castellanos-Rubio et al., 2017; Hrdlickova et al., 2014; Jauregi-Miguel et al., 2019; Plaza-Izurieta et al., 2015; Ricaño-Ponce et al., 2016; Santin et al., 2018). Here we review the lncRNAs identified to be associated with CD and the main findings on the molecular mechanisms by which some of these lncRNAs are involved in disease development (Table 1).

2. *AC104820.2* and CD altered genes

The two GWAS and the Immunochip study performed in CD helped identify new disease-associated variants. However, most of these associated variants map to intergenic regions pointing to more than one potential candidate gene or to unknown regulatory elements that could be affected by the SNP genotypes. Overall 66 candidate genes have been proposed based on their localization under association peaks and further studies such as

Table 1 CD-associated lncRNAs.

Name	References	Chr location	Associated SNP	eQTL	Cell type	Differentially expressed	Function
AC104820.2	Plaza-Izurieta et al. (2015)	GRCh37: Chr2q31	rs1018326	Trans eQTL	Human distal duodenal biopsy	Upregulated in active CD patients	Putative regulator of CD-associated candidate genes.
Lnc13	Castellanos-Rubio et al. (2016)	GRCh37: Chr2q12	rs917997	Trans eQTL	iBMMs, human U937 cell line, Human distal duodenal biopsy	Downregulated in CD patients	Forms chromatin complex with hnRNPD and HDAC1 to regulate inflammatory genes.
LncRNA *Carlr*	Castellanos-Rubio et al. (2017)	GRCm38: Chr2qA3 GRCh37: Chr10p12.2	–	–	iBMMs, human THP1 cell line, intestinal murine C26 cell line, intestinal human Caco2 cell line, Human distal duodenal biopsy	More cytoplasmic in CD patients	Cytoplasmic upon LPS and NFkB activator. Putative role in signaling between macrophages and intestinal cells.
LncRNA *HCG14*	Santin et al. (2018)	GRCh38: Chr6p22	rs3135316	Trans eQTL	Human T84 cell line, Human epithelial cells from duodenal biopsy	Downregulated in CD patients' epithelial cells	Putative NOD1 transcription regulator.
RP4-587D13.2	Jauregi-Miguel et al. (2019)	GRCh38: Chr7q21	rs6962966	–	Human Caco2 cell line, Human intestinal biopsies	Downregulated in CD patients	Putative MAGI2 transcription regulator.

RP3–395M20.9	Hrdlickova et al. (2014)	GRCh37: Chr1p36	–	–	High expression in B and T cells	Proposed candidate in TNF pathway.
IL21-AS1	Hrdlickova et al. (2014)	GRCh38: Chr4q27	–	–	Th1 lymphocytes	Proposed regulator of *IL21* and immune response mediated by T lymphocytes.
AP002954.4	Ricaño-Ponce et al. (2016)	GRCh37: Chr1q23	rs10892258	–	Monocytes	Proposed in immune response to fungal infection and regulation of cytokine response.

Functionally characterized lncRNAs in bold; lncRNAs predicted to be involved in CD in gray.

eQTL analysis have been performed, where the effect of associated SNPs in the expression pattern of candidate genes has been interrogated. It has been proved that eQTL analyses need to be done in the disease-specific tissues and follow up functional studies have to be performed in order to verify the real implication of the genes and the associated SNPs in the disease development (Gutierrez-Arcelus et al., 2015; Ongen et al., 2017). In a work published by Plaza-Izurieta et al. they performed an expression analysis of previously proposed CD candidate genes in distal duodenal biopsies (the disease target tissue) and studied the implication of the genotype of CD-associated variants in the expression of those genes (Plaza-Izurieta et al., 2015).

The expression of the candidate genes was quantified in CD active patients, CD patients in gluten free diet (GFD) and non-celiac controls. GO term analysis showed that, as expected, differentially expressed genes were enriched in immune-related pathways. eQTL analysis revealed that many SNPs affected gene expression in a stimuli dependent manner, as the relationships between SNP genotype and gene expression were different among the 3 analyzed groups. Moreover, most of the eQTLs found were acting in trans and were located in non-coding regions. These non-coding regions harboring the eQTL SNPs included different regulatory elements (open chromatin regions, microRNAs, lincRNAs, …) which could be affecting gene expression of the associated target genes. Further co-expression analysis of the proposed CD candidate genes in each group (active CD, GFD CD and non-celiac controls) led to discover a subset of 18 genes that seemed to be influenced by the genotype of 3 SNPs in CD patients. One of these SNPs (rs1018326) was located within the sequence of lncRNA AC104820.2 and affected correlation of genes such as *ICOSLG, ZMIZ1, ATXN2* or *HCFC1* (all involved in diverse immune response processes). They found that this lncRNA is upregulated in active CD patients compared with treated patients, so rs1018326 genotype could be regulating these associated co-expressed genes by affecting the functionality of lncRNA *AC104820.2* (Fig. 2A).

This work showed that disease-associated SNP mediated gene regulation goes far beyond the simplistic idea of a SNP affecting the expression of a coding gene. In this study, it was seen that several differentially expressed genes may be regulated by same non-coding SNPs located in regulatory elements, such as lncRNAs. This approach identified *AC104820.2* lncRNA that harbors the CD-associated SNP rs1018326 and is differentially expressed in active patients.

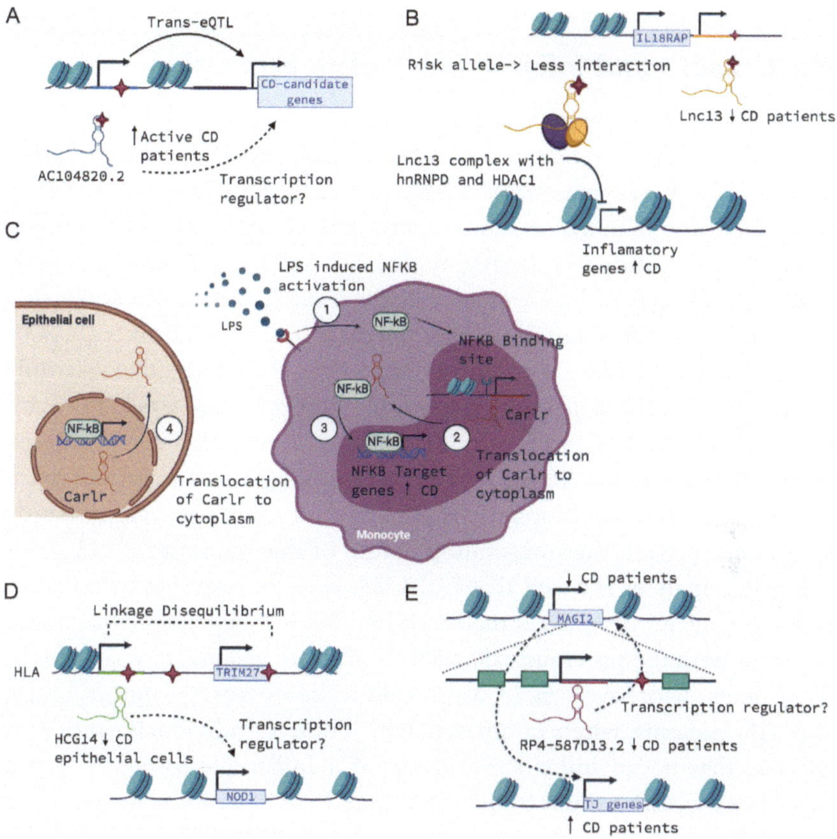

Fig. 2 Proposed mechanisms for CD-associated lncRNAs. Violet stars represent CD-associated SNPs. (A) lncRNA *AC104820.2* is upregulated in CD patients and harbors a CD-associated SNP that seem to regulate CD candidate genes in trans. (B) *Lnc13* is downregulated in CD patients and harbors a CD-associated SNP that modifies the secondary structure and alters *lnc13* function. *Lnc13* interacts with hnRNPD and HDAC1 in the chromatin to regulate the expression of disease related inflammatory genes. (C) *Carlr* lncRNA is induced and translocated to the cytoplasm in response to LPS. Once in the cytoplasm it interacts with the active form of NFKB facilitating its translocation to the nucleus and the subsequent expression of CD related genes. LncRNA *Carlr* seems to participate in the crosstalk between human macrophages and enterocytes. (D) *HCG14* is located in the HLA region and is downregulated in CD intestinal cells, harbors an intronic SNP associated to CD. *HGC14* is able to regulate the levels of *NOD1*. (E) *RP4-587D13.2* lncRNA is located in an intronic region of *MAGI2* coding gene close to a CD-associated SNP and influences *MAGI2* expression. *RP4-587D13.2* lncRNA and *MAGI2* are downregulated in CD patients, while Tight Junction (TJ) genes regulated by *MAGI2* are upregulated.

3. *Lnc13* and inflammatory gene expression

Lnc13 is the only lncRNA so far whose functional characteristics are described to be affected by the genotype of a CD-associated intergenic SNP. As previously mentioned, many disease-associated SNPs are located in non-coding regions which makes it difficult to assess their effects. The intergenic SNP rs917997, located in the region 2q12, 1.5 kb away from the 3′ end of *IL18RAP* coding gene had been repeatedly pointed, but never confirmed, as the functional candidate of the region (Dema et al., 2009; Hunt et al., 2008). In a work published by Castellanos-Rubio et al. they were able to confirm the existence of an independent transcript in this region (Castellanos-Rubio et al., 2016). The low expression profile among different tissues, the predominantly nuclear localization of the transcript (both in human and murine cell lines and human biopsies), together with additional computational experiments proved the non-coding nature of this transcript.

Further analyses showed that *lnc13* decreases in response to inflammation, and that this decrease is mediated by a NFKB dependent mechanism, consistent with the previously described constitutive activation of NFKB in CD patients. Moreover, *lnc13* was found to be downregulated in biopsies from CD patients when compared with control individuals, and it was observed that it can influence a group of 4 inflammatory genes that are upregulated in CD patient biopsies. As lncRNAs can modulate gene regulation at different stages, they wanted to elucidate the mechanisms by which *lnc13* regulates these group of inflammatory genes. *Lnc13* showed chromatin localization, and it was found that it acts as a linker between hnRNPD and HDAC1 proteins and chromatin in the transcription start site of the inflammatory genes, repressing their expression. Interestingly, the CD-associated SNP genotype was able to alter the secondary structure of *lnc13* and they demonstrated that disease-associated genotype has lower ability to bind hnRNPD, HDAC1 and chromatin, affecting the repression of the target inflammatory genes and predisposing to disease. These results showed that the previously unknown lncRNA *lnc13* regulates CD related inflammatory genes by forming a repressing complex in the chromatin in a genotype specific manner (Fig. 2B).

It is noteworthy that *lnc13* was differentially expressed in the tissues analyzed, pointing to a possible cell type specific function. Moreover, rs917997, like many other inflammation-associated SNPs, is also associated to other autoimmune diseases. Interestingly, while T allele is the risk allele for

CD, it is the protective allele for type 1 diabetes (T1D). A recent work published by Gonzalez-Moro et al. described *lnc13* function in the context of T1D and pancreatic beta cell inflammation, finding a totally independent mechanism for *lnc13* in this autoimmune disease and highlighting the importance of the cell context for lncRNA function (Gonzalez-Moro et al., 2020).

4. *Carlr* and NFKB signaling

The ability of lncRNAs to rapidly respond to external stimuli highlights the potential role of these molecules in the development of diverse diseases. Among other signaling networks, lncRNAs have been involved in diverse inflammatory pathways, such as NFKB activation (Liu et al., 2020; Lu et al., 2017; Shang et al., 2019). In CD, when NFKB pathway is activated proinflammatory cytokines are secreted producing the development of intestinal inflammation, where NFKB transcription factor is a key factor in the integration of inflammatory signals.

In a study by Castellanos-Rubio et al., alterations of lncRNAs in response to lipopolysaccharide (LPS), a NFKB activator, were assessed in mouse macrophages (Castellanos-Rubio et al., 2017). Macrophages are important cells for the maintenance of the mucosal homeostasis and the protective immunity in the intestine. LncRNA *Carlr*, previously related with cardiac apoptosis, was identified to be induced upon LPS stimulation and further analyses to characterize its role in CD development were performed.

Carlr is an intergenic lncRNA located in mouse chromosome 2 that showed to be upregulated upon LPS stimulation. NFKB binding sites in the *Carlr* promoter and the decreased response to LPS in the presence of NFKB inhibitor confirmed that *Carlr* increase in response to LPS is NFKB dependent. The implication of *Carlr* in the NFKB pathway was further demonstrated as *Carlr* knockdown in mouse macrophages led to lower NFKB activation and downstream gene repression. Interestingly, *Carlr* showed a primarily nuclear localization in basal conditions but was found to be translocated to the cytoplasm upon LPS stimulation, suggesting that it may work as a posttranscriptional regulator.

As NFKB is known to be altered in CD patients, the presence of *Carlr* human homolog was evaluated. Human *Carlr* expression was confirmed in several human cell lines as well as among different tissues. Silencing of the lncRNA in human monocytic THP-1 cells also led to a downregulation of the NFKB target genes. It is well known that interaction between different cell types is essential in a range of biological processes. Indeed, the appropriate interaction between macrophages and epithelial cells in the intestine is key in the inflammatory response. It is known that macrophage activation, which release proinflammatory cytokines, disrupts the enterocyte monolayer and contribute to inflammatory disease development (Na et al., 2019). Therefore, the role of lncRNA *Carlr* in the crosstalk between human macrophages and enterocytes was evaluated. Using cocultures of LPS activated human monocytic THP-1 cells and Caco2 intestinal cells, it was observed that in the presence of activated monocytes *Carlr* was augmented and translocated to the cytoplasm in the intestinal cells, pointing to the cytoplasmic localization of *Carlr* as a sign of inflammation in the intestinal epithelia. Additionally, *Carlr* silenced THP-1 cells were able to decrease proliferation and NFKB activation in the cocultured Caco2 cells, suggesting that *Carlr* has an important role in the signaling between macrophages and intestinal cells. Accordingly, CD patients showed cytoplasmic localization of *Carlr* together with an increased expression of the *Carlr* regulated genes. Moreover, cytoplasmic *Carlr* was shown to interact with active NFKB preferentially in CD patients (Fig. 2C).

Taken together, these findings indicate that an elevated cytoplasmic *Carlr* activity is associated with the disease characteristic activation of the NFKB signaling in the intestinal epithelia of CD patients and implicate *Carlr* as a potential novel player of the NFKB induced inflammation in CD.

5. *HCG14* and innate response

As previously mentioned, even if MHC region can explain a big part of CD-associated genetics, it is not enough to explain the totality of the genetic predisposition to the disease. It has been described that conserved extended haplotypes (CEH), large genomic regions that are maintained unaltered and have been inherited during evolution, are characteristic of the MHC region and that the CD-associated HLA-B8-DR3-DQ2 CEH is highly conserved. Some studies have suggested that the B8-DR3-DQ2 CEH harbors additional CD predisposing variants independent of HLA classical risk variants (DQ2 and DQ8) that could increase CD predisposition (Bilbao et al., 2006). In order to identify additional HLA-independent CD risk variants in the MHC Santin et al. performed a high-resolution SNP genotyping of the MHC, selecting CD patients and control individuals homozygous for the HLA-DR3-DQ2 genotype and harboring a single copy of the B8-DR3-DQ2 CEH (Santin et al., 2018).

One of the SNPs identified by this approach is located in an intergenic region between *TRIM27* and the lncRNA *HCG14*. This SNP is in absolute Linkage Disequilibrium (LD) with two other SNPs located in an intron of the *TRIM27* gene (rs3135293) and in an intronic region of lncRNA *HCG14* (rs3135316). Expression of *TRIM27* and lncRNA *HCG14* was quantified in duodenal biopsies from CD patients and control groups and no significant differences were observed between the two groups in neither of the genes. However, considering that lncRNAs have generally cell type specific functions, expression was quantified in immune and epithelial fractions purified from duodenal biopsies. *HCG14* was found to be slightly downregulated in the CD patient group in epithelial cells indicating that this lncRNA is only affected in the intestinal epithelial cell compartment of CD patients. These results further confirm that lncRNAs have cell type specific functions and highlight the importance of performing the studies in the appropriate tissues and cell types.

Analysis of lncRNA localization in T84 intestinal cells showed that this lncRNA is located in the nucleus, pointing to a possible transcriptional regulatory role. Using eQTL analysis in duodenal epithelial cells, it was found that *NOD1* expression is altered based on the genotype of a group of 8 associated SNPs located in *HCG14* region, suggesting that CD-associated polymorphisms in *HCG14* region may alter its function, affecting its capacity to regulate *NOD1* expression. Additionally, *HCG14* silenced cells showed decreased expression levels of *NOD1* (Fig. 2D). NOD1 is an innate immune receptor able to recognize bacterial ligands and induce activation of different pro-inflammatory pathways. Interestingly, the implication of the innate immune system in the initiation of CD is gaining evidence and it has been shown that gluten peptides can act as ligands for several innate receptors, inducing innate immune activation (Caruso et al., 2014; Robertson et al., 2013).

All together, these data suggest that CD-associated variants in *HCG14* lncRNA, located in the MHC region, can alter *NOD1* expression in intestinal cells, influencing the innate response in CD patients. However, the exact molecular mechanism underlying this effect and its impact on CD pathogenesis still remains to be clarified.

6. *RP4-587D13.2* lncRNA and tight junction network

Epithelial barrier is known to be disrupted in CD patients, and as a consequence small intestinal permeability is increased. Tight junction (TJ) proteins, such as claudins and occludins, are involved in the formation of the barrier, forming a belt-like structure in the lateral side of the epithelial cell, physically separating the lumen from the basal lamina (Shen, 2012). The alteration of the TJ network is postulated to be responsible of the misfunction of the epithelial barrier observed in CD pathogenesis (Schulzke et al., 1998).

MAGI proteins are membrane associated scaffolding proteins with protein-protein interaction domains that are involved in the regulation of the TJ network (Guillemot et al., 2008). MAGI-2 protein is one of the three MAGI proteins that regulates the epithelial barrier formation in the intestine (Nagashima et al., 2015).

Jauregi-Miguel et al. studied a CD-associated SNP (rs6962966) on chromosome 7q21.11, that is located in an intronic region of *MAGI2* gene and 4 kb downstream of a non-characterized long RNA transcript named *RP4-587D13.2* (Jauregi-Miguel et al., 2019). This transcript was confirmed to be a lncRNA with a primarily nuclear localization, suggesting a possible role in gene regulation. It was observed that *MAGI2* and *RP4-587D13.2* are positively correlated across tissues and silencing of *RP4-587D13.2* in intestinal epithelial cells, resulted in reduced MAGI-2 protein levels. In a downstream analysis, it was confirmed that *MAGI2* is implicated in the regulation of other TJ related genes, as *MAGI2* silencing in intestinal epithelial cells altered the expression of TJ related genes *CLDN1, PARD6A* and *ZAK*, suggesting an implication of the lncRNA in the regulation of TJ genes mediated by MAGI-2.

In vitro results were confirmed in human intestinal biopsies, as *RP4-587D13.2* and *MAGI2* expression was constitutively downregulated in CD patients compared to controls. Correlation between *RP4-587D13.2* and *MAGI2* was also observed among CD patients. An increased expression of MAGI-2 regulated TJ genes in active CD patients was also confirmed. Although the risk allele of the associated SNP did not seem to be responsible for the lower expression of *MAGI2* and *RP4-587D13.2*, it did affect the correlation of both genes, showing stronger correlation in the presence of the risk allele which could in turn contribute to the regulation of downstream TJ genes (Fig. 2E).

Although direct evidence is still needed, this work suggests that *RP4-587D13.2* is involved in the increased intestinal permeability observed in CD patients by reducing MAGI-2 amounts in intestinal cells, which will ultimately result in an alteration of genes involved in the formation of TJ provoking an impairment of the epithelial barrier.

7. LncRNAs predicted to be involved in CD

RNA sequencing and co-expression analyses have helped identify candidate lncRNAs that could be involved in CD pathogenesis. It has been observed that lncRNA transcripts are enriched in autoimmune disease (AID)-associated loci (Hrdlickova et al., 2014), suggesting they may play a role in such disorders. Some lncRNAs have been associated to CD using these approaches, but there is no experimental evidence of their implication in the development of this disease yet.

7.1 *RP3-395 M20.9* and TNF signaling pathway

In order to characterize the coding and non-coding transcripts on different AID-associated loci, RNA sequencing-based co-expression analyses using different immune cell type and subtypes were performed by Hrdlickova et al. (2014). 240 autoimmunity related lncRNAs were found in autoimmunity associated regions. One of these lncRNAs, *RP3–395M20.9*, is located on chromosome 1p36.32 in a locus shared by CD, rheumatoid arthritis, ulcerative colitis and primary sclerosing cholangitis. This lncRNA showed high expression in B and T cells, similar to other protein coding genes in the same locus. Indeed, co-expression analysis revealed that expression of *RP3–395M20.9* lncRNA correlates with that of genes known to be involved in T lymphocyte-related and B cell biology processes, as proliferation and negative regulation of activation of T lymphocytes, regulation of the TCR signaling pathway or TNF-mediated signaling pathway.

RP3–395M20.9 lncRNA was proposed to take part in TNF signaling pathway via *cis* regulation of *TNFRS14/HVEM* and trans regulation of *TNFβ/LTA*, *UBASH3A* and *LINC00158* lncRNA. TNF signaling pathway is known to be altered in CD, with an increased TNF proinflammatory cytokine production in active patients (Marafini et al., 2015). Moreover, *TNFRS14/HVEM* is known to activate NFKB pathway (Shui et al., 2011), which is known to be constitutively active in CD patients. Nonetheless, the implication of this lncRNA in the regulation of TNF signaling has not been confirmed and expression pattern of this lncRNA in context of disease-associated SNPs has not been analyzed. Functional studies are needed in order to decipher the implication of *RP3–395M20.9* lncRNA in TNF signaling pathway and to confirm its implication in the pathogenesis of CD.

7.2 *IL21-AS1* and IL21 expression

Beyond the HLA region, 4q27 locus showed the strongest association with CD in the GWAS and Immunochip studies (Dubois et al., 2010; Trynka et al., 2011; van Heel et al., 2007). This locus has been also associated to other autoimmune disorders as juvenile idiopathic arthritis or inflammatory bowel disease. Although the candidate genes in the region have been the genes encoding IL-2 and IL-21 cytokines, there is also a lncRNA, *IL21-AS1*, located within this locus and 30 kb away from the CD-associated SNP.

Co-expression analyses have shown a strong correlation between *IL21-AS1* and *IL21* in Th1 lymphocytes (Hrdlickova et al., 2014). *IL21* is known to induce expression of cytokines and transcription factors such as IFNγ and T-bet, which are involved in innate immune response activation and Th1 cell differentiation, respectively (Fina et al., 2008). Therefore, the strong correlation observed suggests this lncRNA may play a role in the regulation of *IL21* and in the immune response mediated by T lymphocytes. Besides, *IL21* is upregulated in duodenal biopsies of untreated CD patients compared to non-celiac controls (Strengell et al., 2002) which could point to an involvement of the lncRNA in the disease. Further studies on the regulation of *IL21* by its antisense lncRNA are needed in order to confirm the implication of *IL21-AS1* in the pathogenesis of CD.

7.3 AP002954.4

LncRNA *AP002954.4* was found via cis-eQTL mapping of 640 AID-associated SNPs in a study performed by Ricaño-Ponce et al. (2016). This lncRNA is located on chromosome 11q23.3 and it harbors a SNP (rs10892258) associated to celiac disease, rheumatoid arthritis and multiple sclerosis. The CD risk allele has been related with reduced expression of the lncRNA, showing a stronger reduction in the homozygous form.

Among different immune cell types, *AP002954.4* is predominantly expressed in monocytes and it might play a role in the immune response to fungal infection and the regulation of cytokine response (Ricaño-Ponce et al., 2016). Co-expression analyses were carried out in order to predict lncRNA function and downstream pathway regulation. *AP002954.4* was predicted to act on different immune-related processes such as negative regulation of IL-1 production, cellular defense response, T cell tolerance induction and myeloid cell and neutrophil activation, suggesting it can play a role in the development of CD. IL-1 family members IL-1α and IL-1β are proinflammatory cytokines that take part in immune response by modulating T cells activation and differentiation (Santarlasci et al., 2013), and their increased production in CD has also been described (Manavalan et al., 2010). Furthermore, induction of tolerance by T cells is highly important in CD, as the lack of tolerance to gluten is an important triggering factor of the disease (Hardy and Tye-Din, 2016). Thus, *AP002954.4* implication may be relevant in celiac disease, even though functional studies are needed in order to confirm this hypothesis.

In brief, the function of most of the lncRNAs located in AID-associated loci that are affected by celiac disease-associated SNPs is still unknown, and their mechanism of action is far to be understood. Co-expression analyses are a good approach to get a sense of which pathways that will ultimately influence disease pathogenesis can be affected by lncRNA changes. Although the implication of *RP3–395M20.9*, *IL21-AS1* and *AP002954.4* lncRNAs in CD has not been functionally confirmed, they are undoubtedly interesting candidates for future research.

8. Concluding remarks

Celiac disease is a complex genetic autoimmune disorder that affects the small intestine and develops in response to dietary gluten. In the last two decades, a lot of effort has been focused on the understanding of the genetic contribution to its development which could open the door to new non-dietary therapies (Castellanos-Rubio and Bilbao, 2018; de Haas et al., 2014). GWAS and Immunochip studies have helped in the identification of new disease-associated regions and more interestingly, have shown that more than 80% of autoimmune disease-associated SNPs locate into non-coding regions (Dubois et al., 2010; Lee et al., 2018; Plaza-Izurieta et al., 2015; Trynka et al., 2011; van Heel et al., 2007). In the near past, protein coding genes were considered as the only functional regions of the genome, but nowadays it is widely known that the non-coding regions can also harbor regulatory elements that play crucial roles in gene regulation. Among the newly identified non-coding transcripts, an increasing number of long non-coding RNAs are being annotated thanks to high throughput techniques such as RNAseq that do not rely only in already known transcripts (Castellanos-Rubio and Bilbao, 2018; Mowel et al., 2018; Robinson et al., 2020). Some of the best characterized lncRNAs are nuclear *XIST* and *HOTAIR* which form complexes with the chromatin remodeling proteins guiding them to their target genomic loci and regulating the expression of both proximal and distant genes (McHugh et al., 2015; Rinn et al., 2007;

Wutz, 2011). Other lncRNA with an interesting mechanism of action is lncRNA *Malat1*, which forms complexes with pre-mRNAs and binds splicing and mRNA processing proteins, controlling alternative splicing (Tripathi et al., 2010). In turn, lncRNA *NORAD* interacts with mRNA translation proteins in the cytoplasm, interfering in the translation process of some target mRNAs (Lee et al., 2016). In addition, some lncRNAs have been recently described to encode for small peptides, increasing the broad spectrum of lncRNA mechanisms and functions (Anderson et al., 2015; Matsumoto et al., 2017; Nelson et al., 2016). Even if there are hundreds of lncRNAs described, only a substantial amount of them have been functionally characterized and their implication in disease development is generally unknown. Mostly expression and co-expression analyses have been performed in order to elucidate their putative functions and their contribution to disease. Using these approaches, some disease-associated lncRNAs have been identified, but further molecular biology based functional experiments are needed to describe their mechanisms (de Haas et al., 2014; Ricaño-Ponce et al., 2016).

LncRNAs have shown to be cell-type and stimuli-dependent transcripts that present a wide variety of functions (Castellanos-Rubio and Ghosh, 2019; Hadjicharalambous and Lindsay, 2019). Some lncRNAs are known to be expressed in immune cells and showed important roles controlling gene expression as *lincRNA-Cox2, PACER* and *LincRNA-EPS* among others (Atianand et al., 2016; Carpenter et al., 2013; Krawczyk and Emerson, 2014). In addition, some lncRNAs have been associated with a protective role in inflammatory diseases, while others seem to increase susceptibility to inflammation or autoimmunity (Castellanos-Rubio et al., 2016; Chen et al., 2017; Mowel et al., 2018). In turn, lncRNA *Morrbid* is significantly up-regulated in eosinophils from hypereosinophilic syndrome patients, suggesting Morrbid-BCL2L11 association with this disease. LncRNA *linc-MAF-4* has been linked to susceptibility to multiple sclerosis, and the lncRNA *rmrp* has been described to play a role in the development of cartilage-hair hypoplasia (Huang et al., 2015; Kotzin et al., 2016; Zhang et al., 2017). All these works highlight the importance of lncRNAs in the immune system and stress the necessity of a comprehensive analytical approach combining genomic, transcriptomic and epigenomic data with molecular biology based functional experiments to characterize these lncRNAs and describe their involvement in immune disease development (Lee et al., 2018; Robinson et al., 2020). Here we review the actual knowledge about several lncRNAs that have been suggested to be involved in CD related pathways. Although the real functional implication of most of these lncRNAs in disease development remains unclear, these molecules have shown to be very interesting therapeutic targets, and thus, further studies on celiac disease-associated lncRNAs will for sure improve our understanding of this disease and help in the development of novel treatments.

Acknowledgments

The lncRNA work in the author's laboratory has been funded by the Spanish Ministry of Science Innovation and Universities (PGC2018-097573-A-I00) to A.C.R. A.O-G. and M.S.dlC. are funded by predoctoral fellowships from the Basque Government and University of the Basque Country, respectively. Figures were created using Biorender.com.

References

Abadie, V., Sollid, L.M., Barreiro, L.B., Jabri, B., 2011. Integration of genetic and immunological insights into a model of celiac disease pathogenesis. Annu. Rev. Immunol. 29, 493–525. https://doi.org/10.1146/annurev-immunol-040210-092915.

Anderson, D.M., Anderson, K.M., Chang, C.L., Makarewich, C.A., Nelson, B.R., McAnally, J.R., Kasaragod, P., Shelton, J.M., Liou, J., Bassel-Duby, R., Olson, E.N., 2015. A micropeptide encoded by a putative long noncoding RNA regulates muscle performance. Cell 160, 595–606. https://doi.org/10.1016/j.cell.2015.01.009.

Atianand, M.K., Hu, W., Satpathy, A.T., Shen, Y., Ricci, E.P., Alvarez-Dominguez, J.R., Bhatta, A., Schattgen, S.A., McGowan, J.D., Blin, J., Braun, J.E., Gandhi, P., Moore, M.J., Chang, H.Y., Lodish, H.F., Caffrey, D.R., Fitzgerald, K.A., 2016. A long noncoding RNA lincRNA-EPS acts as a transcriptional brake to restrain inflammation. Cell 165, 1672–1685. https://doi.org/10.1016/j.cell.2016.05.075.

Bilbao, J.R., Calvo, B., Aransay, A.M., Martin-Pagola, A., Pérez de Nanclares, G., Aly, T.A., Rica, I., Vitoria, J.C., Gaztambide, S., Noble, J., Fain, P.R., Awdeh, Z.L., Alper, C.A., Castano, L., 2006. Conserved extended haplotypes discriminate HLA-DR3-homozygous Basque patients with type 1 diabetes mellitus and celiac disease. Genes Immun. 7, 550–554. https://doi.org/10.1038/sj.gene.6364328.

Carlevaro-Fita, J., Rahim, A., Guigó, R., Vardy, L.A., Johnson, R., 2016. Cytoplasmic long noncoding RNAs are frequently bound to and degraded at ribosomes in human cells. RNA 22, 867–882. https://doi.org/10.1261/rna.053561.115.

Carpenter, S., Aiello, D., Atianand, M.K., Ricci, E.P., Gandhi, P., Hall, L.L., Byron, M., Monks, B., Henry-Bezy, M., Lawrence, J.B., O'Neill, L.A.J., Moore, M.J., Caffrey, D.R., Fitzgerald, K.A., 2013. A long noncoding RNA mediates both activation and repression of immune response genes. Science 341, 789–792. https://doi.org/10.1126/science.1240925.

Caruso, R., Warner, N., Inohara, N., Núñez, G., 2014. NOD1 and NOD2: signaling, host defense, and inflammatory disease. Immunity. https://doi.org/10.1016/j.immuni.2014.12.010.

Castellanos-Rubio, A., Bilbao, J.R., 2018. Profiling celiac disease-related transcriptional changes. In: Galluzzi, L. (Ed.), International Review of Cell and Molecular Biology. Elsevier Inc, pp. 149–174. https://doi.org/10.1016/bs.ircmb.2017.07.003.

Castellanos-Rubio, A., Ghosh, S., 2019. Disease-associated SNPs in inflammation-related lncRNAs. Front. Immunol. 10, 1–9. https://doi.org/10.3389/fimmu.2019.00420.

Castellanos-Rubio, A., Fernandez-Jimenez, N., Kratchmarov, R., Luo, X., Bhagat, G., Green, P.H.R., Schneider, R., Kiledjian, M., Bilbao, J.R., Ghosh, S., 2016. A long noncoding RNA associated with susceptibility to celiac disease. Science 352, 91–95. https://doi.org/10.1126/science.aad0467.

Castellanos-Rubio, A., Kratchmarov, R., Sebastian, M., Garcia-Etxebarria, K., Garcia, L., Irastorza, I., Ghosh, S., 2017. Cytoplasmic form of Carlr lncRNA facilitates inflammatory gene expression upon NF-κB activation. J. Immunol. 199, 581–588. https://doi.org/10.4049/jimmunol.1700023.

Chen, Y.G., Satpathy, A.T., Chang, H.Y., 2017. Gene regulation in the immune system by long noncoding RNAs. Nat. Immunol. 18, 962–972. https://doi.org/10.1038/ni.3771.

de Haas, E.C., Kumar, V., Wijmenga, C., 2014. Immunogenetics of Celiac Disease. pp. 53–66. https://doi.org/10.1007/978-1-4614-8560-5_5.

Dema, B., Martínez, A., Fernández-Arquero, M., Maluenda, C., Polanco, I., de La Concha, E.G., Urcelay, E., Núñez, C., 2009. Association of IL18RAP and CCR3 with coeliac disease in the Spanish population. J. Med. Genet. 46, 617–619. https://doi.org/10.1136/jmg.2009.067041.

Dieli-Crimi, R., Cénit, M.C., Núñez, C., 2015. The genetics of celiac disease: a comprehensive review of clinical implications. J. Autoimmun. 64, 26–41. https://doi.org/10.1016/j.jaut.2015.07.003.

Dubois, P.C.A., Trynka, G., Franke, L., Hunt, K.A., Romanos, J., Curtotti, A., Zhernakova, A., Heap, G.A.R., Ádány, R., Aromaa, A., Bardella, M.T., van den Berg, L.H., Bockett, N.A., de La Concha, E.G., Dema, B., Fehrmann, R.S.N., Fernández-Arquero, M., Fiatal, S., Grandone, E., Green, P.M., Groen, H.J.M., Gwilliam, R., Houwen, R.H.J., Hunt, S.E., Kaukinen, K., Kelleher, D., Korponay-Szabo, I., Kurppa, K., MacMathuna, P., Mäki, M., Mazzilli, M.C., McCann, O.T., Mearin, M.L., Mein, C.A., Mirza, M.M., Mistry, V., Mora, B., Morley, K.I., Mulder, C.J., Murray, J.A., Ñez, C., Oosterom, E., Ophoff, R.A., Polanco, I., Peltonen, L., Platteel, M., Rybak, A., Salomaa, V., Schweizer, J.J., Sperandeo, M.P., Tack, G.J., Turner, G., Veldink, J.H., Verbeek, W.H.M., Weersma, R.K., Wolters, V.M., Urcelay, E., Cukrowska, B., Greco, L., Neuhausen, S.L., McManus, R., Barisani, D., Deloukas, P., Barrett, J.C., Saavalainen, P., Wijmenga, C., van Heel, D.A., 2010. Multiple common variants for celiac disease influencing immune gene expression. Nat. Genet. 42, 295–302. https://doi.org/10.1038/ng.543.

Fina, D., Sarra, M., Caruso, R., del Vecchio Blanco, G., Pallone, F., MacDonald, T.T., Monteleone, G., 2008. Interleukin 21 contributes to the mucosal T helper cell type 1 response in coeliac disease. Gut 57, 887–892. https://doi.org/10.1136/gut.2007.129882.

Gonzalez-Moro, I., Olazagoitia-Garmendia, A., Colli, M.L., Cobo-Vuilleumier, N., Postler, T.S., Marselli, L., Marchetti, P., Ghosh, S., Gauthier, B.R., Eizirik, D.L., Castellanos-Rubio, A., Santin, I., 2020. The T1D-associated lncRNA Lnc13 modulates human pancreatic β cell inflammation by allele-specific stabilization of STAT1 mRNA. Proc. Natl. Acad. Sci. U. S. A. 117 (16), 9022–9031. https://doi.org/10.1073/pnas.1914353117.

Guillemot, L., Paschoud, S., Pulimeno, P., Foglia, A., Citi, S., 2008. The cytoplasmic plaque of tight junctions: a scaffolding and signalling center. Biochim. Biophys. Acta Biomembr. 1778 (3), 601–613. https://doi.org/10.1016/j.bbamem.2007.09.032.

Gutierrez-Achury, J., Zhernakova, A., Pulit, S.L., Trynka, G., Hunt, K.A., Romanos, J., Raychaudhuri, S., van Heel, D.A., Wijmenga, C., de Bakker, P.I.W., 2015. Fine mapping in the MHC region accounts for 18% additional genetic risk for celiac disease. Nat. Genet. 47, 577–578. https://doi.org/10.1038/ng.3268.

Gutierrez-Arcelus, M., Ongen, H., Lappalainen, T., Montgomery, S.B., Buil, A., Yurovsky, A., Bryois, J., Padioleau, I., Romano, L., Planchon, A., Falconnet, E., Bielser, D., Gagnebin, M., Giger, T., Borel, C., Letourneau, A., Makrythanasis, P.,

Guipponi, M., Gehrig, C., Antonarakis, S.E., Dermitzakis, E.T., 2015. Tissue-specific effects of genetic and epigenetic variation on gene regulation and splicing. PLoS Genet. 11, 1004958. https://doi.org/10.1371/journal.pgen.1004958.

Hadjicharalambous, M.R., Lindsay, M.A., 2019. Long non-coding RNAs and the innate immune response. Non-coding RNA 5 (2), 34. https://doi.org/10.3390/NCRNA5020034.

Hardy, M.Y., Tye-Din, J.A., 2016. Coeliac disease: a unique model for investigating broken tolerance in autoimmunity. Clin. Transl. Immunol. 5, e112. https://doi.org/10.1038/cti.2016.58.

Hrdlickova, B., Kumar, V., Kanduri, K., Zhernakova, D.v., Tripathi, S., Karjalainen, J., Lund, R.J., Li, Y., Ullah, U., Modderman, R., Abdulahad, W., Lähdesmäki, H., Franke, L., Lahesmaa, R., Wijmenga, C., Withoff, S., 2014. Expression profiles of long non-coding RNAs located in autoimmune disease-associated regions reveal immune cell-type specificity. Genome Med. 6, 88. https://doi.org/10.1186/s13073-014-0088-0.

Huang, W., Thomas, B., Flynn, R.A., Gavzy, S.J., Wu, L., Kim, S.v., Hall, J.A., Miraldi, E.R., Ng, C.P., Rigo, F.W., Meadows, S., Montoya, N.R., Herrera, N.G., Domingos, A.I., Rastinejad, F., Myers, R.M., Fuller-Pace, F.v., Bonneau, R., Chang, H.Y., Acuto, O., Littman, D.R., 2015. DDX5 and its associated lncRNA Rmrp modulate TH17 cell effector functions. Nature 528, 517–522. https://doi.org/10.1038/nature16193.

Hunt, K.A., Zhernakova, A., Turner, G., Heap, G.A.R., Franke, L., Bruinenberg, M., Romanos, J., Dinesen, L.C., Ryan, A.W., Panesar, D., Gwilliam, R., Takeuchi, F., McLaren, W.M., Holmes, G.K.T., Howdle, P.D., Walters, J.R.F., Sanders, D.S., Playford, R.J., Trynka, G., Mulder, C.J.J., Mearin, M.L., Verbeek, W.H.M., Trimble, V., Stevens, F.M., O'Morain, C., Kennedy, N.P., Kelleher, D., Pennington, D.J., Strachan, D.P., McArdle, W.L., Mein, C.A., Wapenaar, M.C., Deloukas, P., McGinnis, R., McManus, R., Wijmenga, C., van Heel, D.A., 2008. Newly identified genetic risk variants for celiac disease related to the immune response. Nat. Genet. 40, 395–402. https://doi.org/10.1038/ng.102.

Husby, S., Koletzko, S., Korponay-Szabó, I.R., Mearin, M.L., Phillips, A., Shamir, R., Troncone, R., Giersiepen, K., Branski, D., Catassi, C., Lelgeman, M., Mäki, M., Ribes-Koninckx, C., Ventura, A., Zimmer, K.P., 2012. European society for pediatric gastroenterology, hepatology, and nutrition guidelines for the diagnosis of coeliac disease. J. Pediatr. Gastroenterol. Nutr. 54, 136–160. https://doi.org/10.1097/MPG.0b013e31821a23d0.

Jauregi-Miguel, A., Santin, I., Garcia-Etxebarria, K., Olazagoitia-Garmendia, A., Romero-Garmendia, I., Sebastian-delaCruz, M., Irastorza, I., Castellanos-Rubio, A., Bilbao, J.R., 2019. MAGI2 gene region and celiac disease. Front. Nutr. 6, 187. https://doi.org/10.3389/fnut.2019.00187.

Kotzin, J.J., Spencer, S.P., McCright, S.J., Kumar, D.B.U., Collet, M.A., Mowel, W.K., Elliott, E.N., Uyar, A., Makiya, M.A., Dunagin, M.C., Harman, C.C.D., Virtue, A.T., Zhu, S., Bailis, W., Stein, J., Hughes, C., Raj, A., Wherry, E.J., Goff, L.A., Klion, A.D., Rinn, J.L., Williams, A., Flavell, R.A., Henao-Mejia, J., 2016. The long non-coding RNA Morrbid regulates Bim and short-lived myeloid cell lifespan. Nature 537, 239–243. https://doi.org/10.1038/nature19346.

Krawczyk, M., Emerson, B.M., 2014. P50-associated COX-2 extragenic RNA (pacer) activates human COX-2 gene expression by occluding repressive NF-κB p50 complexes. eLife 3, e01776. https://doi.org/10.7554/eLife.01776.

Lee, S., Kopp, F., Chang, T.C., Sataluri, A., Chen, B., Sivakumar, S., Yu, H., Xie, Y., Mendell, J.T., 2016. Noncoding RNA NORAD regulates genomic stability by sequestering PUMILIO proteins. Cell 164, 69–80. https://doi.org/10.1016/j.cell.2015.12.017.

Lee, P.H., Lee, C., Li, X., Wee, B., Dwivedi, T., Daly, M., 2018. Principles and methods of in-silico prioritization of non-coding regulatory variants. Hum. Genet. 137, 15–30. https://doi.org/10.1007/s00439-017-1861-0.

Liu, X., Shao, Y., Zhou, J., Qian, G., Ma, Z., 2020. Nuclear factor κB signaling and its related non-coding RNAs in cancer therapy. Mol. Ther. Nucleic Acids 19, 208–217. https://doi.org/10.1016/j.omtn.2019.11.007.

Lu, Y., Liu, X., Xie, M., Liu, M., Ye, M., Li, M., Chen, X.-M., Li, X., Zhou, R., 2017. The NF-κB-responsive long noncoding RNA FIRRE regulates posttranscriptional regulation of inflammatory gene expression through interacting with hnRNPU. J Immunol. 199 (10), 3571–3582. https://doi.org/10.4049/jimmunol.1700091.

Manavalan, J.S., Hernandez, L., Shah, J.G., Konikkara, J., Naiyer, A.J., Lee, A.R., Ciaccio, E., Minaya, M.T., Green, P.H.R., Bhagat, G., 2010. Serum cytokine elevations in celiac disease: association with disease presentation. Hum. Immunol. 71, 50–57. https://doi.org/10.1016/j.humimm.2009.09.351.

Marafini, I., Monteleone, I., di Fusco, D., Cupi, M.L., Paoluzi, O.A., Colantoni, A., Ortenzi, A., Izzo, R., Vita, S., de Luca, E., Sica, G., Pallone, F., Monteleone, G., 2015. TNF-α producing innate lymphoid cells (ILCs) are increased in active celiac disease and contribute to promote intestinal atrophy in mice. PLoS One 10, e0126291. https://doi.org/10.1371/journal.pone.0126291.

Matsumoto, A., Pasut, A., Matsumoto, M., Yamashita, R., Fung, J., Monteleone, E., Saghatelian, A., Nakayama, K.I., Clohessy, J.G., Pandolfi, P.P., 2017. MTORC1 and muscle regeneration are regulated by the LINC00961-encoded SPAR polypeptide. Nature 541, 228–232. https://doi.org/10.1038/nature21034.

McHugh, C.A., Chen, C.K., Chow, A., Surka, C.F., Tran, C., McDonel, P., Pandya-Jones, A., Blanco, M., Burghard, C., Moradian, A., Sweredoski, M.J., Shishkin, A.A., Su, J., Lander, E.S., Hess, S., Plath, K., Guttman, M., 2015. The Xist lncRNA interacts directly with SHARP to silence transcription through HDAC3. Nature 521, 232–236. https://doi.org/10.1038/nature14443.

Mowel, W.K., Kotzin, J.J., McCright, S.J., Neal, V.D., Henao-Mejia, J., 2018. Control of immune cell homeostasis and function by lncRNAs. Trends Immunol. 39 (1), 55–69. https://doi.org/10.1016/j.it.2017.08.009.

Na, Y.R., Stakenborg, M., Seok, S.H., Matteoli, G., 2019. Macrophages in intestinal inflammation and resolution: a potential therapeutic target in IBD. Nat. Rev. Gastroenterol. Hepatol. 16, 531–543. https://doi.org/10.1038/s41575-019-0172-4.

Nagashima, S., Kodaka, M., Iwasa, H., Hata, Y., 2015. MAGI2/S-SCAM outside brain. J. Biochem. 157 (4), 177–184. https://doi.org/10.1093/jb/mvv009.

Nelson, B.R., Makarewich, C.A., Anderson, D.M., Winders, B.R., Troupes, C.D., Wu, F., Reese, A.L., McAnally, J.R., Chen, X., Kavalali, E.T., Cannon, S.C., Houser, S.R., Bassel-Duby, R., Olson, E.N., 2016. Muscle physiology: a peptide encoded by a transcript annotated as long noncoding RNA enhances SERCA activity in muscle. Science 351, 271–275. https://doi.org/10.1126/science.aad4076.

Ongen, H., Brown, A.A., Delaneau, O., Panousis, N.I., Nica, A.C., Dermitzakis, E.T., 2017. Estimating the causal tissues for complex traits and diseases. Nat. Genet. 49, 1676–1683. https://doi.org/10.1038/ng.3981.

Patil, D.P., Pickering, B.F., Jaffrey, S.R., 2018. Reading m6A in the transcriptome: m6A-binding proteins. Trends Cell Biol. 28 (2), 113–127. https://doi.org/10.1016/j.tcb.2017.10.001.

Plaza-Izurieta, L., Fernandez-Jimenez, N., Irastorza, I., Jauregi-Miguel, A., Romero-Garmendia, I., Vitoria, J.C., Bilbao, J.R., 2015. Expression analysis in intestinal mucosa reveals complex relations among genes under the association peaks in celiac disease. Eur. J. Hum. Genet. 23, 1100–1105. https://doi.org/10.1038/ejhg.2014.244.

Ricaño-Ponce, I., Zhernakova, D.v., Deelen, P., Luo, O., Li, X., Isaacs, A., Karjalainen, J., di Tommaso, J., Borek, Z.A., Zorro, M.M., Gutierrez-Achury, J., Uitterlinden, A.G., Hofman, A., van Meurs, J., Netea, M.G., Jonkers, I.H., Withoff, S., van Duijn, C.M., Li, Y., Ruan, Y., Franke, L., Wijmenga, C., Kumar, V., BIOS Consortium, 2016. Refined mapping of autoimmune disease associated genetic variants with gene expression suggests an important role for non-coding RNAs. J. Autoimmun. 68, 62–74. https://doi.org/10.1016/j.jaut.2016.01.002.

Rinn, J.L., Kertesz, M., Wang, J.K., Squazzo, S.L., Xu, X., Brugmann, S.A., Goodnough, L.H., Helms, J.A., Farnham, P.J., Segal, E., Chang, H.Y., 2007. Functional demarcation of active and silent chromatin domains in human HOX loci by noncoding RNAs. Cell 129, 1311–1323. https://doi.org/10.1016/j.cell.2007.05.022.

Robertson, S.J., Yu Zhou, J., Geddes, K., Rubino, S.J., Ho Cho, J., Girardin, S.E., Philpott, D.J., 2013. Nod1 and Nod2 signaling does not alter the composition of intestinal bacterial communities at homeostasis. Gut Microbes 4 (3), 222–231. https://doi.org/10.4161/gmic.24373.

Robinson, E.K., Covarrubias, S., Carpenter, S., 2020. The how and why of lncRNA function: an innate immune perspective. Biochim. Biophys. Acta, Gene Regul. Mech. 1863 (4), 194419. https://doi.org/10.1016/j.bbagrm.2019.194419.

Santarlasci, V., Cosmi, L., Maggi, L., Liotta, F., Annunziato, F., 2013. IL-1 and T helper immune responses. Front. Immunol. 4, 182. https://doi.org/10.3389/fimmu.2013.00182.

Santin, I., Jauregi-Miguel, A., Velayos, T., Castellanos-Rubio, A., Garcia-Etxebarria, K., Romero-Garmendia, I., Fernandez-Jimenez, N., Irastorza, I., Castaño, L., Bilbao, J.R., 2018. Celiac diasease–associated lncRNA named HCG14 regulates NOD1 expression in intestinal cells. J. Pediatr. Gastroenterol. Nutr. 67, 225–231. https://doi.org/10.1097/MPG.0000000000001970.

Schulzke, J.D., Bentzel, C.J., Schulzke, I., Riecken, E.O., Fromm, M., 1998. Epithelial tight junction structure in the jejunum of children with acute and treated celiac sprue. Pediatr. Res. 43, 435–441. https://doi.org/10.1203/00006450-199804000-00001.

Shang, Z., Yu, J., Sun, L., Tian, J., Zhu, S., Zhang, B., Dong, Q., Jiang, N., Flores-Morales, A., Chang, C., Niu, Y., 2019. LncRNA PCAT1 activates AKT and NF-B signaling in castration-resistant prostate cancer by regulating the PHLPP/FKBP51/IKK complex. Nucleic Acids Res. 47, 4211–4225. https://doi.org/10.1093/nar/gkz108.

Shen, L., 2012. Tight junctions on the move: molecular mechanisms for epithelial barrier regulation. Ann. N. Y. Acad. Sci. 1258, 9–18. https://doi.org/10.1111/j.1749-6632.2012.06613.x.

Shui, J.-W., Steinberg, M.W., Kronenberg, M., 2011. Regulation of inflammation, autoimmunity, and infection immunity by HVEM-BTLA signaling. J. Leukoc. Biol. 89, 517–523. https://doi.org/10.1189/jlb.0910528.

Strengell, M., Sareneva, T., Foster, D., Julkunen, I., Matikainen, S., 2002. IL-21 up-regulates the expression of genes associated with innate immunity and Th1 response. J. Immunol. 169, 3600–3605. https://doi.org/10.4049/jimmunol.169.7.3600.

Tripathi, V., Ellis, J.D., Shen, Z., Song, D.Y., Pan, Q., Watt, A.T., Freier, S.M., Bennett, C.F., Sharma, A., Bubulya, P.A., Blencowe, B.J., Prasanth, S.G., Prasanth, K.v., 2010. The nuclear-retained noncoding RNA MALAT1 regulates alternative splicing by modulating SR splicing factor phosphorylation. Mol. Cell 39, 925–938. https://doi.org/10.1016/j.molcel.2010.08.011.

Trynka, G., Hunt, K.A., Bockett, N.A., Romanos, J., Mistry, V., Szperl, A., Bakker, S.F., Bardella, M.T., Bhaw-Rosun, L., Castillejo, G., de La Concha, E.G., de Almeida, R.C., Dias, K.R.M., van Diemen, C.C., Dubois, P.C.A., Duerr, R.H., Edkins, S., Franke, L., Fransen, K., Gutierrez, J., Heap, G.A.R., Hrdlickova, B., Hunt, S., Izurieta, L.P.,

Izzo, V., Joosten, L.A.B., Langford, C., Mazzilli, M.C., Mein, C.A., Midah, V., Mitrovic, M., Mora, B., Morelli, M., Nutland, S., Núñez, C., Onengut-Gumuscu, S., Pearce, K., Platteel, M., Polanco, I., Potter, S., Ribes-Koninckx, C., Ricaño-Ponce, I., Rich, S.S., Rybak, A., Santiago, J.L., Senapati, S., Sood, A., Szajewska, H., Troncone, R., Varadé, J., Wallace, C., Wolters, V.M., Zhernakova, A., Thelma, B.K., Cukrowska, B., Urcelay, E., Bilbao, J.R., Mearin, M.L., Barisani, D., Barrett, J.C., Plagnol, V., Deloukas, P., Wijmenga, C., van Heel, D.A., 2011. Dense genotyping identifies and localizes multiple common and rare variant association signals in celiac disease. Nat. Genet. 43, 1193–1201. https://doi.org/10.1038/ng.998.

van Heel, D.A., Franke, L., Hunt, K.A., Gwilliam, R., Zhernakova, A., Inouye, M., Wapenaar, M.C., Barnardo, M.C.N.M., Bethel, G., Holmes, G.K.T., Feighery, C., Jewell, D., Kelleher, D., Kumar, P., Travis, S., Walters, J.R.F., Sanders, D.S., Howdle, P., Swift, J., Playford, R.J., McLaren, W.M., Mearin, M.L., Mulder, C.J., McManus, R., McGinnis, R., Cardon, L.R., Deloukas, P., Wijmenga, C., 2007. A genome-wide association study for celiac disease identifies risk variants in the region harboring IL2 and IL21. Nat. Genet. 39, 827–829. https://doi.org/10.1038/ng2058.

Wutz, A., 2011. Gene silencing in X-chromosome inactivation: advances in understanding facultative heterochromatin formation. Nat. Rev. Genet. 12, 542–553. https://doi.org/10.1038/nrg3035.

Zhang, F., Liu, G., Wei, C., Gao, C., Hao, J., 2017. Linc-MAF-4 regulates Th1/Th2 differentiation and is associated with the pathogenesis of multiple sclerosis by targeting MAF. FASEB J. 31, 519–525. https://doi.org/10.1096/fj.201600838R.

Zhang, Z., Luo, K., Zou, Z., Qiu, M., Tian, J., Sieh, L., Shi, H., Zou, Y., Wang, G., Morrison, J., Zhu, A.C., Qiao, M., Li, Z., Stephens, M., He, X., He, C., 2020. Genetic analyses support the contribution of mRNA N6-methyladenosine (m6A) modification to human disease heritability. Nat. Genet. 52, 939–949. https://doi.org/10.1038/s41588-020-0644-z.

FSC
www.fsc.org
MIX
Papier | Fördert gute Waldnutzung
FSC® C083411

Kolibri 360 GmbH
Ferdinand-Jühlke-Straße 7
99095 Erfurt, Deutschland
produktsicherheit@kolibri360.de